Smart Farming Technologies for Sustainable Agricultural Development

Ramesh C. Poonia
Amity University Jaipur, India

Xiao-Zhi Gao
University of Eastern Finland, Finland

Linesh Raja
Amity University Jaipur, India

Sugam Sharma
Iowa State University, USA

Sonali Vyas
Amity University Jaipur, India

A volume in the Advances in Environmental Engineering and Green Technologies (AEEGT) Book Series

IGI Global
DISSEMINATOR OF KNOWLEDGE

Published in the United States of America by
IGI Global
Engineering Science Reference (an imprint of IGI Global)
701 E. Chocolate Avenue
Hershey PA, USA 17033
Tel: 717-533-8845
Fax: 717-533-8661
E-mail: cust@igi-global.com
Web site: http://www.igi-global.com

Library of Congress Cataloging-in-Publication Data

Names: Poonia, Ramesh C., 1979- editor.
Title: Smart farming technologies for sustainable agricultural development / Ramesh C. Poonia, Xiao-Zhi Gao, Linesh Raja, Sugam Sharma, and Sonali Vyas, editors.
Description: Hershey, PA : Engineering Science Reference (an imprint of IGI Global), 2018.
Identifiers: LCCN 2017060577| ISBN 9781522559092 (hardcover) | ISBN 9781522559108 (ebook)
Subjects: LCSH: Agricultural innovations. | Agricultural informatics. | Agricultural productivity.
Classification: LCC S494.5.I5 H362 2018 | DDC 338.1/6--dc23 LC record available at https://lccn.loc.gov/2017060577

This book is published in the IGI Global book series Advances in Environmental Engineering and Green Technologies (AEEGT) (ISSN: 2326-9162; eISSN: 2326-9170)

British Cataloguing in Publication Data
A Cataloguing in Publication record for this book is available from the British Library.

All work contributed to this book is new, previously-unpublished material.
The views expressed in this book are those of the authors, but not necessarily of the publisher.

For electronic access to this publication, please contact: eresources@igi-global.com.

Advances in Environmental Engineering and Green Technologies (AEEGT) Book Series

ISSN:2326-9162
EISSN:2326-9170

Editor-in-Chief: Sang-Bing Tsai, University of Electronic Science and Technology of China Zhongshan Institute, China & Ming-Lang Tseng, Lunghwa University of Science and Technology, Taiwan & Yuchi Wang, University of Electronic Science and Technology of China Zhongshan Institute, China

MISSION

Growing awareness and an increased focus on environmental issues such as climate change, energy use, and loss of non-renewable resources have brought about a greater need for research that provides potential solutions to these problems. Research in environmental science and engineering continues to play a vital role in uncovering new opportunities for a "green" future.

The **Advances in Environmental Engineering and Green Technologies (AEEGT)** book series is a mouthpiece for research in all aspects of environmental science, earth science, and green initiatives. This series supports the ongoing research in this field through publishing books that discuss topics within environmental engineering or that deal with the interdisciplinary field of green technologies.

COVERAGE

- Alternative Power Sources
- Waste Management
- Radioactive Waste Treatment
- Sustainable Communities
- Green Transportation
- Green Technology
- Cleantech
- Renewable energy
- Policies Involving Green Technologies and Environmental Engineering
- Air Quality

IGI Global is currently accepting manuscripts for publication within this series. To submit a proposal for a volume in this series, please contact our Acquisition Editors at Acquisitions@igi-global.com or visit: http://www.igi-global.com/publish/.

The Advances in Environmental Engineering and Green Technologies (AEEGT) Book Series (ISSN 2326-9162) is published by IGI Global, 701 E. Chocolate Avenue, Hershey, PA 17033-1240, USA, www.igi-global.com. This series is composed of titles available for purchase individually; each title is edited to be contextually exclusive from any other title within the series. For pricing and ordering information please visit http://www.igi-global.com/book-series/advances-environmental-engineering-green-technologies/73679. Postmaster: Send all address changes to above address.

Titles in this Series

For a list of additional titles in this series, please visit:
https://www.igi-global.com/book-series/advances-environmental-engineering-green-technologies/73679

Effective Solutions to Pollution Mitigation for Public Welfare
Ahmet Ozan Gezerman (Yildiz Technical University, Turkey) Burcu Didem Corbacioglu (Yildiz Technical University, Turkey) and Bhola R. Gurjar (Indian Institute of Technology Roorke, India)
Engineering Science Reference • ©2018 • 274pp • H/C (ISBN: 9781522533795) • US $180.00

Handbook of Research on Microbial Tools for Environmental Waste Management
Vinay Mohan Pathak (Gurukul Kangri University, India) and Navneet (Gurukul Kangri University, India)
Engineering Science Reference • ©2018 • 335pp • H/C (ISBN: 9781522535409) • US $215.00

Food Systems Sustainability and Environmental Policies in Modern Economies
Abiodun Elijah Obayelu (Federal University of Agriculture – Abeokuta (FUNAAB), Nigeria)
Engineering Science Reference • ©2018 • 371pp • H/C (ISBN: 9781522536314) • US $195.00

Microbial Biotechnology in Environmental Monitoring and Cleanup
Pankaj (G. B. Pant University of Agriculture and Technology, India) and Anita Sharma (G. B. Pant University of Agriculture and Technology, India)
Engineering Science Reference • ©2018 • 427pp • H/C (ISBN: 9781522531265) • US $235.00

Handbook of Research on Renewable Energy and Electric Resources for Sustainable ...
Valeriy Kharchenko (Federal Scientific Agroengineering Center VIM, Russia) and Pandian Vasant (Universiti Teknologi PETRONAS, Malaysia)
Engineering Science Reference • ©2018 • 672pp • H/C (ISBN: 9781522538677) • US $345.00

Green Production Strategies for Sustainability
Sang-Binge Tsai (University of Electronic Science and Technology of China (Zhongshan Institute), China & Civil Aviation University of China, China) Bin Liu (Shanghai Maritime University, China) and Yongian Li (Nankai University, China)
Engineering Science Reference • ©2018 • 325pp • H/C (ISBN: 9781522535379) • US $235.00

For an entire list of titles in this series, please visit:
https://www.igi-global.com/book-series/advances-environmental-engineering-green-technologies/73679

IGI Global
DISSEMINATOR OF KNOWLEDGE

701 East Chocolate Avenue, Hershey, PA 17033, USA
Tel: 717-533-8845 x100 • Fax: 717-533-8661
E-Mail: cust@igi-global.com • www.igi-global.com

Editorial Advisory Board

Table of Contents

Detailed Table of Contents

improve the performance of various applications in smart environments. These smart environments include smart home, smart agriculture, smart office, and smart hospital. In various countries throughout the world, agriculture is regarded as the backbone of the economy. To date, many families in Africa are employed in the agriculture industry. The dawn of information communication technologies (ICT) has changed the conventional way in which farming is conducted. However, quite a large number of farmers in Africa are still stagnant and highly dependent on the traditional ways of farming which started hundreds of years ago. This chapter has presented the current problems in farming and irrigation systems and suggests solutions through proposing the intelligent WSNs in a smart environment.

Vaibhav Bhatnagar, Amity University Jaipur, India
Ramesh C. Poonia, Amity University Jaipur, India

Agriculture is the prime source of livelihood for human beings, animals, and all living beings. Agriculture also plays a vital role in the economy of India. This chapter describes the importance of agriculture and factors affecting the development of agriculture. The international scenario of agriculture, current status of Indian agriculture, and position of Rajasthan (state in India) in agriculture are described in this chapter. The total production, total imports and exports, method of irrigation, net area of irrigation, types of crops, fertilization consumption, and highlight of Union Budget 2018-19 of Indian agriculture are described in this chapter. The geography of Rajasthan according to agriculture, production of crop, and consumption of fertilization are also elaborated in this chapter. This chapter is concluded with future perspectives of India agriculture.

S. Umamaheswari, Dr. G. R. Damodaran College of Science, India

Agriculture is an important economic sector that leads to the overall economic growth of a nation. This sector has to meet the increasing consumption needs of the world population, which is expected to increase by 70% by 2050, according to the Food and Agriculture Organization of the United Nations. The advancements in the technology in the field of agriculture will facilitate the growth in farming activities. The internet of things (IoT) industry enables the farmers to face these challenges. IoT sensors provide the information about the crop yields, soil moisture, rain fall, livestock, etc. to the farmers that can be used for decision making. The information can be accessed by the farmers with the aid of smart phones. This chapter focuses on the smart agriculture practices with IoT that helps the farmer to analyze the operational data to improve decision making. The implementation platform for the IoT and the communication technologies for smart farming are discussed. A prototype for a smart irrigation system has also been proposed.

By combining the different monitoring and automation techniques available today, we can develop cutting-edge internet of things (IoT) systems that can support sustainable development through smart agriculture. Systems are able to monitor the farming areas and react to the parameters being monitored on their own without the presence of human beings. This automation can result in a more precise way of maintaining the aspects that affect the growth of plants, leading to an increase in the food production on farmlands. This chapter focuses on IOT for automation in smart agriculture and provides a pathway to develop automation system in the smart environment.

In the European Union (EU), the use of fossil fuels brings several disadvantages, as they are the main culprits responsible for pollutants and GHG emissions. The increasing demand for sustainable fuels leads to the research of alternative technologies, such as biogas production from lignocellulosic materials. Therefore, the acquisition of biomass from marginal areas under Danish conditions has been evaluated in terms of alternative harvesting equipment: an automated robot (Grassbot) versus a regular tractor for key grass materials used for biogas plants (chopped, unchopped, and baled grass) and compared regarding operational, economical, and environmental performances. The evaluation uses two operations models (IRIS and DRIFT) to consider the field characteristics, machinery characteristics, etc. Selected results show that in terms of fuel consumption, chopping, and mowing are the most demanding operations, and that there is no significant difference between the harvesting equipment regarding CO2 emissions.

Agriculture and allied activities play a vital role in a country's economic prosperity. The conventional methods in agricultural practices have become insufficient to cater to the increasing needs. To fulfill the demands, new technologies are to be introduced to raise agricultural standards. Over the past few years, there has been significant interest in designing smart agricultural systems. The manageability of agricultural frameworks has turned into a noteworthy concentration for discusses about future human survival. A significant part of the contention seems to depend on shortsighted understanding of biological models and flops enough to define what maintainability goals are being looked for. To adapt to the undeniably multifaceted nature and between connectedness of current cultivating frameworks with regards to globalization and potential bothers like environmental change, we require a pluralistic way to deal with strategy, which can adapt to the abnormal amounts of vulnerability in these territories and which enables most extreme flexibility of reaction to evolving conditions.

We always say agriculture is a backbone of Indian economy. In this chapter, the authors have proposed a novel approach based on priority-driven scheduling-based irrigation model (for home garden) which supplies optimum and good quality water to the crops. Technology used in this proposed system is wireless sensor network. Proposed system will be very useful as it immediately irrigates the plants if the moisture level decreases to avoid the future losses. In this process, soil moisture values will be sensed and compared to find out the lowest value. Such systems will start a new era in agriculture and will prove this could be the major requirement in future due to many critical factors such as irregularity of monsoon, less availability of water, etc. Subtopics enlisted in this chapter such as literature survey, design and modeling, technical specification, sampling, results, and analysis will elaborate depth of proposed system.

Rising temperatures and increased frequency of extreme events will have direct and negative impacts on natural resources. Water resources are limited on earth; hence, there is a need to manage the utilization techniques of water. The irrigation system improvement using the wireless network is a solution to accomplish water conservation goal as well as improvement in irrigation practices. Smart farming enhances the capacity of the agricultural systems to support food security. The need for adaptation and the potential for mitigation into sustainable agriculture development strategies can be incorporated into such system. The smart farming system includes different techniques of agricultural practices to conserve different resources including water. Solar powered smart irrigation systems are a part of the smart irrigation system. Smart irrigation system includes temperature, moisture, and humidity sensors system. Different smart irrigation systems which are used all over the world will be discussed in this chapter.

Agriculture is important for people all over the world in order to obtain food to sustain the ever-growing population. However, the current practices for obtaining more and more food has several environmental challenges. Hence, new environmentally friendly fertilizers, herbicides, and pesticides have been developed that enhance crop yield by facilitating maximum nutrient uptake by the application of nanotechnology that will help in promoting sustainable agriculture by the slow or controlled release fertilizers. This slow discharge encourages improved delivery of nutrients to the plants that further speeds up early germination, fast growth, and high nutritional level. The current study is aimed to review nano-chemicals used in agriculture that have been developed by the researchers all over the world.

Foreword

Smart farming technology is a combination of agriculture and computation used in the modern times for the sustainability of agriculture. Introduction of this new technology will provide the synergy between agriculture and computing. Internet of Things (IoT) and wireless networks are some of the areas which could be combined with agriculture to make it smart enough to monitor various decisions in terms of irrigation, fertilizer, photoperiod and insect pests.

The book entitled "Smart Farming Technologies for Sustainable Development" provides valuable information about smart farming technologies for sustainable agricultural development. The chapters are written in well organized manner and cover all the important aspects of the book theme.

The first few chapters provide the basic information about the technological development in the field of agriculture. It also provides their past and present scenario of the technology. Wireless sensor network and their application with respect to smart farming are discussed in well précised manner. The other chapters discuss the various technological developments, in smart farming. It provides information about IoT based practices, automation in agriculture and smart farming framework. The last few chapters elaborate the various technique based smart farming methods, smart irrigation system, water utilization and use of nano chemicals etc.

Finally we can conclude that this book is a good attempt to discuss overall aspects of smart farming technologies and their sustainable development. The UG & PG students, researchers and peoples from agricultural organization may find this book quite useful to get better idea about smart farming in relation with their sustainable development.

S. L. Kothari
Amity University Jaipur, India

Jagdish Prasad
Amity University Jaipur, India

Preface

Agriculture is considered as the back bone of most of the country's economy. Globally we consume almost 12 million pounds of food in every one minute. So to fulfill such a huge amount of need, we required smart farming. Introduction of new technologies-based farming will form a bridge between agriculture and Internet of Thing's (IoT's). It will modernize the agriculture activities and addresses the various issues of smart farming. IoT will help the farmers to collect huge amount of information from crop yields, soils, weather forecasting and fertilizers. The sensor-based monitoring system will improve the overall agriculture production.

Smart Farming Technologies for Sustainable Agricultural Development discussed the latest farming advancements in terms of informatics and communication. This helps farming smart enough in decision making. It improves productivity by introducing sensors, automated machines, smart phones; yield meters and drones etc. This book covers the evolution of communication technologies which help the farmers in more accurate prediction for decision making. Smart farming is still very expensive. But the discussion of upgraded enhancement and awareness regarding the smart farming will make it profitable.

The topics include, but are not limited to the following:

- Smart Agriculture/Farming
- Sensor System
- Intelligent Sensor
- Wireless Communication
- IoT
- Agriculture Task Automation
- Soil Moisture Content
- Irrigation
- Moisture Measurement
- Energy Consumption
- Automation
- Microprocessor

- pH
- Communication Standards

TARGET AUDIENCE

This book is ideally designed for farmers, academicians, researchers, PG students, government agency and technology developers to understand the smart farming technologies for sustainable agricultural development.

CHAPTER DESCRIPTIONS

Chapter 1

Agriculture is the back bone of any developing country for their sustainable development. So this is our responsibility to educate the society regarding the sustainable development of agriculture. From last 10-15 years technology are developing at rapid speed. In recent time various researchers are giving more emphasis on applying technology on agriculture. This is termed as smart farming. Smart farming uses computer technology and communication for greater yield and production of crops. This chapter studies the various technological developments in the field of smart farming. Few of them are related to Internet of Things (IOT), Wireless Communication, Irrigation System and Agriculture Automation etc. This chapter helps the new researchers in the field of smart farming to understand the current technological development in their respective scope.

Chapter 2

In recent years, there has been increased use of Wireless Sensor Networks (WSNs) which consists of sensor nodes and the intelligent Machine Learning algorithms to improve the performance of various applications in Smart Environments. These Smart Environments include Smart Home, Smart Agriculture, Smart Office and Smart Hospital etc. In various countries throughout the world, agriculture is regarded as the main backbone of the economy. To date, many families in Africa are employed in the agriculture industry. The dawn of Information Communication Technologies (ICT) has changed the conventional way in which farming is conducted. However, quite a large number of farmers in Africa are still stagnant and highly dependent on the traditional ways of farming which started hundreds of years ago. This chapter has presented the current problems in farming and irrigation systems and suggests solutions through proposing the intelligent WSNs in a smart environment.

Chapter 3

Agriculture is the prime source of livelihood for human being, animals and all living beings. Agriculture also plays a vital role in economy of India. This chapter describes the importance of agriculture and factors affecting the development of agriculture. International scenario of agriculture, current status of Indian agriculture and position of Rajasthan (state in India) in agriculture is described in this chapter. Total production, total imports and exports, method of irrigation, net area of irrigation, types of crops, fertilization consumption and highlight of Union Budget 2018-19 of Indian agriculture is described in this paper. Geography of Rajasthan according to agriculture, production of crop and consumption of fertlization is also elaborated in this paper. This chapter is concluded with future perspectives of India agriculture.

Chapter 4

Agriculture is an important economic sector that leads to the overall economic growth of a nation. This sector has to meet the increasing consumption needs of the world population which is expected to increase by 70% by 2050, according to the Food and Agriculture Organization of the United Nations. The advancements in the technology in the field of agriculture will facilitate the growth in farming activities. The Internet of Things (IoT) industry enables the farmers to face these challenges. IoT sensors provide the information about the crop yields, soil moisture, rain fall, livestock, etc. to the farmers that can be used for decision making. The information can be accessed by the farmers with the aid of smart phones. This chapter focuses on the Smart Agriculture practices with IoT that helps the farmer to analyze the operational data to improve decision making. The implementation platform for the IoT and the Communication technologies for Smart farming are discussed. A prototype for Smart Irrigation System has also been proposed.

Chapter 5

These days by combining the different monitoring and automation techniques available today, we can develop cutting-edge Internet of Things (IoT) systems that can support sustainable development through Smart Agriculture. Systems are able to monitor the farming areas and react to the parameters being monitored on their own without the presence of human being. This automation can result in a more precise way of maintaining the aspects that affect the growth of plants, leading to an increase in the food production on farmlands. This chapter focuses on IOT for Automation in Smart Agriculture and provides a pathway to develop automation system in the smart environment.

Chapter 6

Finding sustainable alternatives for fossil fuels have never been so high. In the European Union (EU), the use of fossil fuels brings several disadvantages, being the main responsible for pollutants and GHG emissions. The increasing demand for sustainable fuels leads to the research of alternative technologies, such as biogas production from lignocellulosic materials. Therefore the acquisition of biomass from marginal areas under Danish conditions has been evaluated in terms of alternative harvesting equipment: an automated robot (Grassbot) versus a regular tractor for key grass materials used for biogas plants (chopped, unchopped and baled grass) and compared regarding operational, economical, and environmental performances. The evaluation uses two operations models (IRIS and DRIFT) to consider the field characteristics, machinery characteristics, etc. Selected results show that in terms of fuel consumption, chopping and mowing are the most demanding operations; and that there is no significant difference between the two harvesting equipment regarding CO2 emissions.

Chapter 7

Agriculture and allied activities play a vital role in a country's economic prosperity. The conventional methods in agricultural practices have become insufficient to cater the increasing needs. To fulfill the demands new technologies are to be introduced to raise agricultural standards. Over the past few years, there has been significant interest in designing smart agricultural systems. The manageability of agricultural frameworks has turned into a noteworthy concentration for discusses about future human survival. A significant part of the contention seems to depend on shortsighted understanding of biological models and flops enough to define what maintainability goals are being looked for. To adapt to the undeniably multifaceted nature and between connectedness of current cultivating frameworks with regards to globalization and potential bothers like environmental change, we require a pluralistic way to deal with strategy, which can adapt to the abnormal amounts of vulnerability in these territories and which enables most extreme flexibility of reaction to evolving conditions.

Chapter 8

We always say agriculture is a backbone of Indian economy. In this chapter, we have proposed a novel approach based on priority driven scheduling based irrigation model (for home garden) which supplies optimum and good quality water to the crops. Technology used in this proposed system is Wireless Sensor Network. Proposed system will be very useful as it immediately irrigates the plants if the moisture level decreases to avoid the future losses. In this process soil moisture values will be sensed and compared to find out the lowest value. Such systems will start new era in agriculture and will prove this could be the major requirement in future due to many critical factors such as irregularity of monsoon, less availability of water etc. Sub topics enlisted in this chapter such as literature survey, design and modeling, technical specification, sampling, results and analysis will elaborate depth of proposed system.

Chapter 9

Rising temperatures and increased frequency of extreme events will have direct and negative impacts on natural resources. Water resources are limited on earth hence there is a need to manage the utilization techniques of water. The irrigation system improvement using the wireless network is a solution to accomplish water conservation goal as well as improvement in irrigation practices. Smart Farming enhances the capacity of the agricultural systems to support food security. The need for adaptation and the potential for mitigation into sustainable agriculture development strategies can be incorporated into such system. The smart farming system includes different techniques of agricultural practices to conserve different resources including water. Solar powered smart irrigation systems are a part of the smart irrigation system. Smart irrigation system includes temperature, moisture and humidity sensors system. Different smart irrigation systems which are used all over the world will be discussed in this chapter.

Chapter 10

Agriculture is an important for people all over the world in order to obtain food to sustain the ever-growing population of the world. However, the current practices for obtaining more and more food has several environmental challenges. Hence, new environmental friendly fertilizers, herbicides and pesticides have been developed that enhances crop yield by facilitating maximum nutrient uptake by the application of nanotechnology that will help in promoting sustainable agriculture by the slow or controlled release fertilizers. This slow discharge encourages improved delivery of nutrients to the plants that further speed up early germination, fast growth and high nutritional level. The current study is aimed to review nano-chemicals used in agriculture that have been developed by the researchers all over the world with the pros and cons.

BOOK IMPACT

Smart Farming Technologies for Sustainable Agricultural Development helps us to understand the latest farming advancements in terms of informatics and communication. This helps farming smart enough in decision making. It improves productivity by introducing sensors, automated machines, smart phones yield meters, drones, etc.

Acknowledgment

The editors would like to acknowledge the help of all the people involved in publication of *Smart Farming Technologies for Sustainable Agricultural Development.*

However this acknowledges is more specifically, to the authors and reviewers that took part in the review process. Without their support, this book would not have become a reality.

First of all editors would like to express their gratitude to each one of the authors for their contributions. Our sincere appreciation goes to the chapter's authors who contributed their time and expertise to this book. Editors wish to acknowledge the valuable contributions of the reviewers regarding the improvement of quality, rationality, and content presentation of chapters. Few of the authors also served as referees; we extremely grateful for their double task. We would also like to acknowledge our Editorial Advisory Board for their timely suggestions.

This acknowledgement remains incomplete without showing editors gratitude towards Ms. Jordan Tepper, Assistant Development Editor, IGI Global, USA. Her timely and sincere guidance motivated us to complete this book.

Finally thank you IGI Global for providing a platform where we can render our work. Editors keep motivated by their commitment towards highest quality standards and excellent service. Hope in years to come we will able to work together to achieve highest standard of innovative research.

At the end editors once again acknowledge their sincere thanks to all who at some stage supported this book.

Ramesh C. Poonia
Amity University Jaipur, India

Xiao-Zhi Gao
University of Eastern Finland, Finland

Linesh Raja
Amity University Jaipur, India

Sugam Sharma
Iowa State University, USA

Sonali Vyas
Amity University Jaipur, India

Chapter 1
The Study of Technological Development in the Field of Smart Farming

Linesh Raja
Amity University Jaipur, India

Sonali Vyas
Amity University Jaipur, India

ABSTRACT

Agriculture is the backbone of any developing country for their sustainable development. So, it is our responsibility to educate the society regarding the sustainable development of agriculture. In the last 10-15 years, technology has been developing at a rapid speed. Various researchers are giving more emphasis to applying technology to agriculture. This is called smart farming. Smart farming uses computer technology and communication for greater yield and production of crops. This chapter studies the various technological developments in the field of smart farming. A few of them are related to internet of things (IOT), wireless communication, irrigation system, and agriculture automation. This chapter helps the new researchers in the field of smart farming to understand the current technological developments.

DOI: 10.4018/978-1-5225-5909-2.ch001

WHAT IS AGRICULTURE?

Introduction

Agriculture is obtained from Latin words Ager means land and Cultura means cultivation. So agriculture is known as cultivation of land. It is also defined as the science of crops production and farm animals from the reserves of the earth. The basic goal of agriculture is to make land produce more quantity of crops ("Using APA," n.d.).

Agriculture is defined as a venture developed and managed by humans in which natural ecosystem is transformed into food production. As population is increasing day by day agriculture is becoming so important for fulfilling needs of people. If the increase in production of food is not done than that day is not so far when there will be nothing to eat. Food production totally depends on the requirement of land, water and other resources ("Using APA," n.d.).

Agriculture encompasses modifying species of plants and animals and also manipulates resources available. As there is advancement in technology, agricultural processes also became more efficient and productive by increasing resources for development of plants with increased capacity for conversion of resources in harvesting forms. With the advancement in agricultural practices some problems also aroused like bio-diversity reduction, fragmentation of natural ecosystem, fresh water pollution, alteration in nutrient balance etc. However these limitations must be compared with increasing human population, also 40% of human population is having agriculture as their primary living necessity. In some countries 80% of population depends on farming for their living, therefore increase in agricultural production makes food availability and also enhances incomes and standard of living. The future of agriculture depends on factors like world food demand, resource availability and cost required for supporting high production demands and advancement in technology for making agriculture more efficient ("Using APA," n.d.; Gu, 2017).

NEED FOR AGRICULTURE

Types with Example

- **Subsistence Agriculture:** Mostly farmers practice this type of agriculture. It is exemplified as dispersed lands and using primal tools for farming. As it is done by poor farmers so there is less use of fertilizers and yielding various types of crops in their fields. There is scarcity of facilities like electricity, etc in such type of farming.

- **Shifting Agriculture:** In this agriculture a forest land is made clear and then farming is done for few years and then again land is deserted as the soil's fertility is lost so the farmers then shift to new land and again the same procedure is repeated. Crops like millets, paddy, vegetables etc are mainly farmed in this agriculture.
- **Plantation Agriculture:** It is also known as tree farming or bush farming. It is termed as single-crop farming of crops like tea, coffee, spices, oranges etc. It is money concentric and needs good managing ability, technology knowledge, fertilizers, and irrigation facilities. It is export oriented and in it mostly crops have two years or more life span. This type of agriculture is practiced in tropical areas.
- **Intensive Farming:** It is an escalation of agricultural practices which promises maximum yield on available land by using various means of pesticides and fertilizers. It is a system of faming by using enormous labor and capital related to area of land. It makes use of capital for acquisition and maintaining efficient machines for plantation and also irrigation facilities.
- **Dry Agriculture:** This type of agriculture can be understood as a procedure of cultivating crops without water in regions receiving annual rainfall less than 500mm. It is practiced for drought-resistant crops or by making use of moisture increasing techniques like growing seeds deeply in land or making use of surface tilt or for delaying evaporation.
- **Mixed and Multiple Agriculture:** Mixed agriculture refers to plantation of crops and animal rising simultaneously. Multiple agriculture refers to cultivation of more than one crop together. This type of agriculture is practiced in regions having abundant rainfall or has irrigation facilities.
- **Crop Rotation:** This type of agriculture refers to cultivating crops one after another in fixed rotation for maintaining soil fertility. After growing cereals pulses are grown because legumes have ability to fix the nitrogen in soil. The rotation is completed in a year in some regions while in some it may be more than a year. Crops like sugarcane or pulses are rotated with cereals. The rotation crops are selected on the basis of soil conditions and farmer's experience.
- **Terrace Cultivation:** For this type of agriculture mountain slopes are made as terraces and its land is utilized for cultivation of crops. Soil erosion is likely to occur because of slopes so preventive measures are taken for this. Mainly terraces are made on slightly grades so that water can be stored and moved in channels slowly towards the crops ("Using APA," n.d.).

Agriculture in Respect to Technology

Smart agriculture is combination of various technologies with agriculture. Smart agriculture replaces traditional farming with more efficient, reliable and consistent agriculture techniques. Examples of such paradigm include context aware farming, remote monitoring, security control, environmental monitoring, etc. Smart agriculture can be understood as Smart farms, Smart Farmers, Smart Consumers. A system is said smart if it follows various intelligence levels: Adaptation, Sensation, Inferring, Learning, Anticipation and Self- Organization. Smart agriculture signifies applying modern Information and Communication Technologies (ICT) for performing agricultural activities. The smart agriculture includes various technologies which can be used individually or together for efficient farming (Patil & Kale, 2016; Prathibha et al., 2017).

Smart agricultural practices also known as Third Green Revolution are overcoming the traditional farming based on applications such as precision equipments, Internet of Things (IOT), sensors, Big Data, Unmanned Aerial Vehicles (UAVs), robotics etc (Walter et al., 2017).

By opting smart farming one can cultivate more efficient and productive agricultural production based on precise and sustainable approach. According to the farmer's, smart agriculture should give farmers a way for better decision making and more efficient management ways. Management Information Systems (MIS) helps in providing planned systems useful in data collection, processing, storage and decision making for conducting agricultural operations. Precision Agriculture helps in improving economy and utilization of resources and helps in reducing environmental impacts. It provides Decision Support Systems (DSS) for managing farm activities for optimization of outcomes while preserving resources and using GPS, drones and hyper spectral images by Sentinel satellites. Agricultural automation is done using robotics, auto- control and artificial intelligence processes at various levels of agriculture which includes farmbots and farmdrones (Walter et al., 2017; APA, n. d).

Smart agriculture not only targets large, conventional farms but also proves to be a boost for growing agricultural trends like family farming, organic farming and also enhances production in transparent farming processes. It also provides advantages in environmental issues like efficient utilization of water, etc (Chandra et al., 2018).

SMART FARMING

The FAO predicts that by 2050 agricultural sector is going to face a massive challenge in order to feed the world population. As per the survey the food production must increase by 70% by 2050. Achieving this target is a challenge for us due to degradation of available resources. The high demand of fresh water and impact of climate change is a big challenge for the upcoming generation. This also leads to the change in life cycle of plant and animals (Guerrini, 2015, “APA,” n. d.; “APA,” n. d.).

One way to address these issues and increase the quality and quantity of agricultural production is using sensing technology to make farms more “intelligent” and more connected through the so-called “precision agriculture” also known as ‘smart farming’.

It’s something that’s already happening, as corporations and farm offices collect vast amounts of information from crop yields, soil-mapping, fertilizer applications, weather data, machinery, and animal health. In a subset of smart farming, Precision Livestock Farming (PLF), sensors are used for monitoring and early detection of reproduction events and health disorders in animals.

Typical monitored data are the body temperature, the animal activity, tissues resistivity, pulse and the GPS position. SMS alerts can be sent to the breeder based on predefined events, say, if a cow is ready for reproduction.

Although the cost of smart farming is still high for any but the largest farms (this, by the way, helps explain why the USA, with its vast territories, is at the forefront of this new paradigm), this doesn’t mean precision agriculture can’t be done in small places. Actually, there are quite a few applications in small-field farming too.

Helpful and sought after as it might be, smart farming has still to overcome many hurdles before it becomes more widespread. “One is that the agricultural sector is extremely low margin. Therefore, investments in innovation are difficult,” the researcher says. Then there’s also what we might call an ‘image problem’ that is causing a hemorrhage of labor. “Being a farmer is not cool because agriculture is perceived as something that belong to history, to the grandfathers,”

Access to real time information about harvesting, planting and yields could also help corporations predict the property value of farms better than anyone else and have unparalleled insight into the commodities market (Guerrini, 2015, “APA,” n. d.; “APA,” n. d.).

Another problem that could slow down IoT in agriculture is the issue of communicating with farmers, who could often not understand the technicalities. “If we tell them that you can do this and that with IoT, they will not understand. The language of the IoT industry has to change dramatically (Yoe, 2017; Zie et al., 2017, Jarchow et al., 2012).

AREA OF STUDY

The smart farming and their sustainable development can be possible by means of technology and technique based study. The below section provide the details about technological based development and technique based development. In respect to technological enhancement, authors provide the details of Internet of Things (IOT) along with wireless technologies. As these two forms the basic requirement of smart farming. At the same time technique based enhancement is also essential for smart farming. The author provides the detail study of irrigation and soil. At the end agricultural automation was discussed.

INTERNET OF THINGS (IOT)

Introduction

The Internet of Things (IOT) is gaining wide interest among the research community, professionals and student due to their diversity. It is gaining their importance in different sectors. It includes industry, agriculture, transportation, education etc (Atzori et al., 2010; Gubbi et al., 2013).

IOT is a collection of various physical devices which connected to each other in order to share data. These physical devices may include home appliances and mobile phones. Once these devices are configured with some smart sensors, it can sense data from the environment followed with their storage to various devices (Weber, 2010; Xia et al., 2012; Al-Fuqaha et al., 2015). Once done the data can be used for personal and business activities. Some of their practical aspects are listed below (Chen et al., 2011; Miorandi et al., 2012):

- Self tracking of personal health
- Automatic ordering of groceries
- Instant warning/safety messaging
- Location based tracking and
- Data collection regarding smart farming

The various sectors can also get benefited by implementing IOT in their day to day activities. They are developing various applications for their business activities. The various sectors are listed below (Bandyopadhyay & Sen, 2011; Ferreira et al., 2017):

- Healthcare
- Retail
- Agriculture
- Smart home/city
- Intelligent transportation system

IOT Architecture

In order to get a better idea of IOT and their working, it is necessary to understand the core architecture of IOT system. The architecture is divided into 4 different categories. The categories are listed below (Wu et al., 2017; Khan et al., 2012; "APA," n. d.; "APA," 2017):

- Sensors
- Gateways
- Server and Cloud
- End User Application

Figure 1 shows the 4 basic categories of IOT architecture.

The first category consists of various sensor devices. They are used to collect data from various objects. Sensors are the basic building block of IOT system architecture. Their objects are used to measure the raw data. Based on the collected data, useful information can be rendered by the system. There are various types of sensors used in different perspective. Few of their broad categories are (Wu et al., 2017; Khan et al., 2012; "APA," n. d.):

- **Accelerometer Sensor:** They are used to detect acceleration (motion). It may be due to the effects of gravity, vibration and movement. It is the measurement of change in speed per unit time. They are widely in cellular devices, laptops, aircraft and consumer electronics etc. to detect motion.
- **Proximity Sensor:** They are used to sense the nearby object presence. It generates signal which can be read by the electromagnetic device to detect their presence. They are widely used to avoid collision detection for making transport system more intelligent. They are also used in retail sector to detect nearby product in a retail shop.

Figure 1. Data Flow and Control for the IOT Architecture

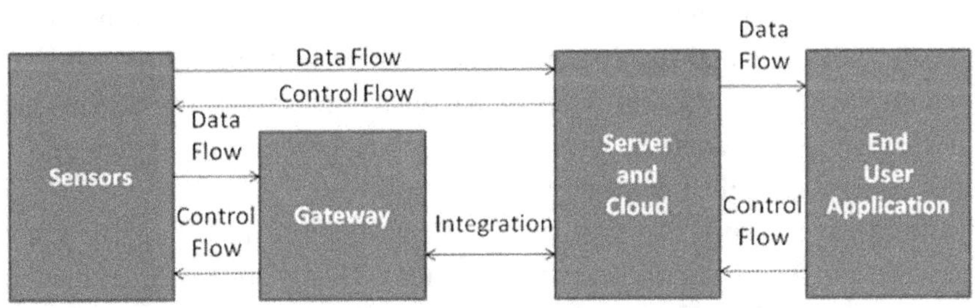

- **IR Sensor:** The infrared (IR) sensors are used to sense the environment characteristics. These sensor can emits or detect infrared radiation for data manipulation. They are widely used in healthcare industry, smart watch, home appliances and remote control etc. Blood pressure measuring devices uses IR sensor to calculate the BP.
- **Temperature Sensor:** They are used to measure the heat energy in the surrounding environment or inside any physical devices. These sensors are mainly used inside refrigerator, AC etc. physical devices. At the same time they are widely used in agriculture to detect the environmental temperature. These measurement increases crop yield to great extent.
- **Chemical Sensor:** They are used to detect state of change in liquid. These sensor devices are mostly used in chemical industry to monitor the environment, process control, chemical detection, radioactive detection inside chemical industry.
- **Gas Sensor:** They are used to detect state of change in air. These sensor devices are mostly used to monitor air quality. They also detect the presence of various gases. Their various application areas cover agriculture, healthcare and manufacturing industries etc.
- **Level Sensor:** They are used to detect the level of fluids inside close or open system. They are used to monitor sea level. Various water reservoirs use this sensor to measure their water level.
- **Motion Detector Sensor:** They are used to detect the continuous motion in a specific area. There application area mostly covers the security aspect of various industries. It converts motion of objects into electric pulses. These pulses are read by the physical devices to make useful analysis. They are widely used in smart camera, automatic door, toll plaza, parking system etc.

Once data is been generated from the various sensors as per the requirement, then it is to be sent to the server or cloud for further processing. It is the responsibility of the internet gateway to collect data from the sensors and route it over wired or wireless medium to server. The various network communication devices, i.e. router, switch, hubs etc are used to transmit data from sensor to server via gateway. Gateway acts as an intermediate between sensors and server. Sometime they need to interact with the cloud for data processing. Gateway helps us to perform data acquisition, malware protection and data management services. Intelligent gateway also performs real time data analytics in order to filter variation in data due to the effect of temperature, dust, humidity and vibration etc.

The third stage consists of server and cloud. Data coming from the earlier category get forwarded to the server or cloud based system. This is done in order to manage, analyze and secure data. This category processes the sensor data to perform some critical analysis regarding various aspects as per the requirement. Sometime sensor data are combined with some external data sources to perform in depth analysis.

The final category of IOT architecture consists of end user application. It utilizes filtered sensor data inside some applications to perform the various tasks. In different cases, applications process the data in real time to develop machine learning capabilities. Learning capability helps the applications to meet industry specific need and requirement ("APA," n. d.).

The architecture of Internet of Things (IOT) may be further classified on the basis of various layers. The below figure 2 shows the same ("APA," 2017; Sethi, 2017):

- **Physical Layer:** It contains various sensors which are used to sense and collect data from the environment or some physical objects.
- **Transport Layer:** Once data is been sensed by the sensors, the transport layer transfer the data to the next upper layer, i.e. processing layer. It is transmitted via network communication devices. Few of them may be wireless or wired network, 4G technology, Bluetooth and infrared etc.
- **Processing Layer:** In order to get some desired information, data is to be redefined. This is done in order to store and analyze the sensor data. The physical layer is responsible for the above task. It manages the data by

Figure 2. 5 Layer Architecture of IOT

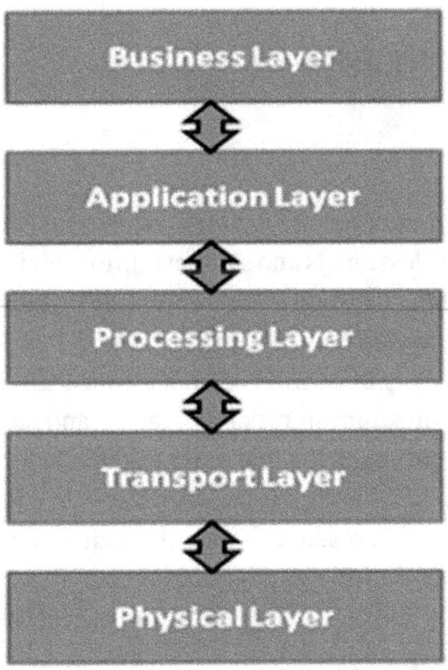

providing various services to their intermediate layers. The services include database operations, cloud computing and big data etc.

- **Application Layer:** This layer utilizes filtered sensor data inside some applications (Smart Health System, Intelligent Transportation System and Smart City etc.) to perform the various tasks. In different cases, applications process the data in real time to develop machine learning capabilities. Learning capability helps the applications to meet industry specific need and requirement.
- **Business Layer:** This layer took care of entire IOT system. It includes business application, user's authentication and privacy, Business profit model etc.

Sometime these 5 layers may be sub classified into three layers. The 3 layer IOT architecture consists of following points (Sethi, 2017):

- Physical layer remains the same in both the architecture.
- Transport and processing layer combined together to form a network layer. Network layer collectively perform the task assigned to transport layer and processing layer respectively.
- Application layer remains the same in both the architecture.
- Business layer is removed from the 3 layer IOT architecture. Their task is now performed by application layer.

WIRELESS TECHNOLOGY

Introduction

Wireless technology is another area which is gaining wide interest among the research community, professionals and student due to their infrastructure less nature. It is gaining their importance in different sectors. It includes industry, agriculture, transportation, education etc (Kim, 2015; Tse, 2005).

Wireless technology is a collection of nodes which are connected to each other through some wireless medium in order to receive and send data for information interchange. It uses electromagnetic and radio frequency for data transmission. The data is sent and received over radio waves with the help of various modulation and demodulation techniques. The sender sends the modulated data over the wireless medium. At the receiver end the data is demodulated to their original form. The wireless network mostly consists of infrastructure less networks. However in some situation it also uses infrastructure for data communication. Wireless network posses various advantages over a wired network. Few of them are discussed below (Rappaport, 1996).

- **Mobility:** It helps in data communication without using infrastructure network. The communication is done with the help of mobile nodes.
- **Accessibility:** The mobile node can be accessed, regardless of their location from where they are operating. It helps people to be stay connected.
- **Simplicity:** It can be deploy with ease. This makes it simple as compared to wired network. Only the initial cost of implementation is high.
- **Maintenance:** Once the implementation is been done. Next course of maintenance charge is low as compared to wired network.
- **Roaming:** It can provide service irrespective of any geographical location where their network exits.

In recent time wireless networks are used in variety of application areas for smart communication. Such as Intelligent Transportation System, Smart City/ Home, Mobile Communication etc. One of the promising area which uses wireless technology is smart farming. Wireless communication can be used to control the farm power distribution and irrigation system. It also helps us to collect/monitor various environmental factors which help in the sustainable development of smart farming, such as temperature, humidity, water, soil etc. It works in relation with Internet of Things (IOT) to get the best result (Dhaka et al., 2014; Dar et al., 2010).

The above figure 3 shows that how wireless network plays an important role in smart farming. It shows that the agriculture field is deployed with smart sensor to get the required data. The data is stored to the server database via gateway and internet respectively. Once stored, the data can be analyzed to get the best result required for the sustainable development of smart farming. These data may be monitor by the end user with the help of cell phones or some other mobile devices (Dhaka et al., 2013; Zhou et al., 2011).

Wireless Network Standards

Each technology uses some standards for their operations. In the similar way in 1997 the Institute of Electrical and Electronics Engineer (IEEE) formed the standards for the wireless networks. Till data various IEEE standards are developed for their smooth functioning. Few of their important standards are discussed below (Cooklev, 2004; Lee et al., 2007):

Figure 3. Wireless Technology in respect to smart farming

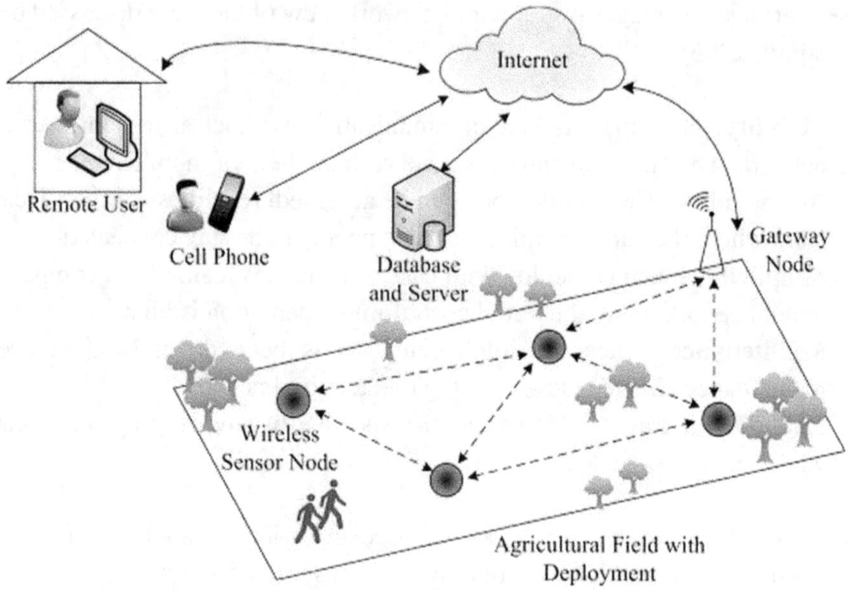

- **IEEE 802.11:** This is the WLAN standard developed in the year 1997. It supports 2.4 GHz frequency and bandwidth of 2 Mbps. It uses DSSS, FHSS and OFDM modulation techniques.
- **IEEE 802.11a:** This is the extension of IEEE 802.11. It was developed in the year of 1999. It supports 5 GHz frequency and bandwidth up to 54 Mbps. It uses OFDM modulation technique. The higher frequency leads more difficulty in penetrating walls. It supports fast speed at the high cost.
- **IEEE 802.11b:** This is the WLAN standard developed in the year 1999. It supports 2.4 GHz frequency and bandwidth up to 11 Mbps. It uses DSSS modulation technique. It supports good signal range at low cost.
- **IEEE 802.11g:** This is the WLAN standard developed in the year 2003. It supports 2.4 GHz frequency and bandwidth up to 54 Mbps. It uses OFDM modulation technique. It supports fast speed at the high cost.
- **IEEE 802.11n:** This is the WLAN standard developed in the year 2009. It supports 2.4/5 GHz frequency and bandwidth up to 300 Mbps. It allows MIMO steaming of data which uses MIMO-OFDM modulation technique. It supports fast speed and best signal range.
- In between above wireless standards there exist various other standards also. Few of them are: IEEE 802.11ac, IEEE 802.11 ad, IEEE 802.11ah, IEEE 802.11aj, IEEE 802.11ax and IEEE 802.11ay.

The below table 1 provide the summary of above information.

RELEVANCE WITH SMART FARMING

Introduction

Smart farming is one of the significant areas where Internet of Things (IOT) along wireless technology plays an important role in their sustainable development. The ultimate goal of relevance with smart farming is to maximize the crop yield with the minimum cost. This is possible once we start getting smart information about various resources such as soil, water, fertilizers and seeds. This can be made possible by deploying various sensors along with some wireless technology. It assists farmers to major crop yield production prediction by monitoring crop life cycle under various climatic conditions. More reliable information for fertilizes, watering etc can improve crop yield (Prathibha et al., 2015; Jarchow et al., 2012). The below section discuss about various wireless agriculture sensor which helps in crop monitoring and enhance crop yield. At the same time the importance of smart farming is discussed. Finally a brief idea of agricultural drones is provided.

Agricultural Sensors

The smart agriculture uses various sensors for their operation (Schribe, n. d.; Andrews & Raja, 2017; Mainetti et al., 2011). Few of them are discussed below:

- **Location based Sensors:** These sensors uses GPS technology to accurately measure the latitude and longitude position.

Table 1. IEEE Wireless Standards Summary

Sr. No.	IEEE Standard	Year	Frequency (GHz)	Bandwidth (Mbps)	MIMO	Modulation
1	802.11	1997	2	2.4	No	DSSS, FHSS and OFDM
2	802.11a	1999	5	54	No	OFDM
3	802.11b	1999	2.4	11	No	DSSS
4	802.11.g	2003	2.4	54	No	OFDM
5	802.11n	2009	2.4/5	300/600	Yes	OFDM

- **Optical Sensors:** These sensors are used to major the various characteristics of soil with the help of light waves. Few of their characteristics are: soil moisture, acidity, clay and organic matter etc.
- **Electrochemical Sensors:** These sensors are used to read the pH and soil nutrient level. At the same time it also reads various chemical composition of soil.
- **Mechanical Sensors:** These sensors are used to major the soil resistance against various attacks.

The above agricultural sensors collect and process data to optimize crop yield. These data can help in different ways to achieve smart farming. Few of them are discussed below:

Wireless Sensor Networks

A wireless sensor network is a little or sometimes infrastructure less. A wireless sensor network consists of many sensor nodes together working for monitoring a target area or place to collect the data packet about the application specific system (Mainetti et al., 2011; Raghavendra et al., 2006; Akyildiz et al., 2002). The wireless sensor networks may be classified into two types:

- **Structure Less:** These sensors have a lot of sensors installed in the node. The sensors deployed in the network may be kept constant without disturbing for the sensing and processing of the data.
- **Structured:** In this the sensor nodes are placed in a pre-defined manner. With full of calculation and by following some placing algorithms these sensor nodes are placed.

Table 2. Smart Farming Data Utilities

Sr. No	Category	Remarks
1	Crop Yield Monitoring	Data related to time, distance and location
2	Crop Yield Mapping	Data related to spatial coordinates from GPS
3	Data Rae Fertilizer	Data related to plant health
4	Data variable Spraying	Data related to spray
5	Data Salinity Mapping	Data related to salinity
6	Data Guidance System	Data related to vehicular position

Researchers had developed and used various devices for detecting and collecting soil and environment condition. They had installed the sensor devices and transmission equipment in appropriate area. These sensor devices have used in wireless sensor network (WSN) (Awasthi et al., 2013; Ruiz-Garcia et al., 2009; Jiber et al., 2011).

Smart farming helps us to achieve following objectives:

- Sensor Based Field and Resource Mapping
- Remote Equipment Monitoring
- Remote crop Monitoring
- Predictive Analytics for crops
- Climate Monitoring and Forecasting
- Livestock Tracking
- Stats on livestock feeding and produce
- Smart Logistics and Warehousing

Agriculture Drones

In recent time the sustainable development in smart faming is the result of the invention of agricultural drones. From last decade the demand of agricultural drones is increasing day by day. As per the statistics available, agriculture is the major sector to incorporate drones. The drones are classified on the basis of their grounded and aerial view. It helps in crop yield assessment, irrigation, spraying, monitoring etc. It also helps us to perform soil and field analysis (Abdullahi et al., 2015; Puri et al., 2017). The major advantages are listed below:

- Crop health imaging
- GIS mapping
- Easy to use
- Save time
- Maximize yield
- Real time data collection and processing etc.

SMART IRRIGATION SYSTEM

Introduction

On an average more than 9 billion gallons of water each day are used largely for landscape irrigation. Out of this almost 50% water is wasted due to overwatering. This is the result of inefficiencies in traditional irrigation methods and systems. The solution to this problem is smart irrigation. Smart irrigation systems are equipped with some sensor based controller which results into the optimized utilization of water resources. These controllers drastically improve the water utilization for irrigation system (Maheshwari, 2016; Maina et al., 2014).

The traditional irrigation controllers works with pre defined programmed schedule. However the smart irrigation controllers monitor weather, soil conditions, evaporation and plant water use to automatically. At the same time it adjusts the watering schedule to actual conditions of the site (Kim et al., 2008). For example, as open-air temperatures increase or rainfall decreases, smart irrigation controllers consider on site-specific variables, such as soil type, sprinklers' application rate, etc. to adjust the watering run times or schedules. There are several options for smart irrigation controllers (Xiao et al., 2010).

Smart Irrigation Controller Types

The smart irrigation controllers are broadly divided into two parts. They are discussed in below section: ("APA,"n. d.; Parmenter et al., 2014; Davis et al. 2014).

Weather-Based Controllers

Also termed as evapo transpiration (ET) controllers. It uses local weather data to adjust irrigation schedules. ET is the combination of evaporation from the soil surface and transpiration by plant materials. These controllers gather local weather information and make irrigation run-time adjustments so the landscape receives the appropriate amount of water.

ET weather data utilizes four weather parameters:

- Temperature
- Wind
- Solar radiation and
- Humidity.

Three forms of weather-based ET controllers are used:

- **Signal-Based Controllers:** It uses meteorological data from a publicly available source and the ET value is calculated for a grass surface at the site. The ET data is then sent to the controller by a wireless connection.
- **Historic ET Controllers:** It uses a pre-programmed water use curve, based on historic water use in different regions. The curve can be adjusted for temperature and solar radiation.

- **On-Site Weather Measurement Controllers:** It uses weather data collected on-site to calculate continuous ET measurements and water accordingly.

Soil Moisture Sensor Based

It is a smart irrigation controller uses one of several well-established technologies to measure soil moisture content. When buried in the root zone of turf, trees or shrubs, the sensors accurately determine the moisture level in the soil and transmit this reading to the controller (Davis & Dukes, 2014).

Two different soil moisture sensor-based systems available:

- **Suspended Cycle Irrigation Systems:** They are set like traditional timer controllers, with watering schedules, start times and duration. The difference is that the system will stop the next scheduled irrigation when there is enough moisture in the soil.
- **Water on Demand Irrigation:** It has a user-set lower and upper threshold, which initiates irrigation when the soil moisture level fails to meet those levels.

Relevance With Smart Farming

The smart irrigation systems and controllers are far better than traditional irrigation controllers. It conserve water across a variety of scenarios. The studies specify that smart irrigation optimize the water utilization from 40% to as high as 70%.

Another study tested a prototype controller/receiver system consisting of a traditional irrigation controller modified to receive a signal broadcasted via satellite. Outdoor water savings were calculated based on 2-years of pre-installation usage and were adjusted for weather conditions. The reported average outdoor savings is 16% and it is also reported this represents 85% of potential savings based on reference ET.

A water efficient irrigation study of the Saving Water Partnership, a coalition of 24 water purveyors, was conducted in Washington State's Puget Sound. Water savings were calculated based on historical consumption and adjustments were made for weather conditions. The reported water savings were 20,735 gallons per year per site for sites with rain sensors controllers and 10,071 gallons per year per site for sites using traditional controllers ("APA," n. d.).

AGRICULTURE AUTOMATION

The farmers uses smart phone as a tool for smart farming. From soil-monitoring sensors to sheep-herding drones, the internet of things has reached the countryside. But so have the bots, and automation is requiring rather fewer humans. "Younger generations don't want to be farmers," says Saverio Romeo, principal analyst at Bath-based Beecham Research and co-author of the report "Towards Smart Farming: Agriculture Embracing the IoT Vision". "Robotics enables agricultural systems where you don't need as much labour, and the technology allows you to optimize the land you have through data." Data is gathered through sources ranging from weather stations to infrared cameras. This data allows farmers to refine the amount of feed their livestock require, or remotely control their connected tractors and combine harvesters. "Yo are moving towards more and more autonomous systems (Yoo et al., 2007; de Lima et al., 2010; Li & Yang, 2012; Jahnavi & Ahamed, 2015; Minbo et al., 2013; dhaka & Vyas, 2014; dadhich et al., 2017).

- **Flying Farm:** It uses SenseFly that makes the eBee Ag - a UAV that flies pre-programmed routes over crops. It transmits live data for analysis.
- **Multispectral Cameras:** It is used to analyze the phenotypes of crops using 3D, infrared and thermal imaging cameras.
- **Connected Vehicles:** It is used to connect tractors to farmers and dealers, Help us to monitor inspections and maintenance closely.
- **Orange Picker:** It uses six cameras to locate and cautiously harvest the ripe fruit.
- **Fake Fish:** It is sensor that can check how factors, such as hydropower facilities, impact real fish.
- **Nutritional Lights:** It is used for indoor farming. It helps to optimize crop growth.
- **Milking Robots:** Milking machine keeps the cows in open space to reduce stress.
- **Super Greens:** A vegetable nutritionally comparable to a mix of kale, broccoli and spinach.
- **Bee Trackers:** It is used to study bee behavior.

CONCLUDING REMARKS

The chapter provides the detail information about the smart farming in terms of sustainable technological development. The study of Internet of Things based wireless sensor networks plays an important role in sustainable development of smart farming. The agricultural drones and automation are also the import aspect of smart farming. The technological enhancement in respect to irrigation, soil takes agriculture to the next level.

REFERENCES

Abdullahi, H. S., Mahieddine, F., & Sheriff, R. E. (2015, July). Technology impact on agricultural productivity: a review of precision agriculture using unmanned aerial vehicles. In *International Conference on Wireless and Satellite Systems* (pp. 388-400). Springer. 10.1007/978-3-319-25479-1_29

Akyildiz, I. F., Su, W., Sankarasubramaniam, Y., & Cayirci, E. (2002). Wireless sensor networks: A survey. *Computer Networks*, *38*(4), 393–422. doi:10.1016/S1389-1286(01)00302-4

Al-Fuqaha, A., Guizani, M., Mohammadi, M., Aledhari, M., & Ayyash, M. (2015). Internet of things: A survey on enabling technologies, protocols, and applications. *IEEE Communications Surveys and Tutorials*, *17*(4), 2347–2376. doi:10.1109/COMST.2015.2444095

Andrews, L. J. B., & Raja, L. (2017). A study on m-health inline with the sensors applying for a real time environment. *Journal of Statistics and Management Systems*, *20*(4), 659–667. doi:10.1080/09720510.2017.1395186

Andrews, L. J. B., & Raja, L. (2017, December). Remote based patient monitoring system using wearable sensors through online and offline mode for android based mobile platforms. In *Infocom Technologies and Unmanned Systems (Trends and Future Directions) (ICTUS), 2017 International Conference on* (pp. 602-606). IEEE. 10.1109/ICTUS.2017.8286080

Atzori, L., Iera, A., & Morabito, G. (2010). The internet of things: A survey. *Computer Networks*, *54*(15), 2787–2805. doi:10.1016/j.comnet.2010.05.010

Awasthi, A., & Reddy, S. R. N. (2013). *Monitoring for precision agriculture using wireless sensor network-a review*. Global Journal of Computer Science and Technology.

Bandyopadhyay, D., & Sen, J. (2011). Internet of things: Applications and challenges in technology and standardization. *Wireless Personal Communications*, *58*(1), 49–69. doi:10.100711277-011-0288-5

Chandra, A., McNamara, K. E., & Dargusch, P. (2018). Climate-smart agriculture: Perspectives and framings. *Climate Policy*, *18*(4), 526–541. doi:10.1080/14693062.2017.1316968

Chen, H., Jia, X., & Li, H. (2011, October). A brief introduction to IoT gateway. In *Communication Technology and Application (ICCTA 2011), IET International Conference on* (pp. 610-613). IET.

Cooklev, T. (2004). *Wireless Communication Standards: A Study of IEEE 802.11, 802.15, 802.16*. IEEE Standards Association.

Dadhich, M., Rathore, V. S., & Vyas, S. (2017, December). Comparison and selection of SaaS service providers using ASMAN framework. In Infocom Technologies and Unmanned Systems (Trends and Future Directions) (ICTUS), 2017 International Conference on (pp. 568-572). IEEE.

Dar, K., Bakhouya, M., Gaber, J., Wack, M., & Lorenz, P. (2010). Wireless communication technologies for ITS applications. *IEEE Communications Magazine*, *48*(5), 156–162. doi:10.1109/MCOM.2010.5458377

Davis, S. L., & Dukes, M. D. (2014). Methodologies for successful implementation of smart irrigation controllers. *Journal of Irrigation and Drainage Engineering*, *141*(3), 04014055. doi:10.1061/(ASCE)IR.1943-4774.0000804

de Lima, G. H., Silva, L. C., & Neto, P. F. (2010, March). WSN as a Tool for Supporting Agriculture in the Precision Irrigation. In *Networking and Services (ICNS), 2010 Sixth International Conference on* (pp. 137-142). IEEE 10.1109/ICNS.2010.26

Dhaka, V. S., Poonia, R. C., & Raja, L. (2013). A literature review on dedicated short range communication for intelligent transport. *International Journal of Advanced Research in Computer Science and Software Engineering*, *3*(9), 1066–1071.

Dhaka, V. S., Poonia, R. C., & Raja, L. (2014). The Realistic Mobility Evaluation of Vehicular Ad-Hoc Network for Indian Automotive Networks. International Journal of Ad hoc. *Sensor & Ubiquitous Computing*, *5*(2), 1. doi:10.5121/ijasuc.2014.5201

Dhaka, V. S., & Vyas, S. (2014). Analysis of Server Performance with Different Techniques of Virtual Databases. *Journal of Emerging Trends in Computing and Information Sciences*, *5*(10).

Dukes, M. D., Shedd, M., & Cardenas-Lailhacar, B. (2009). *Smart Irrigation Controllers: How Do Soil Moisture Sensor (SMS) Irrigation Controllers Work?* University of Florida, Institute of Food and Agricultural Sciences.

Ferreira, D., Corista, P., Gião, J., Ghimire, S., Sarraipa, J., & Jardim-Gonçalves, R. (2017, June). Towards smart agriculture using FIWARE enablers. In *Engineering, Technology and Innovation (ICE/ITMC), 2017 International Conference on* (pp. 1544-1551). IEEE. 10.1109/ICE.2017.8280066

Gu, X., Chai, Y., Liu, Y., Shen, J., Huang, Y., & Nan, Y. (2017). A MCIN-based architecture of smart agriculture. *International Journal of Crowd Science*, *1*(3), 237–248. doi:10.1108/IJCS-08-2017-0017

Gubbi, J., Buyya, R., Marusic, S., & Palaniswami, M. (2013). Internet of Things (IoT): A vision, architectural elements, and future directions. *Future Generation Computer Systems*, *29*(7), 1645–1660. doi:10.1016/j.future.2013.01.010

Guerrini, F. (2015). *The Future of Agriculture? Smart Farming*. Retrieved from https://www.forbes.com/

How robotics is transforming 21st century farming. (n.d.). Retrieved from http://www.wired.co.uk/article/farmville-gets-an-upgrade

Jahnavi, V. S., & Ahamed, S. F. (2015). Smart wireless sensor network for automated greenhouse. *Journal of the Institution of Electronics and Telecommunication Engineers*, *61*(2), 180–185. doi:10.1080/03772063.2014.999834

Jarchow, M. E., Kubiszewski, I., Larsen, G. D., Zdorkowski, G., Costanza, R., Gailans, S. R., & Petrehn, M. R. (2012). The future of agriculture and society in Iowa: Four scenarios. *International Journal of Agricultural Sustainability*, *10*(1), 76–92. doi:10.1080/14735903.2012.646730

Jiber, Y., Harroud, H., & Karmouch, A. (2011, July). Precision agriculture monitoring framework based on WSN. In *Wireless Communications and Mobile Computing Conference (IWCMC), 2011 7th International* (pp. 2015-2020). IEEE. 10.1109/IWCMC.2011.5982844

Khan, R., Khan, S. U., Zaheer, R., & Khan, S. (2012, December). Future internet: the internet of things architecture, possible applications and key challenges. In *Frontiers of Information Technology (FIT), 2012 10th International Conference on* (pp. 257-260). IEEE. Retrieved from https://techbeacon.com/4-stages-iot-architecture

Kim, H. (2015). Wireless Communications Systems Design and Considerations. *Wireless Communications Systems Design*, 349-378.

Kim, Y., Evans, R. G., & Iversen, W. M. (2008). Remote sensing and control of an irrigation system using a distributed wireless sensor network. *IEEE Transactions on Instrumentation and Measurement*, *57*(7), 1379–1387. doi:10.1109/TIM.2008.917198

Lee, J. S., Su, Y. W., & Shen, C. C. (2007, November). A comparative study of wireless protocols: Bluetooth, UWB, ZigBee, and Wi-Fi. In *Industrial Electronics Society, 2007. IECON 2007. 33rd Annual Conference of the IEEE* (pp. 46-51). IEEE.

Maheshwari, B. (2016). Understanding the performance of irrigation systems around homes. *Journal of Environmental Engineering and Landscape Management*, *24*(4), 278–292. doi:10.3846/16486897.2016.1176575

Maina, M. M., Amin, M. S. M., & Yazid, M. A. (2014). Web geographic information system decision support system for irrigation water management: A review. Acta Agriculturae Scandinavica, Section B—Soil &. *Plant Science*, *64*(4), 283–293.

Mainetti, L., Patrono, L., & Vilei, A. (2011, September). Evolution of wireless sensor networks towards the internet of things: A survey. In *Software, Telecommunications and Computer Networks (SoftCOM), 2011 19th International Conference on* (pp. 1-6). IEEE.

Minbo, L., Zhu, Z., & Guangyu, C. (2013). Information service system of agriculture IoT. *Automatika, 54*(4), 415-426.

Miorandi, D., Sicari, S., De Pellegrini, F., & Chlamtac, I. (2012). Internet of things: Vision, applications and research challenges. *Ad Hoc Networks*, *10*(7), 1497–1516. doi:10.1016/j.adhoc.2012.02.016

Parmenter, J., Jensen, A. N., & Chiu, S. (2014, June). Smart irrigation controller. In *Electro/Information Technology (EIT), 2014 IEEE International Conference on* (pp. 394-398). IEEE 10.1109/EIT.2014.6871796

Patil, K. A., & Kale, N. R. (2016, December). A model for smart agriculture using IoT. In *Global Trends in Signal Processing, Information Computing and Communication (ICGTSPICC), 2016 International Conference on* (pp. 543-545). IEEE 10.1109/ICGTSPICC.2016.7955360

Prathibha, S. R., Hongal, A., & Jyothi, M. P. (2017, March). IOT Based Monitoring System in Smart Agriculture. In *Recent Advances in Electronics and Communication Technology (ICRAECT), 2017 International Conference on* (pp. 81-84). IEEE 10.1109/ICRAECT.2017.52

Puri, V., Nayyar, A., & Raja, L. (2017). Agriculture drones: A modern breakthrough in precision agriculture. *Journal of Statistics and Management Systems, 20*(4), 507–518. doi:10.1080/09720510.2017.1395171

Raghavendra, C. S., Sivalingam, K. M., & Znati, T. (Eds.). (2006). *Wireless sensor networks*. Springer.

Rappaport, T. S. (1996). *Wireless communications: principles and practice* (Vol. 2). Prentice Hall.

Report: Global Agriculture Towards 2050. (n.d.). Retrieved from http://www.fao.org/

Report: How to Feed the World in 2050. (n.d.). Retrieved from http://www.fao.org/

Ruiz-Garcia, L., Lunadei, L., Barreiro, P., & Robla, I. (2009). A review of wireless sensor technologies and applications in agriculture and food industry: State of the art and current trends. *Sensors (Basel), 9*(6), 4728–4750. doi:10.339090604728 PMID:22408551

Schribe, S. (n.d.). *Smart Agriculture Sensors: Helping Small Farmers and Positively Impacting Global Issues, Too.* Retrieved from https://www.mouser.in/applications/smart-agriculture-sensors/

Sethi, P., & Sarangi, S. R. (2017). Internet of things: Architectures, protocols, and applications. *Journal of Electrical and Computer Engineering*.

Smart Irrigation. (n.d.). Retrieved from https://www.hydropoint.com/what-is-smart-irrigation/

The essential building blocks of an IoT architecture. (2017). Retrieved from https://mobilebusinessinsights.com/2017/09/the-essential-building-blocks-of-an-iot-architecture/

Tse, D., & Viswanath, P. (2005). *Fundamentals of wireless communication*. Cambridge University Press.

Walter, A., Finger, R., Huber, R., & Buchmann, N. (2017). Opinion: Smart farming is key to developing sustainable agriculture. *Proceedings of the National Academy of Sciences of the United States of America, 114*(24), 6148–6150. doi:10.1073/pnas.1707462114 PMID:28611194

Weber, R. H. (2010). Internet of Things–New security and privacy challenges. *Computer Law & Security Review*, *26*(1), 23–30. doi:10.1016/j.clsr.2009.11.008

Wu, M., Lu, T. J., Ling, F. Y., Sun, J., & Du, H. Y. (2010, August). Research on the architecture of Internet of things. In *Advanced Computer Theory and Engineering (ICACTE), 2010 3rd International Conference on* (Vol. 5, pp. V5-484). IEEE.

Xia, F., Yang, L. T., Wang, L., & Vinel, A. (2012). Internet of things. *International Journal of Communication Systems*, *25*(9), 1101–1102. doi:10.1002/dac.2417

Xiao, K., Xiao, D., & Luo, X. (2010). Smart water-saving irrigation system in precision agriculture based on wireless sensor network. *Nongye Gongcheng Xuebao (Beijing)*, *26*(11), 170–175.

Xie, T., Huang, Z., Chi, Z., & Zhu, T. (2017, April). Minimizing amortized cost of the on-demand irrigation system in smart farms. In *Proceedings of the 3rd International Workshop on Cyber-Physical Systems for Smart Water Networks* (pp. 43-46). ACM 10.1145/3055366.3055370

Yoe, H. (2017, June). Smart farming: ICT based agriculture: Keynote address. In *Software Engineering, Artificial Intelligence, Networking and Parallel/Distributed Computing (SNPD), 2017 18th IEEE/ACIS International Conference on* (pp. 3-3). IEEE.

Yoo, S. E., Kim, J. E., Kim, T., Ahn, S., Sung, J., & Kim, D. (2007, June). A 2 S: automated agriculture system based on WSN. In *Consumer Electronics, 2007. ISCE 2007. IEEE International Symposium on* (pp. 1-5). IEEE

Zhou, Y., Evans, G. H. Jr, Chowdhury, M., Wang, K. C., & Fries, R. (2011). Wireless communication alternatives for intelligent transportation systems: A case study. *Journal of Intelligent Transport Systems*, *15*(3), 147–160. doi:10.1080/15472450.2011.594681

Chapter 2
Wireless Sensor Networks Technologies and Applications for Smart Farming

Dharm Singh Jat
Namibia University of Science and Technology, Namibia

Chucknorris Garikayi Madamombe
Namibia University of Science and Technology, Namibia

ABSTRACT

In recent years, there has been increased use of wireless sensor networks (WSNs) which consists of sensor nodes and the intelligent machine learning algorithms to improve the performance of various applications in smart environments. These smart environments include smart home, smart agriculture, smart office, and smart hospital. In various countries throughout the world, agriculture is regarded as the backbone of the economy. To date, many families in Africa are employed in the agriculture industry. The dawn of information communication technologies (ICT) has changed the conventional way in which farming is conducted. However, quite a large number of farmers in Africa are still stagnant and highly dependent on the traditional ways of farming which started hundreds of years ago. This chapter has presented the current problems in farming and irrigation systems and suggests solutions through proposing the intelligent WSNs in a smart environment.

DOI: 10.4018/978-1-5225-5909-2.ch002

INTRODUCTION

In various countries throughout the world, agriculture is regarded as the main backbone of the economy (Ruby & Jawahar, 2017). Various studies have shown that the main occupation of many families in Africa is agriculture (Ruby & Jawahar, 2017). In most African countries, more than 70% of the land is suitable for agricultural uses especially cultivating of various crops. The most well-known crops that are planted in Africa are maize, sweet potatoes, potatoes, sorghum, wheat, beans, cotton, tomatoes, onions, rice, sugar cane and cereals (Lokesh, Silver & Anuradha, 2017).

In this modern day and age ICT technologies can be used to improve any kind of business in any industry (Spijker, 2014). ICT technologies can be adopted to reduce human intervention. The power of ICT can be used to reduce costs as well as to increase efficiency. According to Spijker (2014), whenever ICT is implemented in any industry, there are significant improvements that are noticed. It is against this background that agriculture is also trying as much as possible to implement ICT technologies so that all the agricultural tasks can be automated. This is important because the costs would be reduced and production would be increased significantly (Verdouw, Beulens & Vorst 2013).

In recent years, farmers are beginning to realise the importance of implementing ICT technologies in their daily operations (Verdouw, Beulens, Reijers & Vorst, 2015). According to Verhoosel, Bekkum, and Verwaart (2016), most farmers are now using latest technologies in order to improve the way in which they do their farming. For instance, the latest technology is now being considered by farmers to switch on and off irrigation systems, monitoring the crop production, monitoring harvest volumes, detecting moisture content of the soil, monitoring the temperature and humidity conditions as well as monitoring the health of the crops and plants.

To date, the technologies that are being implemented by farmers are so intelligent and they are able to provide farmers with the most valuable information that could help them to achieve good harvests. Furthermore, these technologies help farmers in making sound decisions. This type of farming is referred to as smart farming (Vogt, 2013). According to Wamba and Wicks (2010), smart farming is sometimes referred to as precision agriculture. However, Wamba and Wicks (2010) noted that smart farming and precision agriculture are sometimes used interchangeably to mean the same but they are actually different.

Precision agriculture mainly focuses on decision support system (DSS) for the entire farming system (Welte, Ault, Bowman, Ellis, Buckmaster, Ess & Krogmeier, 2013). On the other hand, smart farming mainly focus on the monitoring of the agricultural fields as well as performing some intelligent operations (Wigboldus, Klerkx, Leeuwis, Schut, Muilerman & Jochemsen, 2016). Therefore, smart farming generally focuses on the monitoring of the outside environment whilst precision agriculture focuses on the data analysis so that the farmers are informed with valuable information in a real-time manner (Williamson, 2016). Various studies have concluded that smart farming technologies joined with precision agriculture would allow farming to be done smoothly and reduce costs significantly (Wolfert, Sørensen & Goense, 2014; Yan, Shi & Huang, 2013; Yang, 2014). Furthermore, crop yields as well as production is significantly increased (Young, 2016).

The advent of ICT technologies has changed the traditional way in which farming is conducted throughout the whole world (Lokesh, Silver & Anuradha, 2017). However, quite a large number of farmers in Africa are still stagnant and highly dependent on the traditional ways of farming which started hundreds of years ago (Zhu, Dai & Huang, 2012). As a result, most crops are not yielding optimum produce that should be harvested if farmers adopt the new ways of farming (Zong, Fares, Romoser & Wood, 2014). In addition, the low yield is also as a result of lack of proper soil management, moisture monitoring, crop protection and adequate resources by farmers (Gutiérrez, Villa-Medina & Porta-Gándara, 2014). It has been noted that most farmers experience great losses due to improper machinery for irrigation, crop pests, plant diseases, incorrect use of pesticides and insecticides and failure to predict accurate weather conditions (Tong, Hong & JingHua, 2015). In order to get a bumper harvest, farmers are required to monitor their crops on a regular basis (Van Rijmenam, 2015). As a result of vast limitations that affect agriculture in the present day, there is a need to develop systems that help the growing of crops and yield better results (Luo, Yang, Li & Xu, 2016).

In the past years, the irrigation systems that have been used by farmers were using normal timers and switching mechanism (Noyes, 2014). These timers and switches were used to trigger irrigation systems at preset time periods without considering weather conditions or the moisture present in the soil (Noyes, 2014). In irrigation systems which are equipped with intelligent sensors, water will only be turned on from the irrigation system when the moisture level in the soil has gone lower than the required (Ruby & Jawahar, 2017). This technique saves a lot of water and also help crops to grow better since the required soil moisture levels is always monitored and maintained (Nuthall, 2017). As a result of including various innovative sensing and controlling mechanisms, the crop yield has been increased. At the same time, the cost of labour has been significantly reduced.

Nevertheless, the main disadvantage of these mechanisms is that they are expensive and also they are difficult to design (Osterwalder, 2014). Consequently, there is a need for alternative technologies which are affordable and yet perform complex operations. Therefore, wireless technologies and automation in the agriculture industry are needed. In the past, wireless technologies such as remote sensors, a global positioning system (GPS) and geographical information system (GIS) have been used (Plume, 2014). Nikesh and Kawitkar (2016) noted that whenever automation of agricultural tasks is implemented, efficiency increases and subsequently the crop yield is improved significantly. According to Kavita and Chaudhari (2016), a Wireless Sensor Network is defined as numerous sensor devices that are intersected together to monitor various environmental conditions. Poppe, Wolfert, Verdouw and Renwick (2015) noted that these WSNs are powerful technologies that can automate agricultural tasks at a relatively lower cost and lower power consumption. In addition, these WSNs can also be used to monitor crops through the available sensor nodes which are placed in the agricultural field (Karan, 2015). The data about the crops can also be sent automatically via the wireless communication system.

The Need for IOT in Agriculture

According to a study which was conducted by United Nations (UN) and the Food and Agriculture Organisations (FAO) (2016), the food which is being generated from the agriculture should be increased by approximately 70% by 2050 in order to feed the growing population worldwide. This means that worldwide, the population is increasing at a faster rate as compared to the rate at which farming is growing. If this continues, it is a major challenge as there might be lack of food to feed people around the globe. As indicated in the previous sections, agriculture is the main source of food worldwide and also most economies of countries are dependent on agricultural productions (Ritaban, Ahsan, Jagannath, D'Este & Aruneema, 2014). Furthermore, most people in many countries worldwide are employed in the agriculture industry (Royse, 2014).

In order to increase productivity in the agriculture industry, there is need for farmers to adopt new technologies (Semantic Community, 2015). Otherwise, food supply will be scarce in the future. As a result of using IoT, agricultural productivity is expected to increase significantly because most of the activities would be automated and costs will also be reduced. Using the traditional way of farming coupled by implementing IoT has resulted in smart farming (2014).

The dawn of modern technologies has made farming activities easier to perform. Thus the need for smart farming in this day and age is inevitable. In recent times, Internet of Things (IoT) has transformed the way in which business is done in different industries across the world. IoT consists of voluminous things cooperating together to produce useful information (Atzori and Morabito, 2010). The IoT has allowed farmers to connect various devices on the internet resulting in better crop yield, a significant reduction of wastage of resources and better crop management (Remya, 2016).

This chapter has examined the current problems faced in the irrigation systems and provides better solutions through proposing the intelligent WSNs. In this study, the intelligent sensor-based automatic irrigation system with IoT is proposed.

Karan (2015) proposed the sensor based irrigation system in which the sensors are used to automatically send the signals to the microcontroller by collecting various environmental conditions such as temperature and humidity conditions. The microcontroller and the GSM were connected together.

This chapter is structured as follows. Section 2 examines the related works in the field of smart irrigation and farming. Section 3 explains the proposed materials and methods of the proposed conceptual framework. Section 4 concluded the chapter.

RELATED WORKS

Generally, the moisture sensor is designed to detect the moistness and humidity conditions and then send the signals to the mobile phone through the microcontroller. This approach reduces power consumption thereby saving much energy. The agricultural control system based on Arduino proposed by Narayut, Sasimanee and Anupong (2016) is an intelligent farming system that improves productivity in the farming system. This technology comprises of two major parts which are control system and intelligent sensor system. Some sort of filtering is generally used to smoothen the sensed data in scenarios were the weather conditions are forecasted based on a decision tree.

Nikesh and Kawitkar (2016) proposed a research study on IoT which was grounded in smart farming. The paper mainly focused on automating agriculture activities through the various state of the art technologies. In their proposed study, the smart remote controlled robot which is based on the GPS was used to do the operations such as sensing moisture, automatic irrigation, spraying insects and weeding. The automatic irrigation system was equipped with intelligent sensors that could make best decisions based on the actual conditions of the soil. The system also monitors temperature, humidity of the soil as well as theft detection in the agricultural fields. In this smart architecture, all the sensors were integrated together using a raspberry pi as well as the wireless communication module. This paper by Nikesh and Kawitkar (2016) examined all the agricultural information that occurs in the fields such as various field activities, problems in the irrigation system as well as storage problems.

Rajalakshmi and Devi (2016) proposed a study on IoT based crop monitoring and automation of the irrigation system. The paper outlined the methods to monitor soil moisture, temperature, humidity as well as the light intensity in the agricultural fields. Furthermore, the paper outlined the methods to implement a successful sensor-based irrigation system. The paper proposed the methods in which the data collected by the sensors was transmitted to the web server via a wireless communication. The JSON format was used to encode the data and to maintain the database servers. In their proposed model, the moisture and temperature that falls below the minimum required will trigger the irrigation system. The information about the field conditions would be sent periodically to the farmers so that they are able to monitor whatever is happening in the fields. In their paper, the variables used were soil moisture, humidity sensor as well as a temperature sensor. This smart agricultural technology is more convenient in areas with little rainfalls. MySQL database was used to store the data used for the automation of irrigation system.

Tanmay and Pushpendra (2016) proposed smart monitoring agricultural systems. The study focused on the protection of agricultural products from being stolen as well as protection of crops from attacks by rodents and insects. The study proposed the real-time notification of problems to farmers after sensing a specific problem. This means that using such an approach, farmers are always updated of any problem that would be happening in the agricultural fields. In this study, Python programming language was used to design the sensors. The proposed model was tested on an area of 10 square meters. After successful tests, it was found out that the proposed technology may help significantly to enhance security and protect agricultural products as well as to prevent insects from attacking the grains in the storage places.

Nelson and Artur (2015) proposed a wireless sensor based system for smart agriculture. The paper examined technologies in WSNs. In their study, the network was used to perform three main tasks as follows; collecting data, presentation as well as analysis of data. The main variables that were used in this mechanism were temperature and soil moisture. The main benefits of automating the irrigation process are that it reduces wastage of water and reduces labour costs as well.

Mohamed and Ibrahim (2016) proposed a technology called Precision Agriculture (PA). Mohamed and Ibrahim (2016) noted that the WSNs is the most successful way in solving agricultural problems such as optimization of farming resources, decision making support, land monitoring as well as monitoring of various environments. The major advantage of this approach is that it provides information to farmers in a real-time manner so that the farmers can be able to make swift decisions. The precision agriculture systems which are based on the IoT technology mainly examines the hardware, software and network architectures.

Liu (2016) proposed a research paper on greenhouse technology in agriculture. The paper outlined the design and the implementation based on the ZigBee technology using the CC2530 chip. This paper mainly focused on the monitoring of the environment. In their paper, the proposed system provided all the real-time data about agriculture and it used the wireless communication like temperature control and humidity conditions. This is an intelligent monitoring system that is used to control the greenhouse. It is very helpful to provide farmers with optimum information that could help them to improve the farming activities.

Critical Analysis of Related Works

Past related works have shown that the advent of modern technologies has made farming activities easier to perform. Thus the need for smart farming in this day and age is unavoidable. In recent years, Internet of Things has transformed the way in which business is done in different industries across the world. IoT consists of voluminous things cooperating together to produce useful information. The IoT has allowed farmers to connect various devices on the internet resulting in better crop yield, a significant reduction of wastage of resources and better crop management.

METHODOLOGY

In the proposed conceptual smart agriculture system, the irrigation system architecture will be made of the hardware output which is linked to the software. The proposed concept for the smart irrigation system will notify the farmers about the conditions of the agricultural field. The proposed system will work with the following hardware and software; intelligent sensors, Arduino, Wi-Fi module and the GSM module. The overall system is as shown in figure 1 below.

Temperature and Humidity Sensor

The DHT11 sensor will be the temperature and humidity sensor. The sensor will be used to measure the air in the surroundings. The system will also be equipped with the capacitive humidity sensor and thermistor sensor which will be used to output signals on the data pin. The temperature measuring device in the sensor will be used to measure the temperature conditions.

Figure 1. The system overview

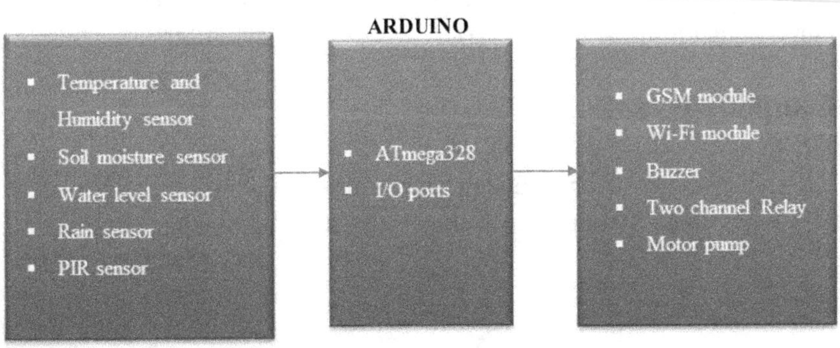

Soil Moisture Sensor

The volume of water in the soil will be measured by the soil moisture sensor. The volumetric water content in the soil is measured using soil moisture sensor. This sensor will consist of the variable resistor which will be used to monitor the level of moisture in the soil and thereby controlling the irrigation system.

Level Sensor

These sensors will be used to measure and detect the level of the fluids in the soil. These type of sensors need to contact with the fluids for them to work. They are comprised of components which are fixed and thus they require less maintenance. Normally, these sensors are fixed on the walls of the tank and the field as well.

Rain Sensor

This sensor is mainly a switching device which is fixed to the smart irrigation system. This sensor is designed to send the alerts to the system with a sound once it has reached the predetermined threshold values during the rainfall.

PIR Sensor

This PIR sensor is a movement based locator sensor which will be used to detect the movement of people and any physical creatures. These are cheap and affordable sensors. These sensors make use of infrared radiation technology in their operations.

Arduino

This is a prototyping platform and it is also an open-source which is a microcontroller board based on ATmega328. The Arduino will be pre-loaded with the software so that it will perform as the IDE. This software is highly portable and it has low power consumption and it has a lot of libraries.

GSM sim900A

This is a mobile communication modem which makes it possible for communication to be done via mobile cellphones. It also allows for data to be transmitted via cellphones.

Proposed Conceptual System Overview

The figure 2 below illustrates how the sensors and the various devices will be connected to the Arduino. Figure 2 is an actual representation of all the devices to make one system. In the proposed system, A Tmega328 microcontroller would be used. The voltage output is transmitted from the sensors as inputs into the Arduino module. The Arduino system will then turn ON or OFF the irrigation system based on the input values that are supplied.

There will be two main functions that will be performed by the suggested system represented in figure 2. These two functions are gardening purposes as well as agricultural purposes. As shown in figure 2, the main backbone of this system is the Arduino. The Arduino board is indicated in the block diagram. The system will make use of intelligent sensors and as a result, the system will also make intelligent decisions. This means that the water will not pump out if there is still enough moisture and water content in the soil. This approach is used to solve wastage of water. In the previous irrigation systems, the water will be pumped out at predefined time intervals even if there is enough water and moisture in the soil.

Figure 2. Representation of the block diagram of the system

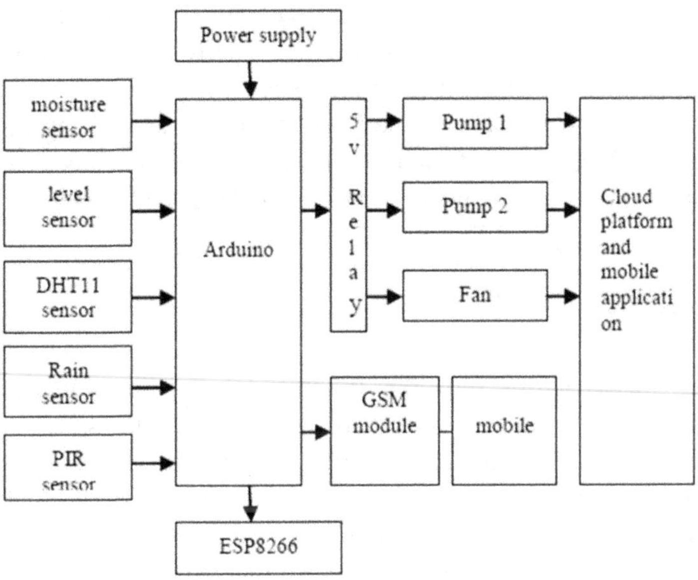

The traditional irrigation systems are usually set to pump out water in the morning and also in the evening. Using the proposed approach, water will be pumped out into the fields whenever it is necessary to do so. This means that during dry and hot conditions, the water will be pumped into the fields more often as compared to during rainy seasons when there is supplementary water from the rainfall. In short, the system will only open the pumps for irrigation when there is not sufficient water in the soil. On the other hand, whenever the water moisture level in the soil decreases significantly, the water will then be pumped out into the fields until such point when the required moisture levels are reached.

As discussed in the previous sections, the moisture sensor is the one which is used to measure the moisture levels. The intelligent sensors will also sense the rainfall. When there is rainfall, the moisture level is very high and the water will be pumped back into the tanks. This water will then be reserved so that it can be used at a later stage. This is very good because water will not be wasted as it is recycled.

This PIR sensor is a movement based locator sensor which will be used to detect the movement of people and any physical creatures. These are cheap and affordable sensors. These sensors make use of infrared radiation technology in their operations. The DHT11 sensor will be the temperature and humidity sensor. The humidity and temperature sensor will be used to measure the air in the surroundings. The system will also be equipped with the capacitive humidity sensor and thermistor sensor which will be used to output signals on the data pin. The temperature measuring device in the sensor will be used to measure the temperature conditions. The moisture sensors will be used to measure and detect the level of the fluids in the soil. These type of sensors need to contact with the fluids for them to work. They are comprised of components which are fixed and thus they require less maintenance. Normally, these sensors are fixed on the walls of the tank and the field as well.

CONCLUSION

The use of Wireless Sensor Networks (WSN) and the intelligent machine learning algorithms together improve the performance of agriculture in a smart environment. This study discussed the significant works in the literature of wireless sensor networks for agriculture in smart environments. The study also suggested conceptual system will concentrate on the advanced ways of improving agriculture and convert conventional irrigation systems to smart irrigation systems through the use of Internet of Things (IoT). The main advantages of the proposed conceptual system are a reduction in wastage of water, reduction in humanitarian efforts and labour improved monitoring of agricultural products by farmers using the sensor networks and GSM. Furthermore, the proposed system can also provide a pathway to reduce theft of crop products by humans. It has been noted that these WSNs are powerful technologies that can automate agricultural tasks at a relatively low cost and low power consumption. In addition, these WSNs can also be used to monitor crops through the available sensor nodes which are placed in the agricultural field. The data about the crops can also be sent automatically via the wireless communication system.

REFERENCES

Atzori, A. I., & Morabito, G. (2010). The Internet of Things: A survey. *Computer Networks*, *54*(15), 2787–2805. doi:10.1016/j.comnet.2010.05.010

Gutiérrez, J. F., Villa-Medina, A., & Porta-Gándara, M. A. (2014). Automated irrigation system using a wireless sensor network and GPRS Module. *IEEE Transactions on Instrumentation and Measurement*, *63*(1), 166–176. doi:10.1109/TIM.2013.2276487

Karan, K. (2015). Sensor-based automated irrigation system with IoT: A technical review. *International Journal of Computer Science and Information Technologies*, *6*, 520–524.

Kavita, B., & Chaudhari, D. (2016). Solar powered sensor base irrigation system. *Internationnal Research Journal of Engineering and Technology*, *3*, 678–682.

Liu, D. (2016). Wireless Sensor Network in Precision agriculture application. *International conference on computer, Information and telecommunication systems (CITS)*, *8*, 345-350.

Lokesh, K. K., Silver, O., Malende, W. F., & Anuradha, K. (2017). Internet of things application for implementation of smart agriculture system. *International conference on I-SMAC (IoT in Social, Mobile, Analytics and Cloud)*, 54-59.

Luo, H., Yang, P. L., Li, Y. K., & Xu, F. P. (2016). An intelligent controlling system for greenhouse environment based on the architecture of the internet of things. *Sensor Letters*, *10*(1), 514–522. doi:10.1166l.2012.1850

Miller, H. G., & Mork, P. (2013). From data to decisions: A value chain for Big Data. *IT Professional*, *15*, 57–59.

Mohamed, R., K., & Ibrahim, M. (2016). *Wireless Sensor Network in Precision Agriculture Application*. Academic Press.

Narayut, P., Sasimanee, P., & Anupong, C. (2016). A control system in an intelligent farming using Arduino technology. *International Student Project Conference ICT-ISPC, 7*, 345-349.

Natale, F., Gibin, M., Alessandrini, A., Vespe, M., & Paulrud, A. (2015). Mapping Agriculture effort through AIS data. *PLoS One*, 10.

NEC. (2014). *NEC and Dacom collaborate on precision farming solution to maximize yields and reduce costs*. Retrieved from https://www.nec-enterprise.com/news/Latest-press-releases/NECand-Dacom-collaborate-on-precision-farming-solution-to-maximize-yields-andreduce-costs-791

Nelson, S., & Artur, A. (2015). Wireless sensor and actuator system for smart irrigation on the cloud. *2nd World forum on Internet of Things (WF-IoT)*, 150-155.

Nikesh, G., & Kawitkar, R. S. (2016). IoT based smart agriculture. *International Journal of Advanced Research Computer and Communication Engineering, 5*, 450–459.

Noyes, K. (2014). Big Data Poised to Change the Face of Agriculture. *Fortune Data*. Retrieved from http:// fortune.com/2014/05/30/cropping-up-on-every-farm-big-data-technology/

Nuthall, P. L. (2017). *Farm Business Management: Analysis of Farming Systems*. CABI.

Omta, S., Trienekens, J., & Beers, G. (2001). Chain and network science: A research framework. *Journal on Chain and Network Science, 1*, 1–6.

Orts, E., & Spigonardo, J. (2014). *Sustainability in the Age of Big Data. IGEL/Wharton*. Pennsylvania: University of Pennsylvania.

Osterwalder, A. (2014). *The Business Model Ontology: A Proposition in a Design Science Approach, l'Ecole des Hautes Etudes Commerciales*. Université de Lausanne.

Plume, K. (2014). The Big Data Bounty: U.S. Startups Challenge Agribusiness Giants. *CloudTweaks*. Retrieved from http://www.reuters.com/article/us-usa-farming-startupsidUSKCN0HX0C620141008

Poppe, K. J., Wolfert, J., Verdouw, C. N., & Renwick, A. (2015). A European perspective on the economics of Big Data. *Farm Policy Journal, 12*, 11–19.

Porter, M. E. (1985). *Competitive Advantage: Creating and Sustaining Superior Performance. 1985*. New York: Free Press.

Porter, M.E., & Heppelmann, J.E. (2014, November). How smart, connected products are transforming competition. *Harv. Bus. Rev.,* 65–88.

Provan, K. G., & Kenis, P. (2008). Modes of network governance: Structure, management, and effectiveness. *Journal of Public Administration: Research and Theory*, *18*(2), 229–252. doi:10.1093/jopart/mum015

Rajalakshmi, P., & Devi, M. (2016). IOT based crop-field monitoring and irrigation automation. *10th International conference on intelligent systems and control (ISCO)*, *7*, 89 – 95. 10.1109/ISCO.2016.7726900

Remya, K. (2016). Greenhouse monitoring and control based on IoT using WSN. *ITSI-TEEE*, *4*, 2016–2020.

Ritaban, D., Ahsan, M., Jagannath, A., D'Este, C., & Aruneema, D. (2014). Development of an intelligent environmental knowledge system for sustainable agricultural decision support. *Environmental Modelling & Software*, *52*, 264–272. doi:10.1016/j.envsoft.2013.10.004

Royse, R. (2014). The Growing Field of Agriculture Technology. *Vator News*. Retrieved from http://vator.tv/ news/2014-05-14-the-growing-field-of-agriculture-technology

Ruby, R. A., & Jawahar, A. (2017). Smart agro system using wireless sensor networks. *International Conference on Intelligent Computing and Control Systems*, 400-403.

Semantic Community. (2015). *Big Data Science for Precision Farming Business*. Retrieved from http://semanticcommunity.info/Data_Science/Big_Data_Science_for_Precision_Farming_Business

Sonka, S. (2014). Big Data and the Ag sector: More than lots of numbers. *The International Food and Agribusiness Management Review*, *17*, 1.

Sonka, S. (2015). Big Data: From hype to agricultural tool. *Farm Policy Journal*, *12*, 1–9.

Spijker, A. (2014). *The New Oil - Using Innovative Business Models to Turn Data into Profit*. Basking Ridge: Technics Publications.

Sun, J., Zhou, Z., Bu, Y., Zhuo, J., Chen, Y., & Li, D. (2013a). Research and development for potted flowers automated grading system based on internet of things. *Shenyang Nongye Daxue Xuebao*, *44*, 687–691.

Sun, Z., Du, K., Zheng, F., & Yin, S. (2013b). Perspectives of research and application of Big Data on smart agriculture. *J. Agric. Sci. Technol.*, *15*, 63–71.

Sundmaeker, H., Verdouw, C., Wolfert, S., & Pérez Freire, L. (2016). Internet of food and farm 2020. In O. Vermesan & P. Friess (Eds.), *Digitising the Industry. Internet of Things Connecting Phy*.

Tanmay, B., & Pushpendra, K. P. (2016). Development of IoT based Smart Security and Monitoring Devices for Agriculture. *6th International Conference - Cloud System and Big Data Engineering*, 978 – 982.

Tien, J. M. (2013). Big Data: Unleashing information. *Journal of Systems Science and Systems Engineering*, *22*, 127–151.

Tong, L., & Hong, T., & JingHua, Z. (2015). Research on the Big Data-based government decision and public information service model of food safety and nutrition industry. *Shipin Anquan Zhiliang Jiance Xuebao*, *6*, 366–371.

UNECE. (2013). *Classification of types of Big Data*. Retrieved from http://www1.unece.org/stat/platform/ display/bigdata/Classification+of+Types+of+Big+Data

Van der Vorst, J., & Beulens, B. (2005). Innovations in logistics and ICT in food supply chain networks. In W. M. F. Jongen & M. T. G. Meulenberg (Eds.), *Innovations in Agri-food Systems: Product Quality and Consumer Acceptance. Wageningen Academic Publishers.*

Van Rijmenam, M. (2015). *John Deere is revolutionizing farming with Big Data*. Retrieved from https:// datafloq.com/read/john-deere-revolutionizing-farming-big-data/511

Verdouw, C. N., Beulens, A. J. M., Reijers, H. A., & van der Vorst, J. G. A. J. (2015). A control model for object virtualization in supply chain management. *Computers in Industry*, *68*, 116–131. doi:10.1016/j.compind.2014.12.011

Verdouw, C. N., Beulens, A. J. M., & van der Vorst, J. G. A. J. (2013). Virtualisation of floricultural supply chains: A review from an internet of things perspective. *Computers and Electronics in Agriculture*, *99*, 160–175. doi:10.1016/j.compag.2013.09.006

Verhoosel, J., van Bekkum, M., Verwaart, T. (2016). HortiCube: a platform for transparent, trusted data sharing in the food supply chain. *Proceedings in Food System Dynamics*, 384–388.

Vogt, W. (2013). Looking at Big Data One Plant at a Time. *Farm Industry News*. Retrieved from http:// farmindustrynews.com/blog/looking-big-data-one-plant-time

Wamba, S. F., & Wicks, A. (2010). RFID deployment and use in the dairy value chain: applications, current issues and future research directions. *Technology and Society (ISTAS), 2010 IEEE International Symposium on. IEEE,* 172–179.

Welte, J. T., Ault, A. C., Bowman, C., Ellis, S., Buckmaster, D. R., Ess, D., & Krogmeier, J. V. (2013). An Approach to Farm Management Information Systems Using Task-Specific, Collaborative Mobile Apps and Cloud Storage Services. *American Society of Agricultural and Biological Engineers*, 1.

Wigboldus, S., Klerkx, L., Leeuwis, C., Schut, M., Muilerman, S., & Jochemsen, H. (2016). Systemic perspectives on scaling agricultural innovations. A review. *Agronomy for Sustainable Development*, *36*(3), 1–20. doi:10.100713593-016-0380-z

Wikipedia. (2016). *Big Data*. Retrieved from https://en.wikipedia.org/wiki/Big_data

Williamson, O. E. (2016). *The Mechanisms of Governance*. Oxford University Press.

Wolfert, J., Sørensen, C. G., & Goense, D. (2014). A Future Internet Collaboration Platform for Safe and Healthy Food from Farm to Fork. *Global Conference (SRII), 2014 Annual SRII,* 266–273.

Yan, B., Shi, P., & Huang, G. (2013). Development of traceability system of aquatic foods supply chain based on RFID and EPC internet of things. *Nongye Gongcheng Xuebao (Beijing)*, *29*, 172–183.

Yang, C. (2014). Big Data and its potential applications on agricultural production. *Zuowu, Huanjing Yu Shengwu Zixun*, *11*, 51–56.

Young, L. (2016). *Farmobile takes ag data to the next level.* Retrieved from http://www.agrinews.com/news/minnesota_news/farmobile-takes-ag-data-to-the-next-level/article_5074de27-ecdf-5dd3-a89e-e3b28815475c.html

Zhu, W., Dai, C., & Huang, P. (2012). Environmental control system based on IOT for nursery pig house. *Nongye Gongcheng Xuebao (Beijing)*, *28*, 177–182.

Zong, Z. L., Fares, R., Romoser, B., & Wood, J. (2014). FastStor: Improving the performance of a large scale hybrid storage system via caching and prefetching. *Cluster Computing*, *17*, 593–604.

Chapter 3
Sustainable Development in Agriculture:
Past and Present Scenario of Indian Agriculture

Vaibhav Bhatnagar
Amity University Jaipur, India

Ramesh C. Poonia
Amity University Jaipur, India

ABSTRACT

Agriculture is the prime source of livelihood for human beings, animals, and all living beings. Agriculture also plays a vital role in the economy of India. This chapter describes the importance of agriculture and factors affecting the development of agriculture. The international scenario of agriculture, current status of Indian agriculture, and position of Rajasthan (state in India) in agriculture are described in this chapter. The total production, total imports and exports, method of irrigation, net area of irrigation, types of crops, fertilization consumption, and highlight of Union Budget 2018-19 of Indian agriculture are described in this chapter. The geography of Rajasthan according to agriculture, production of crop, and consumption of fertilization are also elaborated in this chapter. This chapter is concluded with future perspectives of India agriculture.

DOI: 10.4018/978-1-5225-5909-2.ch003

INTRODUCTION

Agriculture is a science as well as an occupation which is mainly concerned with plough the field, growing seeds, irrigation of the crop, fertilizing, harvesting and livestock for food and commercial purpose. Agriculture is not only limited to cultivating but also extended to bee farming, poultry and dairy. Agriculture industry plays a vital role in economy of a country (Herrera, 1999). Agriculture also provides a large number of employment opportunities.

IMPORTANCE OF AGRICULTURE

- **Prime source of Livelihood:** Agriculture is the main source of livelihood in most of the countries even if they are developed countries. Agriculture contributes around 70% in an economy, which is a heavy component indeed. India is the country of agriculture, in which agriculture products are directory used at farmer's kitchen rather than in the market.
- **Generate National Income:** Agriculture also use to generate income at national level. In India, many commercial commodities like tobacco, tea and spices etc. are exported which generate heavy national income.
- **Generate Employment:** Agriculture provides large number of employment. In India, there are so many peoples who do not have their own land for agriculture, but they are working on someone's land and they are maintaining their family well.
- **Maintain Ecological Balance:** Greenhouse effect and global farming are the well-known international problems. Plantation is only the solution of these problems. Orchids and gardens not only prevent Greenhouse Effect and Global Farming but also enhance the beauty of the earth.

FACTORS AFFECT AGRICULTURE

As it is seen above, agriculture is very important for livelihood and to maintain economy of a country. There are so many factors that affect the growth of agriculture. Some factors are chemical & Biological in nature whereas some factors are technical and depend upon market demand (Factors Affect Irrigation, 2018). Factors affect agriculture is depicted in Figure 1.

Figure 1. Factors affect Agriculture

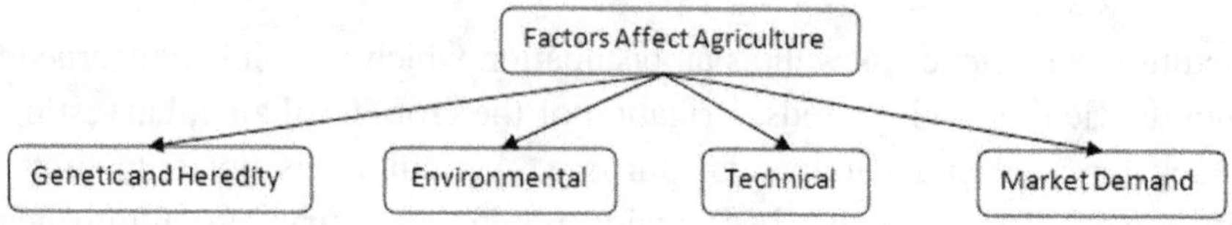

Genetic and Heredity

There are so many genetic and heredity factors that affect the growth of agriculture like chemical compositions, tolerance of disease, early maturity, quality of straw and high yield ability. For example, wheat has several varieties like H1 1531, DBW 16, HD 2888, NIAW 917 and etc. H1 1531 yields 25-27 Q/H and DBW 16 yields 37-39 Q/H. So, if a farmer yields DBW 16, he will yield more amount of wheat.

Environmental

The second category is environmental issues, in which there are large number of factors that affect a crop. All factors are listed below:

- **Climate:** Around 50% yield is influenced by different climate factors. So many sub factors affect the crop production.
- **Temperature:** Crop production heavily depends upon environmental temperature. Every geographical place has its own temperature, which varies time to time even in a single day. Moreover, every crop needs its favorable temperature for production. If a crop say cotton require 20-28 ·C for production, and if farmers are growing crop in Rajasthan, then it is must require to be give additional water to crop otherwise crop will get collapse.
- **Relative Humidity:** Amount of water present in the form of vapors in the air is called as humidity. Relative humidity is defined as ratio between moisture present to the saturation capacity of the air at a specific temperature. It affects the crop water need. If there is high relative humidity at specific temperature then a crop required less water for irrigation. If relative humidity is always high, then there is chance of disease and pest outbreak.
- **Solar Radiations:** Solar radiations are the most vital components responsible for crop production. They are required in the process of photosynthesis. Photosynthetically Active Radiations are responsible for production of biomass.

- **Wind Velocity:** Velocity of wind indirectly affects the growth of crop. They are responsible for soil erosion, increase evaporation, dispersal of pollens and spreading out the pests and diseases. Around 4-6 km/h velocity of wind is suitable for crop yielding, otherwise they can also harm the crop by removing the leaves and flowers.
- **Precipitation:** Precipitation is nothing but the water that fall from natural phenomenon like rainfall, snow and fog. Heavy precipitation may collapse the crop completely, whereas low precipitation required heavy amount of irrigation. In India, crop like *mustard* grows out with precipitation only, they do not require irrigation additionally.

Edaphic Factor

Edaphic factors are those factors that are influenced by soil. Nobody cannot imagine crop production without soil. Edaphic factors are the backbone of crop production. Soil moistures, air, temperature, organism and reaction are the factors that affect the crop production. Fertilizers are applied based on deficiency of nitrogen, potassium and phosphorus present in the soil.

- **Biotic Factors:** Biological organism also affects the growth of a crop. When two crops are yielded simultaneously, high amount of production can be expected. For example, Legumes and Cereals are yielded together and beneficial for are high production because of synergetic effects. Some animals also affect the production of crops like protozoa help in decomposition of organic matter. Honey bees helps in pollination which leads to high production.
- **Physiographic Factors:** Some Physiographic and geographical affect crop indirectly like increase in altitude leads to increase in precipitation but decrease in temperature. A sloppy land is harmful for production because rainfall water always ran out and it carries the nutrition with it.

Technical

Technology is a factor that affects all the systems. Agriculture is also no longer remains untouched with it. Change in technology heavily contributed the high yield production in agriculture. Change in Technology is depicted in Figure 2.

As it is seen above, technology was the backbone of Green Revolution. Figure 2, also depict the same thing. From the Plow to GPS based tractor and from vegetable hawker to mobile-based vegetable selling application, technology has changed the

Figure 2. Change in Technologies in Agriculture

economic status of the farmer. Technology does not only refer the developing the high yield seeds but also developing the chemical fertilizer, using ICT (Information and Communication Technology) and using advance mechanism for irrigation. Precision Agriculture is emerging field of agriculture research that is helping the farmer in real time monitoring and decision support system.

Market Demand and Government Policy

Market demand also influences the production of crops, especially those crops which are exported outside the country. In India, some crops like saffron, tea, tobacco are heavily yielded because these crops give good economic profit to the farmer. Some farmers do not cultivate the routine crops but cultivate that crop which is demanded in the market or there is scarcity in the market. Some farmers always yield coriander in spare when it is demanded high. Some government policies also influence the crop, even at the time of British rule, Indian farmers are forced to yield the Neel. Nowadays, Indian government is doing well in favor of farmers. Abolition of intermediaries, providing subsidies and loan exemptions are some policies which motivate the farmers and indirectly increase the production of a crop.

International Scenario of Agriculture

Agriculture is the only reason for civilization. Before doing agriculture human being were called as nomadic. It is the agriculture that made human being independent and they started to live in a civilized manner. Agriculture was also a medium of exchange in ancient time like 1 kg rice can be exchanged with ½ kg corn (known as barter system). At this present era, not even a single country remains untouched with agriculture. Farming conditions of developing countries have a vital role to play in addressing food needs, combatting with poverty, safeguarding water systems, handling climate change, protecting biodiversity, and many more crucial issues, however agriculture itself is a vibrant and complex sector, having huge diversity. Therefore analysts and practitioners working on these big pictures and global issues can all benefit from a basic understanding of what the global distribution of crops actually looks like. Countries like India, Pakistan and Ethiopia heavily depended upon agriculture. Agriculture contributes heavy component in Gross Domestic Product (GDP). In Table 1 and Figure 3, here depicting contribution of components for GDP of some countries.

It is depicted in the table 1 (Countries with their GDP Components, 2018) that followed by Pakistan and Nigeria, India is the third country in which agriculture contributes in the Gross Domestic Product. It is noticed here that these three countries are developing countries, developed countries like United State, United Kingdom are more depended upon service sector.

National Scenario of Agriculture

At present scenario our country holds second position over worldwide in farming output. Agriculture is the largest economic sector that plays a vital role in socio-economic condition of India. In the year 2013, our country became the seventh largest agriculture exporter over the globe. Country exports to around 120 countries including Southeast Asia, United States, Middle East and SAARC countries. As per the report of FAO 2010, India is the largest producers of fresh vegetables, fruits, spices, jute and castor oil. India is also the second largest wheat and rice producer.

History of Indian Agriculture

(Suryawanshi, 2011), the history of Indian Agriculture has begins with Indus valley civilization. Indian agriculture started around 9000 BC as a result of domestication of animals and crops. With the development of agriculture the concept of settled life (colonizing tradition) was implemented. Favorable climate conditions like double monsoon helped two harvests being reaped in one year. Indian agriculture products

Table 1. Countries with Components of GDP

S.N.	Country Name	Components of GDP in Percentage		
		Agriculture	**Industry**	**Services**
1	Argentina	11.4	30.2	80
2	Australia	3.6	28.2	68.2
3	Bangladesh	15.1	28.6	56.3
4	China	8.6	40.7	50.7
5	Ethiopia	36.2	17	73.6
6	Germany	0.6	30.3	69.1
7	India	16.5	29.8	45.4
8	Indonesia	13.7	40.3	46
9	Israel	2.1	27.3	69
10	Japan	1.2	27.7	71.1
11	Nigeria	21.1	19.4	59.5
12	Pakistan	25.2	19.2	55.6
13	Russia	4.7	33.1	62.2
14	United States	1.1	19.4	79.5
15	United Kingdom	0.6	19.2	80.2

Figure 3. Countries with Components of GDP

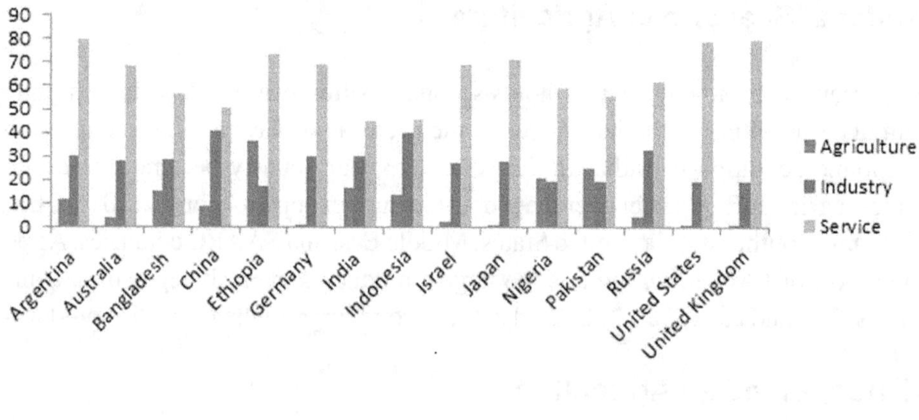

crossed the boundaries of the nation using trading network and foreign crops were also imported in India. At the time of British rule (1757-1947) commercial crops like Indigo, cotton and opium etc. were heavily imported with high value of revenue gained by the British Government. Due to canal based irrigation system, states like Punjab, Andhra Pradesh and Narmda Valley became the prime centers of agriculture reformation. During the time period of First World War (1918-139), the agriculture production stagnated. The price of basic commodities was raised to three times during 1870-1920. After the independence, special programs like Grow More Food Campaign (1940s) and the Integrated Production Programme (1950s) focused on food and cash crops supply. For upgrading the level of Indian agriculture, government introduced 5 year planning system. Before 1960, Indian agriculture was depended upon imports and foreign aids. After the Green Revolution, condition gets improved in Punjab, Haryana and West Bangal of wheat and rice. By the year 2000, farmer was able to yield 6 tons of wheat per hectare. For doing these irrigation methods was changed and tube-well was initiated. But from the year 1970 to 2011 the GDP was fall down from 43% to 17% not because of poor farming but rapid growth in other sectors like industry and services. In order to combat with side effects of chemical fertilizer Kerala and Sikkim are started towards the organic farming.

Green Revolution in India

(Chakravarti, 1973), During the British rule, India faced biggest food disaster in 1943. Surprisingly, 4 million people died because of hunger. For combat this issue, The Green Revolution was started in 1960 for increasing food production especially in Punjab, Haryana and Uttar Pradesh States. The main crop was wheat. Dr. Norman Borlaug and M.S. Swaminathan were the scientists who contributed in this revolution. Chemical fertilizers were used to increase the production of Wheat. Three basic formulae were used of Green Revolution:

- **Continued Expansion of Farming Area:** In this method, farming areas were expanded in order to enhance the yielding of grains.
- **Double Cropping Existing Farmland:** It was the main technique used for Green Revolution. Two Crop Seasons method was introduced in this scheme. In this method, one crop was based on natural mansoon and second was based on artificial mansoon. The artificial mansoon was developed with the help of dams, tube well etc.
- **Using Seeds with Superior Genetics:** Indian Council of Agriculture Research (ICAR) developed High Yield Value Seeds of Wheat and Rice. It was the scientific aspect of Green Revolution in which was used to yield heavy amount of grain. These seeds were of superior quality having immune system to fight from diseases and insects. It played prominent role in yielding the surplus grains not only wheat and rice but also corn and millet.

Green Revolution tremendously increased the production of grains like rice and wheat. Since, chemical fertilizers were used at that time as a result decreased soil fertility and land degradation.

In the year 2003-04, agriculture contributed around 22% in total GDP and provided livelihood to more 52% of Indian population. At this time India is the largest producer of Cashew nuts, coconuts, ginger, turmeric, banana, sapota, pulses and black pepper and second largest producer of groundnut, wheat, vegetables, sugar. Moreover country also produce tobacco and rice in heavy quantity.

Seasons of Crops

In India, there are three different seasons for crop yielding i.e. Kharif, Rabi and Zaid. The season of Kharif is June to September, season of Rabi is October to March and season of Zaid lies between April and June. Table 2 depicts (Singh, Ramesh, 2017) the crops that yield in these seasons:

Different Categories of Crops Yield in the Country

Agriculture is not only the source of food, but it also generates revenue for the farmers to increase the level of standard for living. There are seven categories of crop that can be yield for the source of food as well as good source of money generation. They are food grains, spices, seeds and pulses, vegetables, fruits, dry fruits and cash crops. Major Crops of India is depicted in Table 3.

Indian Agriculture at Current Scenario

In this section, elaborating the current agriculture scenario of country. According to the census report 2011, more than 54% people of India are engaged in agriculture and its allied activities. Still the statistics of share of agriculture and its allied activities are declining from 18.2% (in the year 2012-13) to 17.0% (in the year 2015-16). The total GVA (Gross Value Added) is increased but the component of agriculture and

Table 2. Types of Crop with suitable Seasons.

S.N.	Season	Northern Area	Southern
1	Kharif	Cotton, Rice, Bazra, Toor and Jowar	Maize, Rice, Jowar, Ragi and Groundnut.
2	Rabi	Gram, Wheat, Mustard, Rapeseeds and Barley	Maize, Rice, Groundnut, Ragi and Jowar.
3	Zaid	Fruits, Vegetables, Fodder	Vegetables, Fodder and Rice

Table 3. Different Types of Crops yield in India

Food Grains	Spices	Seeds and Pulses	Vegetables	Fruits	Dry Fruits	Cash Crops
Wheat	Cardamom	Fenne	Lemons	Coconuts	Arecanut	Tobacoo
Rice	Nutmeg and mace	Safflower seeds	Green peas	Banana	Cashew	Saffron
Jowar	Garlic	Castor oil seeds Pulses	Pumpkins	Mangoes	Cocoa	Tea
Ragi	Turmeric	Sesame seeds	Potatoes	Guavas	Betelvine	Coffee
Bazra	Dry chillies and peppers	Cotton seed and lint	Gourds	Squashes		Cotton
Sorghum	Coriande	Lentil	Tomatoes	Sugarcane		Rubber
	Anise		Cabbages			
			Onions			
			Dry beans			
			Limes			
			Pigeon peas			
			Brinja			
			Cauliflowers			
			Okra			
			Chickpeas			

its allied services are decreased. (Government of India. Department of Agriculture, Cooperation and Farmers Welfare, 2017) Component of agriculture and its allied sector is depicted in Table 4 and Figure Number 4.

Area of Production of Different Crops

Area of production refers to the total area of land size used to yield a particular crop. Here, depicting the total area of production of major crops like wheat, rice, pulses and food-grains. Here depicting the facts and figures are from year 2014 to 2016. These commodities are heavily yielded in Rajasthan. Area of Production of different crops is depicted in Table 5 (Government of India. Department of Agriculture, Cooperation and Farmers Welfare, 2017).

Table 4. Component of Agriculture and its Allied Sector

S.N.	Year	Total GVA (in %)	Component of Agriculture and its Allied Sectors (in %)
1	2013	5.4	1.5
2	2014	6.3	4.2
3	2015	7.1	-0.12
4	2016	7.2	1.2

Figure 4. Component of Agriculture and its Allied Sector

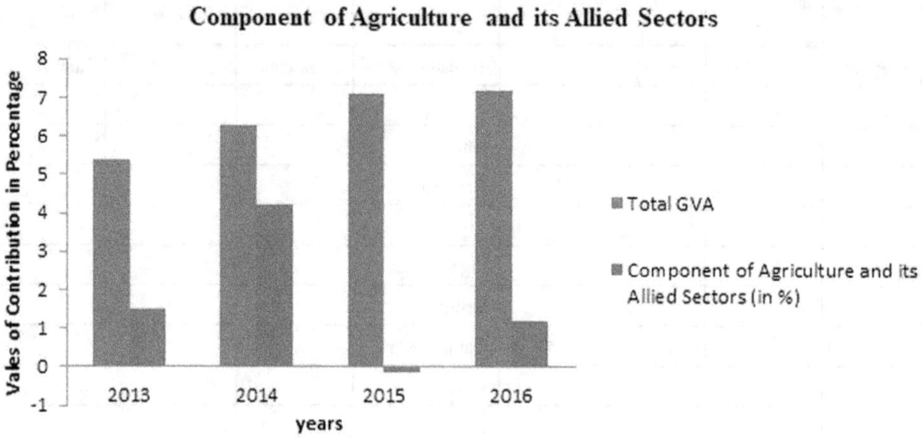

Total Production of Different Crops

In this section, depicting total production of major crops in Million Tonnes. Here also the figures are disappointing, there are less production in the 2015-16 compare to 2014-2015. Here,mdepicting total production of wheat, rice, pulses and food-grains. These commodities are heavily yielded in Rajasthan. Total Production of different crops is depicted in Table 6(Government of India. Department of Agriculture, Cooperation and Farmers Welfare, 2017).

International Scenario of Indian Agriculture

In this section, exploring the international collaborations of Indian agriculture sector. Our country is one of the founder members of Food & Agriculture Organization, and actively participating in different events and projects conducted by FAO. Here is the list of projects undergoing with the collaboration of FAO. Research Project funded

Table 5. Area of Production of Different Crops

S.N.	Area (Lakh Hectare)	2013-2014	2014-2015	2015-2016
1	Wheat	304.73	314.65	302.27
2	Rice	441.36	441.10	433.88
3	Pulses	252.12	235.54	252.59
4	Food Grains	1250.41	1243.00	1226.50

Table 6. Total Production of Different Crops

S.N.	Total Production (Million Tonnes)	2013-2014	2014-2015	2015-2016
1	Wheat	95.85	86.52	93.50
2	Rice	106.65	105.48	104.32
3	Pulses	19.25	17.15	16.47
4	Food Grains	265.04	252.02	252.22

by FAO is depicted in Table 7 (Government of India. Department of Agriculture, Cooperation and Farmers Welfare, 2017-18).

The Department of Agriculture and Farmer Welfare has signed more than 55 Memorandum of Understanding, Maintenance of Certificate and agreements with more than 55 countries, some of these organizations are World Food Program, Trust Fund of FAO for Desert Locust in Eastern Region, Trust Fund for International Desert Locust, Organization for Economic Cooperation & Development (OECD), Asia and Pacific Coconut Community (APCC), Asia-Pacific Plant Protection Convention (APPPC), Global Crops Diversity Trust Fund (GCDT).

EXPORT AND IMPORT OF INDIAN AGRICULTURE

India is the significant exporter of cotton, rice, nuts and castor seed. According to the WTO Indian agriculture contributes 1.74% in import and 2.26% in export of overall in world agriculture trade. Indian export scenario is depicted in the Table 8 given below: (Export and Import of Indian Agriculture Commodities, 2018)

Now, depicting the agriculture products that are exported since 2013-14. The major commodities which heavily exported are rice, spices, cotton and cashew nut. The quantity is shown in 1000 tons in the below Table 9 (Government of India. Department of Agriculture, Cooperation and Farmers Welfare, 2017-17).

Table 7. Research Projects funded by FAO

S.N.	Name of Projects	Date	Amount
1	Development of Extension and Outreach organizational and managerial capacities by state and public institutions in Mizoram	22 June 2015	$ 496.000
2	Global Environment Facility (GEF) 6 Formulation in India	26th June 2015	$ 110.787
3	Strengthening National Forest Inventory and Monitoring Protocols and Capacities in India	1st May 2016	$ 397.000
4	Technical Cooperation Program Facility (TCPF): Strengthening institutional capacities for sustainable mountain development in the Indian Himalayan Region	11th July 2016	$ 99.955.
5	Technical Cooperation Program Facility (TCPF): Promoting Nutrition Education and Communication in India	1st August 2016	$ 99.821
6	Technical Cooperation Program Facility (TCPF): Supporting Project Preparation of FAO India GEF 6 Full Scale Project	1st August 2016	$ 99.821.
7	Strengthening Agricultural Market Information Systems in India using innovative methods and digital technology (Baby Project)	14th September 2015	$ 884.374.
8	Programme Support to Nationally Executed (NEX) Land and Water Programme in India	1st January 2004	$ 4652.611
9	Green-Agriculture: Transforming Indian agriculture for global environmental benefits (PPG)	16th August 2016	$ 300.000
10	Technical Assistance to Farmer Water School Programme and Agricultural Activities under the Uttar Pradesh Water Sector Restructuring Project (UPWSRP)	9th March, 2016	$ 1847.539.

Table 8. Export and Import of India

S.N.	Year	Export (Billion $)	Import (Billion $)
1	2013-14	35.1	22.2
2	2014-15	29.1	22.3
3	2015-16	27.1	23.3
4	2016-17	22.4	23.7

Now, depicting the agriculture products that are imported since 2013-14. The major commodities which heavily imported are pulses, fruits, sugar and coffee. The quantity is shown in 1000 tons in the below Table 10 (Government of India. Department of Agriculture, Cooperation and Farmers Welfare, 2017-18).

Table 9. Exports of Major Crops of India

S.N.	Name of Commodity	Year 2013-14	2014-15	2015-16	2016-17 (Up to Sept.)
1	Rice	10902	11928	10419	5436
2	Spices	897	923	821	475
3	Cotton	1948	1143	1346	210
4	Cashew Nut	121	135	103	37

Table 10. Major Imports of India

S.N.	Name of Commodity	Year 2013-14	2014-15	2015-16	2016-17 (Up to Sept.)
1	Pulses	3178	4585	5798	2014
2	Fruits	769	858	836	451
3	Sugar	881	259	232	333
4	Coffee	60	75	66	37

Irrigation in India

Irrigation is the process of supply of water to a crop for its growth. It also helps to maintain landscape, revegetate disturbed soils, forest production, mitigates weed growth and avoids soil consolidation. In India, canals (major and minor), river, open well, tube well and dams are the prime sources of irrigation which ultimately reduces the dependency on mansoon. These primary sources of irrigation are also available in Rajasthan. Since Indian agriculture system is dependent upon irrigation, here, depicting percentage of Net Area Irrigated of some major states of India. (Types of Irrigation Methods, 2017). Net Area Irrigated is depicted in Table 11 and Figure 5.

Types of Irrigation Practices in India

In this section, describing different types of irrigation system followed in India. (Types of Irrigation Methods, 2017)

Table 11. Net Area Irrigated

S.N.	Name of State	Net Area Irrigated (in %)
1	Andhra Pradesh	38.76
2	Gujarat	31.24
3	Haryana	76.99
4	Madhya Pradesh	30.01
5	Maharashtra	14.33
6	Punjab	92.94
7	Rajasthan	31.56
8	Tamil Nadu	49.14
9	Uttar Pradesh	67.10
10	West Bengal	34.99

Figure 5. Net Area Irrigated of States of Rajasthan

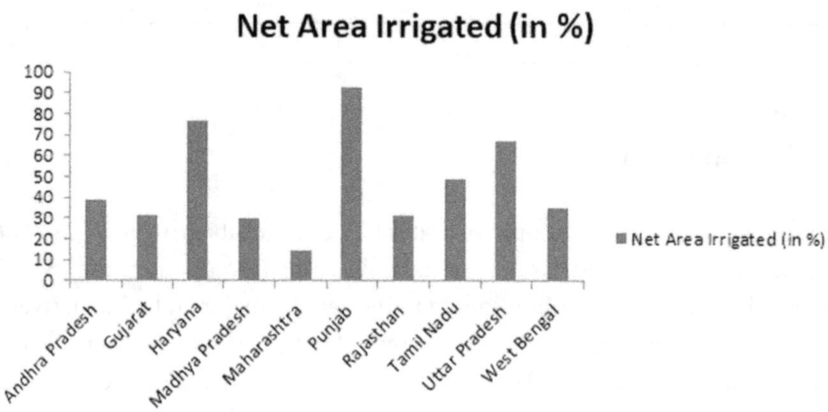

Direct Irrigation Method

In the direct irrigation method, the water is diverted into the canal by developing barrage or weir across the river. It is easy to develop and widely used in India. It also increases the water level of river. The flow of water is controlled by gated structure and combination of this and diversion is known as headwork. If the water is available for the irrigation is throughout the irrigation time then it is known as Perennial Irrigation Scheme, and if water is available for short period of time on the basis of availability of source then it is called as Direct Irrigation, moreover if there is no barricades for controlling the flow of water then it is known as Inundation Irrigation. Direct irrigation is depicted in the Figure 6 below.

Figure 6. Direct Irrigation Method

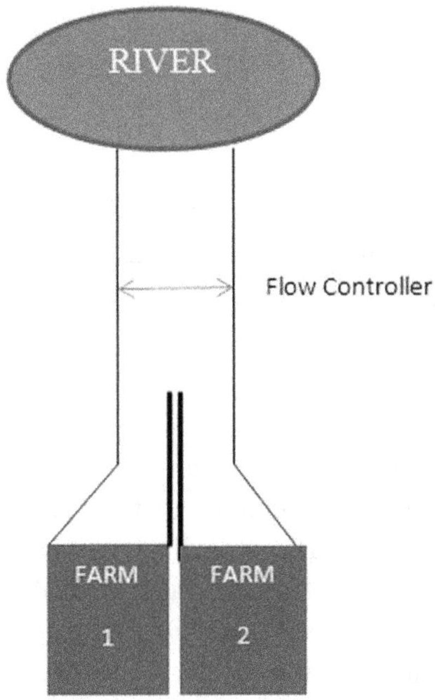

Storage Irrigation Method

Storage Irrigation Method is an efficient model, in which the excess amount of water during mansoon season is stored in a dam or reservoir. Again canal is created from these water stores. Indira Gandhi Canal, Buckingham Canal Satlej and Yamuna Link Canal are some examples of Storage Irrigation Method.

Methods of Field Water Application

In above section, studied that how water is reached up to the farmer's land. Now it will be discussed how this amount of water will be supplied to the crops. These methods are Surface Irrigation Method, Subsurface Irrigation Method, Sprinkler Irrigation Method and Drip Irrigation. Taxonomy of Irrigation is depicted in Figure 7.

Surface Irrigation Method

In surface irrigation method, the water is applied to the soil directly from the channel (like canal). There is a need to develop adequate water distribution system for controlling the distribution of water. The most common Surface Irrigation Method

Figure 7. Taxonomy of Irrigation

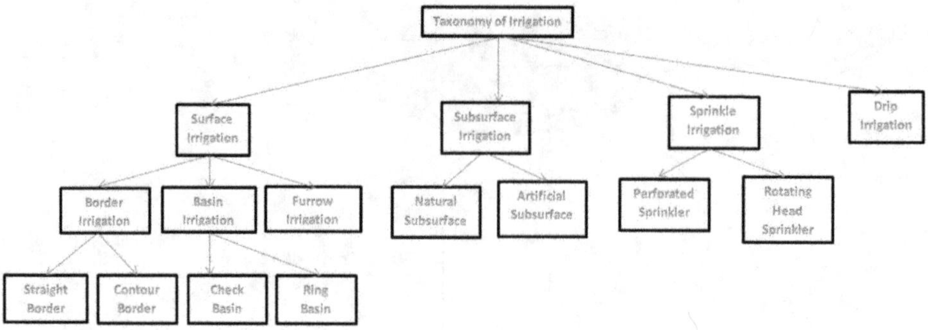

is Flooding Method where continuous water is disseminated to the surface of the land. The main advantage of this method is that soil absorbs the sufficient amount of water as per the requirement. The main dis-advantage of this technique is water distribution is uneven that is why it is known as uncontrolled flooding. There are two ways to make it controlled irrigation are Border Irrigation and Basin Irrigation. In Border Irrigation, the water is supplied to the border of the farm. The Basin Method is again classified into two types that is Check Basin and Ring Basin Method. Check Basin Method is most common type of Irrigation System. Furrow irrigation system is also a type of Surface Irrigation Method in which furrows are created, this type of irrigation is feasible for those crops which are not able to stand in water for longer time. (Bishop et al. 1967).

Subsurface Irrigation Method

In Subsurface Irrigation Method, the water is supplied to the root zone of the crop. It mitigates the loss of water through evapotranspiration. There are two ways to implement the subsurface irrigation which are Natural Subsurface Irrigation Method and Artificial Subsurface Irrigation Method. (Macleod, 1951).

Sprinkler Irrigation System

Sprinkler Irrigation system is similar like the natural rainfall system. The pumps are embedded in the field and water is disseminated through the pumps just like it happens in rainfall. Sprinkler system is suitable for sandy soil, lack of water availability and undulating surface. Sprinkler system is of two types, Perforated Pipe System and Rotating Head System. (Types of Irrigation, 2017).

Drip Irrigation System

Drip Irrigation is water conservation technique but very costly to apply. In this technique, water is supplied in soil with very slow speed (around 15 liters per hour) using the plastic pipes outlet by drippers (similar look of nozzles). This technique is suitable for applying where there is scarcity of water and labor is costly to be hired. (Mehoudar, 1980).

Fertilization in India

(Finck, 1982). Process of Agriculture can't be defined with the vitality of soil. As like a human that needs diet supplements, tonics and power booster, soil also need the additional nutrients. The process of providing the extra nutrients to the soil is known as Fertilizations. As a human body is composed of different vitamins and minerals (Vitamin A, Vitamin D, Iron and Calcium etc.), soil also have 17 types of different elements responsible for successful yielding of the crop. Major soil nutrients are depicted in Table 12.

Out of these 17 soil nutrients, Nitrogen, Phosphorous and Potassium (NPK) are three primary elements present in the soil (Indian Fertilization Pattern, 2017). Since Indian farmers are not so much aware about the chemistry of the fertilization, so they directly purchase the compound of these NPK Fertilizers. NPK Compound is depicted in Table 13.

Now, depicting (Government of India. Department of Fertilization, 2017) the Total Production, Import and Sales of Urea, DAP and Potash latest by 2017 September. These figures are expressed in MT (megaton). Total Production, Import and sales of Fertilizer is depicted in Table 14.

However, the domestic price of Urea is fixed at Rs. 5360 without taxes per Lac Mega Ton. The price of DAP is 21951.22 Megaton and the price of MOP is 12234.70 per Lac Mega Ton.

Major Highlights of Union Budget 2018 With Agriculture Perspectives

(Indian Union Budget, 2018). With the guidance of Hon'ble Prime Minister Mr. Narendra Modi, Finance Minister Mr. Aruj Jaitly presented the Union Budget 2018. This budget is farmer and agriculture oriented, and helps the farmer to raise the living standard. Here, pointing out the major highlights of Budget which is concerned with agriculture and farmers.

Table 12. Major Soil Nutrients

S.N.	Name of Nutrient	Source	Importance of Nutrient
1	Carbon	Air & water	Promote Structure of Crop, Health of soil and provide resistance for harmful substances.
2	Hydrogen	Air & Water	Photosynthesis Process
3	Oxygen	Air & Water	Photosynthesis Process
4	Nitrogen	Soil (Required Fertilization)	Basic Building Block
5	Phosphorus	Soil (Required Fertilization)	Respiration process of Plant
6	Potassium	Soil (Required Fertilization)	Activation of Enzymes
7	Sulfur	Soil (Required Fertilization)	Formation of Chlorophyll
8	Zinc	Soil (Required Fertilization)	Internode Elongation
9	Iron	Soil (Required Fertilization)	Maintain Plant metabolism process
10	Boron	Soil (Required Fertilization)	Cell Division
11	Manganese	Soil (Required Fertilization)	Nitrogen assimilation
12	Copper	Soil (Required Fertilization)	Activation of enzymes
13	Calcium	Soil (Required Fertilization)	Cell wall deposition
14	Nickle	Soil (Required Fertilization)	Maintain toxic level
15	Magnesium	Soil (Required Fertilization)	Building block of Chlorophyll
16	Chlorine	Soil (Required Fertilization)	Opening and closing of stomata
17	Molybdenum	Soil (Required Fertilization)	Associated with nitrogen, help in building block

Table 13. NPK Compounds

S.N.	Name of Elements	Compound	Amount of Content
1	Nitrogen	Urea (carbamide)	1 Unit of Urea contains 46% of Nitrogen
2	Phosphorous	DAP (Diammonium phosphate)/ Super Phosphate	1 Unit of DAP contains 46% of Phosphorous
3	Potassium	Potash/ (Muriate Potash)	1 Unit of Muriate Potash contains 50% of Potassium

Table 14. Total Production, Import and Sales of Fertilizer

S.N.	Name of Fertilizer	Total Production (LMT)	Import (LMT)	Sale (LMT)
1	Urea	18.68	4.54	24.70
2	DAP	3.59	28.77	9.97
3	Complexes	7.81	0.21	9.75

- In the year 2016-17, Total food Production is 275 million Tonnes and 300 million fruits and vegetables.
- Minimum Support Price of Kharif Crops with 1.5 times of production cost.
- Development and Upgradation of 22,000 Gramin Agriculture Markets, for this Rs. 2000 crore is corpus.
- Organic Farming will be motivated, around 1000 hectare will be entertained.
- Amount of Rs 200 Crore will be allocated medicinal and aromatic plants.
- Amount of Rs 14000 Crore is allocated for Prime Minister Krishi Sampada Yojna.
- Amout of Rs 500 Crore is allocated for agri-logistics, professional management and processing facilities.
- Set up of 42 Mega Food Marks.
- Amount of Rs. 1290 Crore is allocated for Bomboo sector promotion.
- Setup of long Term Irrigation Fund in NABARD
- Amount Rs 10,000 is allocated to Dairy Processing Infrastructure Development Fund (DPIDF) and Aquaculture Infrastructure Development Fund (FAIDF).
- Amount of Rs 11 Lakh Crore would be raised for increment of volumes of institutional credit for agriculture.

GEOGRAPHICAL INTRODUCTION ABOUT RAJASTHAN

(Geography of Rajasthan, 2017). Rajasthan is the largest state in terms of geographical boundaries of 342,239 square kilometers of India. There are 32 districts in Rajasthan .It is located at 26.57268°N 73.83902°E. The state was formed on 30 March 1949, and city Jaipur was declared as capital of Rajasthan. Rajasthan is the land of desert and driest state of India. Rajasthan is the last producer of Mustard, Millet, Cluster Bean and Psyllium Husk, Heena and Coriander whereas the state is second largest producer of all pulses and Cumin. Rajasthan also produce Soyabean, Maize and wheat in large amount. In the western Rajasthan, humidity and rainfall both are low, whereas in eastern Rajasthan, both are better in state. The mansoon comes in the month of July to September from southwest side. The average rainfall lies between 200-400 mm. Mostly, sandy soil is available in Rajasthan. This sandy soil is alkaline and calcareous in nature .However, in some areas loamy soil with nitrogenous nature is also found. Major crops yield in Rajasthan are depicted in Table 15.

Types of Soil in Rajasthan

Soil is the upper most layer of earth. As it is mentioned that soil is very important for yielding, that is why here depicting the types of soil distributed in the Rajasthan. There are eight types of soil distributed in Rajasthan

- **Desert Soil:** Desert soil pale brown in color and well drained in nature. Desert soil is not so much suitable for crop yielding because nitrogen presence is very low. It is sandy in nature with high Ph value. Desert soil is generally found in Jodhpur, Naugur Barmer, Jalore, Sriganganagar, Hanumangarh, Sikar, Jhunjhunu and Sikar districts of Rajasthan.
- **Dunes and Associated Soils:** Dunes soil is in yellowish brown in color and well drained in nature. This soil is generally used for pulses in Kharif and Rabi seasons. The texture of Dunes soil is loamy in nature. It is generally found in Bikaner, Barmer, Jaipur, Jaisalmer, Churu and Jodhpur states of Rajasthan.
- **Brown Soil:** As the name indicates, Brown soil is of brown color. It is rich in calcium and requires good fertilization for yielding the crop. The texture of Brown soil is sandy loam to clay loam. Rabi crops are sown in this soil. It is

Table 15. Major Crops of Rajasthan

Kharif	Rabi	Zaid
Milet	Wheat	Moong
Corn	Mustard	Urad
Vigna	Gram	
Radiata	Jou	
Moth	Green Clover	
Lentils	Coriander	
Peanuts	Onion	
Sesame	Cumin	
Soyabeen	Tomato	
Cotton		

generally found in Bundi, Tonk, Udaipur, Bhilwara, Chittorgarh and Udaipur states of Rajasthan.

- **Sierozems:** Sierozems is yellowish brown in color, best for low rainfall area. The texture of Sirozems is sandy loamy to sandy clay. Bajra, pulses, wheat and mustard is grown in this soil. It is generally found in Naguar, Pali, Ajmer and Aravali Hills area of Rajasthan.
- **Red Loam:** As the name indicates, it is of reddish in color, having high concentration of iron-oxide. The texture of Red Loam is Sandy Loam to Sandy, which is suitable for chilies, maize, barley and wheat cultivation. It is generally found in Banswara, Dungarpur and some part of Chittorgarh and Udiapur states of Rajasthan.
- **Hill Soil:** The scientific name of hill soil is Lithosol. It is of reddish, yellow and brownish in color. The texture of this soil is sandy loam to clay. It is also well drained in nature. It is not suitable for crop yielding and also suffers from soil erosion. It is generally found in Hilly area of Rajasthan i.e. Pali, Sirohi, Rajsamand and some part of Udaipur and Chittorgarh.
- **Saline Sodic Soil:** Saline Sodic soil is grey and brownish in color. It contains large component of salt. Cultivation of crop is restricted in this soil excluding some shrubs and some vegetables. It is generally found in Sambhar, Deedwana and some parts of Barmer and Jalore.
- **Black Soil:** The scientific name of black soil is Alluvial soil. The texture of this soil is clay to sandy loam. It is generally found in Hadoti Region of Rajasthan (Kota, Jhalawar, Bundi and Baran), suitable for tobacco, mustard etc. crops.

Area and Production of Kharif Crops

(Government of Rajasthan, Department of Agriculture,, 2016-17) In Rajasthan, Kharif is sown in around of 140-160 Lakh per hectare. In which, grains are sown with 60 to 65%, Oilseeds are sown with 10-15% and rest all are sown with around 20%. Here it is depicting (in Lakh per ton) the total production of Kharif Production of last three years. Production of Kharif Crop of Rajasthan is depicted in Table 16.

Table 16. Production of Kharif crop of Rajasthan

S.N.	Year	Cereals	Pulses	Oilseeds	Cotton
1	2014-15	69.04	9.63	24.22	15.27
2	2015-16	50.70	10.47	22.38	12.44
3	2016-17	63.23	16.50	28.16	15.85

Area and Production of Rabi Crops

(Government of Rajasthan, Department of Agriculture, 2016-17) In Rajasthan, Rabi is sown in around of 70-95 Lakh per hectare. Here depicting (in Lakh per ton) the total production of Rabi Production of last three years. Production of Rabi Crop is depicted in Table 17.

Net Area of Irrigation

(Government of Rajasthan, Department of Agriculture, 2016-17), In the previous section, here exhibited the net area of irrigation of some major states. Now here, depicting the Net Area of Irrigation of Rajasthan latest for three year with the sources of irrigation. These data is in Hectare. The Net Area Irrigated of Rajasthan is depicted in Table 18.

Consumption of Fertilization of Rabi and Kharif Crop

(Government of Rajasthan, Department of Agriculture, 2016-17), In this section, here elaborating the consumption of fertilization (Urea, DAP and MOP) of last three years. Consumption of fertilization of Rajasthan is depicted in Table 19.

Table 17. Production of Rabi crop of Rajasthan

S.N.	Year	Cereals	Pulses	Oilseeds
1	2014-15	69.04	9.63	24.22
2	2015-16	50.70	10.47	22.38
3	2016-17	63.23	16.50	28.16

Table 18. Net Area Irrigated of Rajasthan

S.N.	Year	Canals	Pond	Wells and Tube Wells	Other Sources
1	2013-14	1895107	67461	556102	162037
2	2014-15	1927840	69699	5733278	14993
3	2015-16	1979480	66193	5775257	117067

Table 19. Consumption of fertilization of Rajasthan

S.N.	Year	Kharif			Rabi		
		Urea	DAP	MOP	Urea	DAP	MOP
1	2014	626457	296492	3363	1127384	204502	980
2	2015	782648	521640	8308	1204097	255059	6733
3	2016	733523	319613	7904	1211242	249640	4262

Major Highlights of Rajasthan Budget 2018-19 With Agriculture Perspectives

(Rajasthan Budget, 2018) Chief Minister of Rajasthan Smt. Vasundhra Raje Scindhiya presented the Budget 2018-19. Here elaborating the major points of this budget which is concerned with agriculture sector only. As per expert view, this budget is also farmer oriented budget.

- Amount of Rs. 2,582.70 crore is allotted to the government of Rajasthan.
- Under this, Rs. 758.16 Crore is sanctioned for Seasonal Based Crop Insurance Scheme.
- Under this, Rs. 26 crore is sanctioned for "*Kisan Seva Kendra Sah Village Knowledge Centre*"
- Under this, Rs. 500 crore is sanctioned for National Agriculture Development Scheme.
- Under this, Rs. 79.47 Crore is sanctioned for "Rashtriya Krishi Vistar avam Takniniki Mission" both for (Agriculture Development and Agriculture Engineering)
- Under this, Rs. 219.18 Crore is sanctioned for "*Rashtriya Khadyya Suraksha Mission*" for (Wheat, Pulses and Cereals)
- Under this, Rs. 27.22 Crore is sanctioned for Soil Health management.
- Under this, Rs. 64.43 Crore is sanctioned for National oilseeds and oil Pam Mission
- Under this, Rs. 20 Crore is sanctioned for drip irrigation.
- Under this, Rs 171.80 Crore is sanctioned for five agriculture universities.
- Under this, Rs. 3,126.54 crore is Irrigation and Flood Control.

CONCLUSION

Agriculture is the base line of livelihood for not only our country but also around the globe. Agriculture not only refer to the farming process but also includes animal husbandry, dairy maintenance etc. Agriculture is very important for livelihood, generate income and employment and also help to maintain the ecological balance. The growth of a crop is influenced by genetic and heredity of the crop and soil, environmental and technical issues and also market demand. Countries like Ethopia, Pakistan and Nigeria have high component of agriculture in their GDP, whereas developed countries like Germany, UK and US has negligible contribution of agriculture in their GDP. In India, agriculture has contributed 16.6% of overall GDP. India heavily depended upon of Agriculture, where concept of colonizing was developed only by agriculture. During the time of British rule, condition of agriculture sector was worst, which was enhanced by Green Revolution using chemical fertilizers. In India, there are three types of seasons for agriculture first one is Rabi, second is Kharrif and third one is Zaid. Since Indian economy is dependent upon agriculture but the component of agriculture and its applied sector in Gross Value Added is declining year by year, which is a disappointing situation. Wheat, rice and pulses are the highest yielded crop in the India. India is also integrated with international agriculture research agencies (Food and Agriculture Organization) with million values projects. Statistics of import and export of Indian agriculture sector is also disappointing, imports are increasing year by year and exports are decreasing year by year. Irrigation is vital component of agriculture process, developed states like Punjab and Haryana have the highest net irrigated area in comparison with other states of India. Surface Irrigation Method is the most common method used for irrigation by Indian farmers but now they are switching to high-tech methods like Sprinkler and Drip Irrigation Method. Fertilization also plays important role in agriculture, Indian farmers directly used urea for Nitrogen, DAP for Phosphorus and Muriate of Pottash for Potassium. However, these fertilizers are produced and used in sufficient amount. Union Budget 2018 is also declared by Government of India, which is becoming the boon of agriculture sector by allocation of heavy amount of fund. Rajasthan- India is the largest state of country in terms of geographical parameters. Rajasthan is the heavy producer of rice, wheat and mustard. Statistics of agriculture development in Rajasthan is quite satisfactory. There is increment noticed in the production of Kharif and Rabi crops for last three years of cereals, pulses and oilseeds. Good growth of net irrigation area is also observed from last three years, although Rajasthan is a desert area. To increase the production, fertilization also played a vital role. Fertilization consumption is also increased in very good amount in both Rabi and Kharif crops. At last it is concluded, India is fully depended upon agriculture and the climate conditions, soil conditions and other environmental issues also favor the agriculture development, but the condition of farmers are not good enough, exports are decreasing and imports are increasing in disappointing manner. There are some issues that are responsible for this, first one is farmers are not so much technology friendly and feel hesitate to adopt new technology, second is the government policies for agriculture does not influence the farmers directly and the last reason is that agriculture research is theory based and remain in research papers only, it is not beneficial for farmers at ground level. However, positive hope for agriculture development can be anticipated from the latest budget.

REFERENCES

Balasaheb. (2011). *A study of export marketing management of grapes with special reference to Nasik district* (Doctoral Thesis). Retrieved from http://shodhganga.inflibnet.ac.in/handle/10603/92452

Bishop, A. A., Jensen, M. E., & Hall, W. A. (1967). *Surface irrigation systems*. Academic Press.

Chakravarti, A. K. (1973). Green revolution in India. *Annals of the Association of American Geographers*, *63*(3), 319–330. doi:10.1111/j.1467-8306.1973.tb00929.x

Countries with their GDP Components. (2017). Retrieved from https://www.cia.gov/library/publications/the-world-factbook/fields/2012.html

Export and Import of Indian Agriculture Commodities. (2018). Retrieved from https://www.export.gov/article?id=India-Agricultural-Commodities

Factor Affect Agriculture. (2018). Retrieved from http://eagri.tnau.ac.in/eagri50/AGRO101/lec09.pdf.9

Finck, A. (1982). *Fertilizers and fertilization: introduction and practical guide to crop fertilization*. Verlag Chemie.

Geography of Rajasthan. (2017). Retrieved from http://niti.gov.in/writereaddata/files/Rajasthan_Presentation_0.pdf

Government of India. (2017). *Department of Fertilization*. Monthly Bulletin September.

Government of Rajasthan, Department of Agriculture. (2016-17). *Administrative Progress Report 2016-17*. Author.

Herrera, J. L. (1999). *Handbook of Agriculture* (1st ed.). Taylor And Francis.

Indian Fertilization Pattern. (2017). Retrieved from http://fert.nic.in/sites/default/files/Indian%20Fertilizer%20SCENARIO-2014.pdf

Indian Union Budget. (2018). Retrieved from http://www.indiabudget.gov.in/

Macleod, N. A. (1951). *U.S. Patent No. 2,536,196*. Washington, DC: U.S. Patent and Trademark Office.

Mehoudar, R. (1980). *U.S. Patent No. 4,210,287*. Washington, DC: U.S. Patent and Trademark Office.

Rajasthan Budget. (2018). Retrieved from http://plan.rajasthan.gov.in/

Singh, R. (2017). *Indian Economy*. McGraw-Hill.

Types of Irrigation. (2017). Retrieved from http://nptel.ac.in/courses/105105110/pdf/m3l01.pdf

Chapter 4
Internet of Things Practices for Smart Agriculture

S. Umamaheswari
Dr. G. R. Damodaran College of Science, India

ABSTRACT

Agriculture is an important economic sector that leads to the overall economic growth of a nation. This sector has to meet the increasing consumption needs of the world population, which is expected to increase by 70% by 2050, according to the Food and Agriculture Organization of the United Nations. The advancements in the technology in the field of agriculture will facilitate the growth in farming activities. The internet of things (IoT) industry enables the farmers to face these challenges. IoT sensors provide the information about the crop yields, soil moisture, rain fall, livestock, etc. to the farmers that can be used for decision making. The information can be accessed by the farmers with the aid of smart phones. This chapter focuses on the smart agriculture practices with IoT that helps the farmer to analyze the operational data to improve decision making. The implementation platform for the IoT and the communication technologies for smart farming are discussed. A prototype for a smart irrigation system has also been proposed.

INTRODUCTION

The exponential growth in the technological advancements leads to the automation of processes in all the sectors. The processes in the field of agriculture are also getting automated by implementing the smart environment. This is done by the use of wireless sensor network with sensors deployed at different locations. The data collection is done by the network and the nodes send the data through the wireless

DOI: 10.4018/978-1-5225-5909-2.ch004

protocol. The primary objective in the field of agriculture is to increase the yield of crops. Crop monitoring is not the only solution to achieve the goal. The entire activities in the field have to be automated in order to increase the productivity to the greater extent. The development of a smart environment for automating the activities will benefit the farmers.

BACKGROUND

The Internet of Things (IoT)

The Internet is emerging with lot of smart connected devices today. There is an exponential growth in the number of smart devices connected to the mobile internet. The devices are embedded with intelligence due to the development in technology. These intelligent (smart) devices are capable of interacting with humans and other smart devices, as well, which led to the development of "Internet of Things" (IoT). The term Internet of Things is coined in 1999 by Kevin Ashton of the Massachusetts Institute of Technology (MIT) in the context of supply chain management that describes a technology of the future based on the Internet and involves sharing of information (K.Ashton, 2009). The intercommunication between machines leads to the M2M implementations. IoT makes the information available to the applications, business systems and data warehouses.

The IoT can be viewed as a global infrastructure for the information society, enabling advanced services by interconnecting (physical and virtual) things based on existing and evolving interoperable information and communication technologies (ICT) (Series Y). In 2050, it is expected to have 50 billion devices connected to the internet. Internet of Things is the Future of the internet.

Sense

This phase is responsible of gathering information from captured by any sensing device. The sensing technology will vary depends on the purpose. Biometric, biological, environmental and audio visual are different means of sensing. The device that senses the signals need not be one that gathers the information.

Communicate

The sensed information at the device level has to be transmitted to a cloud-based service for processing. The transmission can be done by means of communication technologies like either WiFi (wireless LAN based communications) or WAN (wide

Figure 1. Steps in the Internet of Things workflow

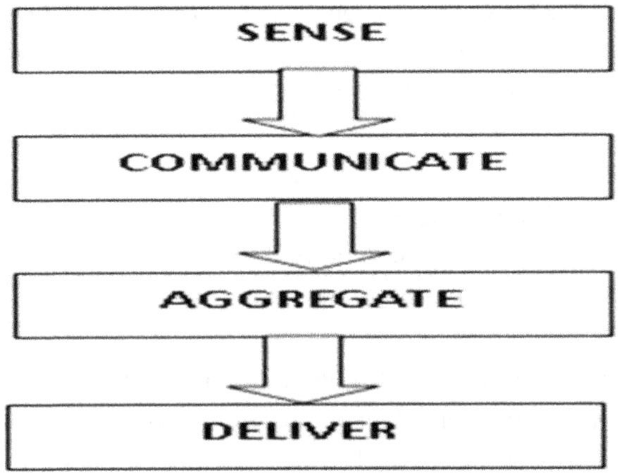

area network… i.e. cellular) communications. Short range communications like Bluetooth, ZigBee, Near-Field may also be needed depending on the requirement. The usage of Global Positioning System (GPS) may also be required.

Aggregate

The sensed data is transmitted to a cloud based service for data processing to get useful information. The data can be gathered from the IoT device and from other internet sources.

Deliver

The processed information which is useful is delivered to the user. The user can be a consumer, a commercial or an industrial user. The main goal is to provide the information as simple as possible. A well designed platform independent user interface is executed to get the processed information.

Applications of IoT

The leading IT companies like IBM, Honeywell, General Electric, Cisco, Schneider Electric, Siemens, Microsoft, NI, etc. are very much involved in IoT to make Smart environments (L. Atzori et al., 2010). The IoT has found its applications in every walk of life. The possible applications of IoT are listed below.

- **Natural Disaster Discovery:** The natural disasters like land-slides will be predicted to take necessary actions in advance with the help of combination of sensors and their autonomous coordination and simulation.
- **Water Management**: The IoT can help to detect the liquid presence outside tanks and pressure variations along pipe for streamlining the water supply. The users can be given alerts if there is any contamination in the water.
- **Smart Health Care:** The IoT can play a major role in medical field for saving lives or improving the quality of life. For e.g. The health parameters and activities can be monitored and also assisted living can help the elderly people at home.
- **Industry applications**: The IoT applications in the industrial sectors make the companies to track the entire supply chain activities with the help of sensors. The sensors embedded in the machineries help in the maintenance activities.
- **Smart Homes:** The Smart Home can be built up with the sensors deployed in the electronic home appliances. The sensors control these devices depending on the environmental factors. For. e.g. The refrigerator can be switched off for a while if the room temperature is too cool that saves the energy.
- **Intelligent transport system:** The advancement in the sensor technology makes the IoT to implement Intelligent transportation system that will provide efficient transportation control and management. The Highways can be made intelligent with warning messages and diversions according to climate conditions and unexpected events like accidents or traffic jams. The intelligent transportation can have many interesting features such as non-stop electronic highway toll, mobile emergency command and scheduling, vehicle rules violation monitoring, reducing environmental pollution, anti-theft system, avoiding traffic jams, reporting traffic incidents, minimizing arrival delays etc.
- **Smart Agriculture**: A network of special type of sensors can be deployed in the agricultural farms that sense data, process the data and inform the farmer through the communication media. Eg. mobile alert stating that a specific plant needs water. Smart Agriculture helps the agriculturists to have better understanding of the plant growth and to have better farming practices by sensing the knowledge of land conditions and climate variations. This improves the agricultural productivity by eliminating improper farming conditions.
- **Smart metering and monitoring:** This application will be helpful in energy consumption by automating the meter reading and collecting the payments. The wind turbine maintenance and monitoring gas, oil and water levels in the storage tanks can also be possible with the help of IoT.

- **Smart City:** Smart Cities can be designed with the help of IoT. For e.g. smart parking, smart lighting, automatic clearing up of waste materials, smart roads etc.
- **Smart Security:** The IoT can be used in the field of security and surveillance systems. For e.g. Alarm at the time of intruders tries to open the doors and windows.

Academicians, industrialists and the government are working towards making everything smart. A few projects initiated by the leading companies are:

- Microsoft's Eye-on-Earth platform.
- European Commission's Cluster of European Research Projects on the Internet of Things (CERP-IoT).
- European FP7 research projects IoT@Work, 'The Internet of Things Initiative' (IoTi) and 'European Research Cluster on the Internet of Things'(IERC).
- HP's Central Nervous System for the Earth initiative.

IoT Enabling Technologies

The key enabling technologies for the IoT can be the network standards, the thing identification mechanisms and the application devices. 6LoWPAN (IPv6 in Low-Power Wireless Personal Area Networks) for Wireless Sensor Networks (IEEE 802.15.4), Bluetooth Low Energy (IEEE 802.15.1) for Wireless Personal Area Network, WiFi Low Power (IEEE 802.11) for WLAN and the Long Tern Evolution – Advanced (LTE-A) for M-to-M Communications in WAN are the standards used for IoT application deployment. The IoT has the capability to collect, investigate and distribute data which can be turned into information or knowledge. The gathering of data from the IoT devices can be done by using the following identification mechanisms.

- **Barcode:** Barcodes are the simplest form used for identifying a smart item. (Eg.Product Identifier). An identification number of an item can be coded by a one- or two-dimensional barcode. The barcode can be available either in the label that is attached to an item or it can be directly printed on the item.
- **Quick Response (QR) or Matrix Barcodes:** QR codes are essentially machine readable two-dimensional barcodes that gives people information about products. The QR codes are used for extended identification of a resource like Uniform Resource Locators (URLs). This codes are the advancements of barcodes (Mellisa Tolentino, 2013).

- **Radio Frequency Identification (RFID):** This technology is used for the digital identification of resources. The components of aa RFID system includes a tag, which is made up of a microchip with an antenna, and an interrogator or reader with an antenna. The reader sends out the electromagnetic waves and the tag antenna is tuned to receive these waves. There are two types of RFID tags namely passive RFID tag and active RFID tag. There are a wide range of applications that use of RFID technologies such as Product Life Cycle Monitoring, Asset Tracking, Monitoring the elderlies, Airport Baggage Tracking logistics etc (Sabita Maharjan (2010). RFID can be used a communication device in the Internet of Things with the help of WSN.
- **Electronic Product Code (EPC):** EPC is one common set of data stored in a tag. EPC's are coded on RFID tags because of which objects can be tracked and identified uniquely. The tag contains a 96-bit string of data. The first eight bits are a header which identifies the version of the protocol. The next 28 bits identify the organization that manages the data for this tag; the organization number is assigned by the EPCGlobal consortium (David L. Brock, 2002).
- **Near field communication (NFC):** NFC is similar to RFID or it can be an integration of RFID reader into a mobile phone, which makes NFC customer-oriented as mobile phone is the most popular personal device worldwide (Vilmos, A. Medaglia et al., 2011). NFC can also be a type of radio communication between NFC enabled mobile devices by touching them together or bring close in the proximity of the other phone. The frequency of range that is utilized by NFC is within the unlicensed radio frequency band of 13.56 MHz.(MEDAGLIA, Carlo Maria et al., 2011). The operating range of NFC device is 20 cm. As the communication cannot be done from a remote location the NFC enabled communication between the smart objects is safe. There will be a significant contribution by the NFC technology to the future development of IoT.

The applications of the IoT can be implemented through the internet technologies and devices such as smart phones, tablets, laptops, industrial technologies, appliances and building automation.

IoT COMMUNICATION STANDARDS

IEEE 802.15.4

The underlying Physical Layer and the Media Access Layer (MAC) for the Internet of Things network is the IEEE 802.15.4 standard. The IEEE 802.15.4-2006 standard which has been released after the IEEE 802.15.4-2003 is the physical layer protocol for low-power and low-data rate Low power Lossy Networks (LLNs). The IEEE 802.15.4 concentrates on the transmission of data between the resource constrained devices. The specifications of IEEE 802.15.4 such as low data rate, low cost, and high throughput are suitable for the IoT, machine-to-machine (M2M) and Wireless Sensor Networks (WSNs).

The High levels of security, encryption, and authentication services are the other important features.This is the basis for the ZigBee protocol as both the protocols offer low data rate services on power constrained devices. The IEEE 802.15.4 is providing support for three frequency channel bands and exploits a direct sequence spreadspectrum (DSSS) method.

- **Note:** The data rates of the three frequency channels are 250 KBPS at 2.4 GHz, 40KBPS at 915 MHz and 20 KPBS at 868 MHz.

High-throughput and low latencies are the advantages of higher frequency and wider bands whereasthe lower frequencies offer better sensitivity and cover larger distances. The IEEE 802.15.4 utilizes the CSMA/CA protocol to reduce potential collisions (IEEE 802.15.4, 2011).

IEEE 802.15.4e

The IEEE 802.15.4e is an extension of the IEEE 802.15.4 protocol to support low power communication. The main features of this standard are time synchronization and channel hopping to enable high reliability, low-cost, and to meet the IoT communications requirements. The specific MAC features are slotframestructure, scheduling, synchronization, channel hopping, and network formation.

The IEEE 802.15.4eslotframestructure schedules and notifies each node what to do. A node in the network may be either in sleep mode or in send/receive mode. In the sleep mode, the node turns off the radio to save power and stores all information it needs to send during the next transmission. Scheduling handles the mobility scenarios with the help of the manager node by informing the schedule to other nodes to follow. The node's connection with their neighbors and the gateways is maintained well by synchronization. This is done in acknowledgment-based mode and frame-based mode. The IEEE802.15.4e launches channel hopping for time slotted access to the wireless medium. Channel hopping needs changing the frequency channel using a predetermined random sequence. This leads to frequency diversity and reduction of the effect of interference and multi-path fading. The network capacity is added to sixteen channels as two frames and the same link can be transmitted on more than one frequency channel at the same time (M. Park, 2015).

6LoWPAN

The flow of IPv6 packets over 802.15.4 networks that support small packets is taken care by IPv6 over Low Power Wireless Personal Area Networks (6LoWPAN). 6LoWPAN acts as an adaptation layer to transport IPv6 packets over 802.15.4 links. The 6LoWPAN standard is defined by RFC 6282 that bridges the gap between the Internet and Low Power Lossy networks (LLNs) by providing IPv6 networking capabilities through special encapsulation and header compression techniques that allow IPv6 packets to be sent over low-power link layer technologies.IPv6 is the most supported IP in 6LoWPAN because of its large addressing space and the built-in support for network auto configuration. 6LoWPAN networks need a gateway which can be either an Ethernet or Wi-Fi to access the Internet. The nodes of the Internet mostly use IPv4. A 6LoWPAN gateway has an IPv6-to-IPv4 conversion protocol. The 6LoWPAN standard is an adaptation layer between the 802.15.4 link layer and a TCP/IP stack. The main advantages include support to large network size, mesh network topology, reliable communication and less energy consumption. 6LoWPAN is well suited for the applications that have internet-connected sensors, less throughput, and battery-powered. A 6LoWPAN device is capable of communicating with any other IP-based server or device on the Internet, including Wi-Fi and Ethernet devices (Z. Shelby and C. Bormann, 2010).

RPL (Routing Protocol for Low Power Lossy Networks)

IETF has come out with a standardized and effective routing protocol known as Routing for Low power and Lossy networks (RPLs), which is capable of quickly building routes, distributing routing knowledge among nodes with little overhead, and adapting topology in an efficient way. RPL is expected to be the standard routing protocol for the majority of the IoT applications including the smart grid. Routing is the primary task of managing the packet forwarding on the efficient route between the sensor nodes. The transceiver part of the nodes which is a short range radio is used to send or receive the packets among the nodes. This radio medium is subject to bit errors and link failures. This makes the LLN nodes difficult in finding the best routes for data delivery. The specific routing requirements of the LLNs have not been satisfied by the existing routing protocols such as OSPF, IS-IS, AODV, and OLSR. The solutions of successful routing should consider the specific application requirements along with the IPv6 behavior and 6LoWPAN mechanisms. The IETF Routing Over Low power and Lossy networks (ROLL) Working Group was formed to design a routing solution for low power and lossy networks (also known as sensor networks). The ROLL working group has proposed the Routing Protocol for Low power and Lossy Networks (RPL), based on a gradient-based approach (T. Winter et al., (2011) and T. Winter, (2012).

CoAP

A Constrained Environment is a workspace with low powered hardware devices, with minimum networking capabilities. The existing web services are extended to the RESTful architecture to provide a service to the IoT devices. The use of web services (web APIs) on the Internet has become ubiquitous in most applications and depends on the fundamental Representational State Transfer [REST] architecture of the Web.

The IoT devices become IP enabled and connected to the Internet and enabled to communicate a language to provide interoperability. Redesigning and optimizations in application protocols are required to implement machine-to-machine (M2M) applications over constrained environments on the IoT.Even though HTTP is widely used with Web Services, it is not the only protocol for M2M communication. The Constrained Application Protocol (CoAP) is designed by IETF Constrained RESTful Environment (Core) working group to provide alightweight RESTful (HTTP) interface and it incorporates optimizations for constrained environments. REST is the standard interface between HTTP client and servers (Z. Shelby et al. (2014). The architecture of CoAP is presented in Figure 2.

The CoAP is a specialized web transfer protocol for the constrained nodes and constrained networks such as LLNs.The nodes often have 8-bit microcontrollers with small amounts of ROM and RAM, while constrained networks such as IPv6 over Low-Power Wireless Personal Area Networks (6LoWPANs) often have high packet error rates and a typical throughput of 10s of Kbit/s.The protocol is designed for machine-to-machine (M2M) applications such as smart energy and building automation.

Figure 2. CoAP Architecture

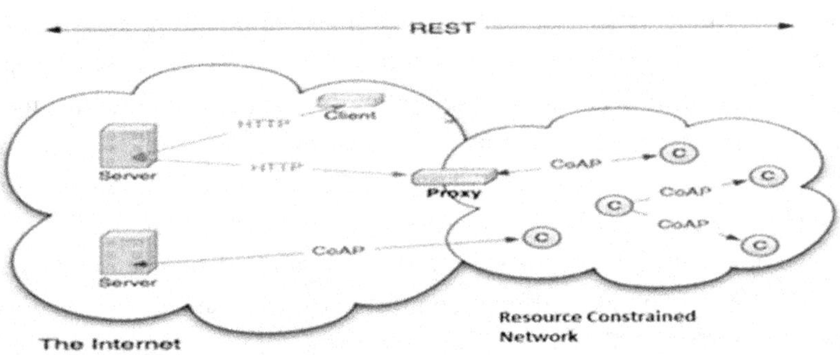

The CoAP provides a request/response interaction model between application endpoints, supportsthebuilt-in discovery of services and resources, and includes key concepts of the Web such as URIs and Internet media types.

The CoAP comprises of some HTTP functionalities which can be suitable for M2M applications over constrained environments on the IoT, meaning it takes into account the less processing power and energy constraints of small embedded devices, such as sensors. REST results in less overhead and energy consumption of lightweight IoT applications. It makes use of UDP, instead of TCP commonly used in HTTP and has a light mechanism to provide reliability. The CoAP enables low-power sensors to apply RESTful services. Messaging layer and request/response layer are the two layers that are part of the CoAP architecture. It uses the Efficient XML Interchanges (EXI) data format that is far better in terms of space when compared to plain text HTML/XML (V. Karagiannis and P. Chatzimisios, 2015).

The CoAP is an application layer service discovery protocol for constrained environments. Built-in header compression, congestion control, resource discovery, IP multicast support, native push model, and asynchronous message exchange are other special features of CoAP that are not offered by HTTP.

Service discovery in the constrained environment is an increasingly important issue as we move towards realizing pervasive systems. The user interaction with the service is an important issue once the service is discovered. The service life span is completed only when the communication is proper. A service layer in the form of middleware resolves this issue. CoAP can be used for service discovery in real-time IoT environments. The Integrating Physical world devices in constrained web environments using Constrained Application Protocol (CoAP) together with an end-to-end IP and RESTful Web Services based architecture can be implemented (S.Umamaheswari and K.Vanitha 2016).

SMART AGRICULTURE

Agriculture plays a vital role in the overall socio-economic framework of the country. Current agricultural practices followed are neither economically nor environmentally sustainable. There are problems in agriculture due to the lack of the use of hi-tech irrigation systems, crop failures due to lack of monsoon, poor seed quality, inefficient farming practices and undue output costs. According to FAO, The agricultural sector is going to face enormous challenges in order to feed 9.6 billion people by 2050. The food production must be increased to 70% and this should be done in the limited available lands. These issues can be resolved by using sensing technology to make farms more “intelligent” and connected to increase quality and quantity of agricultural production. This scenario is referred as Smart Farming. The solutions for the smart farming exploit Artificial Intelligence technology that learns from data, identifies patters and generates valid predictive models automatically. These predictive models can solve the challenges in smart farming like crop maintenance within ideal soil moisture range, reduction of water costs and prediction of the need of water for the crops in the case of irrigation.

The farmers can plan sowing, irrigating, putting fertilizers and pesticides, harvesting, and related crop management paraphernalia based on weather forecasts. The farmers can also manage the farm remotely using smart automated sensors to detect humidity, temperature, risk of frost, etc. Based on humidity, watering requirements can be determined. Using hi-tech irrigation systems, the farmer can make optimum use of water. There are lots more farming related equipment's that are automated. Precision Livestock Farming (PLF) which is a part of smart farming uses sensors for monitoring and detecting the reproduction events and health disorders in animals.

Related Work

An auto irrigation system, temperature control for livestock and farm products and auto sprinkler system and to control through mobile application using GSM(interface) technology are presented. This system consists of IR transmitter and receiver, sensors, GSM modems, keypad, microcontroller, LCD, relay, fan, light, hotter, valve and sprinkler. Automatic lighting, fire and smoke detection, moisture control and feeder control are implemented along with digital lock for animal security. (Dhristi kanjilal et.al., 2014).

The automation of moisture level detection is implemented with the aid of web application, mobile application and sensor network with Zigbee and Beaglebone controller. Python API is used for communicating with the device and the server. (Vaibhavraj S.Roham et.al., 2015). Smart GPS based remote controlled robot to perform task like weeding, spraying, moisture sensing, animal scaring, vigilance, Smart irrigation and Smart Warehouse maintenance are implemented. Smart device (robot), interfacing sensors, WiFi or ZigBee, Camera, microcontrollers and raspberri pi (a system performs small computing and network operations) are the elements of this system . Robot device, Warehouse, Wireless moisture sensor and mobile application are the basic components. This system is a combination of multiple activities of smart farming. (Nikesh Gondchawar and Dr.R.S Kawwitkar, 2014)

Agriculture monitoring system that senses moisture and temperature is developed to automate the irrigation. An Android application is used for controlling the irrigation system. (N.Suma et.al., 2017). A system for automating the moisture level maintenance and animal intrusion is implemented using wireless sensor network and GSM technology. The data received from the sensors of the wireless sensor network are analysed by the control module for decision making. The alerts are sent through the farmers through SMS services. (Akshay S et.al., 2016).

IMPLEMENTATION PLATFORM

Interconnected hardware devices are the backbone of the IoT. These interconnected IoT devices observe the real-world objects known as "things", which include home appliances, industrial environment and people. The Internet of Things has transformed from high-level hype towards solid ideas and products for making the business community to develop suitable hardware prototype or device to implement their ideas. An implementation platform has to be designed to develop a new IoT solution. This implementation platform is designed by designing and prototyping of hardware and software components. An IoT device depicts the hardware component that has been adopted for a specific purpose. It refers the hardware components that comprises of sensors, actuators and prototyping boards. New versions/releases of the devices and device platforms are being released continuously.

The IoT devices can be characterized in terms of Data acquisition, Data storage and processing, Network Connectivity and Power management. *Data acquisition* is the method of measuring real-world conditions and converting into digital readings. This method is used to manipulate raw sensor readings and convert analog sensor readings into digital signals so that they can be processed and analyzed. *Sensors* are the input devices that measure physical variables and translate them to electrical signals. The physical variables include temperature, smoke, gas, humidity, sound, speed, proximity, pressure, light and so on. The output devices include LEDs, speakers, screens and actuators like motors or solenoids. *Actuators* are the components that move or control things in real time applications.

Data storage and processing capabilities are necessary for the IoT devices to perform handling, transformation and analysis of data that is captured. The IoT devices are capable of processing data directly or the data can be transmitted to other devices or cloud services for analysis. Data analysis can be done in real-time on the devices or on a nearby gateway device that the IoT devices are connected to. Edge analytics processes and analyses the data at the edges of the network. Devices that are involved in edge analytics will require more processing capabilities than devices that perform only basic data processing.

Network connectivity is an important characteristic of any IoT device. The IoT devices will communicate with the other devices and generate data to services and applications in the cloud. IEEE 802.11, Bluetooth, RFID and Low Power Wide Area Network (LPWAN) technologies are the communication standards adopted by the IoT devices. Depending on the usage patterns and the power requirements of the IoT devices, a device may be put up into sleep mode or low-power mode periodically to conserve power. This process is done by the Radio Duty Cycles (RDCs).

CHOICE OF HARDWARE

The Internet of things is a set of physical objects that utilize network support to transfer data. These objects include sensors, software, boards, and so on. There are wide range of low-cost, commercially available implementation platforms, hardware development boards and prototyping kits. The implementation of a project can be done in a modular way with flexibility. Alternative components can be substituted and tested with different sensors with different specifications. Data processing, networking and storage modules can be separately upgraded based on the requirements.

The implementation hardware is designed as *System-on-a-chip* (SoC) ICs. This SOC is a package of data processing, networking and storage capabilities. Self-contained and RF-certified modules are created by the manufacturers like Gainspan, Wiznet and TI that have the features of TCP, UDP and IP on chip with built-in security features. There are a variety of devices available with different configuration range. Microcontroller development boards and Single-Board-Computers are the two different widely used platforms for implementing IoT projects.

MICROCONTROLLER DEVELOPMENT BOARDS

A *microcontroller* is a SoC that offers data processing and storage capabilities. Microcontrollers consists of a processor core (or cores), memory (RAM), and *erasable programmable read-onlymemory* (EPROM) for storing the custom programs that run on the microcontroller. Printed Circuit Boards (PCBs) that support the microcontroller to design a convenient prototype with a program is referred as Microcontroller Development Boards. General Purpose Input/Output (GPIO) pins connect the microcontroller with the sensors and actuators. Arduino is an open source device platform used for designing compatible development boards and tooling. I2C and SPI are the standard communication protocols that are used for intra-device communication with the components connected with the bus.

Arduino UNO

Arduino UNO is an open source, microcontroller based prototyping platform to launch with electronics and programming. The UNO is a microcontroller that can be used by anyone to begin learning about programming. The developer has the freedom to create connected environments quickly and easily with the help of UNO's easy-to-use hardware and software. UNO consists of a starter kit and sample projects that makes the developer to begin the programming easily. The data communication between the external devices can be easily handled by Arduini UNO. The software that runs on Arduino-compatible microcontroller can be developed using C or C++ and the Arduino IDE. Arduino UNO Microcontroller Development Board is presented in Figure 3.

Simply connect it to a computer with a USB cable or power it with a AC-to-DC adapter or battery to get started. You can tinker with your UNO without worrying too much about doing something wrong, worst case scenario you can replace the chip for a few dollars and start over again. - Arduino

SINGLE BOARD COMPUTERS

Single board computers (SBCs) are the improved version of microcontrollers, as they allow the user to attach peripheral devices like keyboards, mice, and screens, as well as offering more memory and processing power. The Raspberry Pi, BeagleBone Black, DragonBoard 410c and WyzBee are some of the most widely used SBCs. The SBC device capabilities can be extended by adding stackable expansion boards. Raspberry Pi uses *hats* and BeagleBone Black uses *capes* as their expansion boards. Most of the SBC devices are similar to a mini-PC and run on a streamlined Linux based embedded operating system.

Figure 3. Arduino UNO Microcontroller Development Board

The choice of the development tools and language used for developing embedded applications that work with the sensors and actuators are more for the developers on the SBC devices than on microcontrolled boards. The setting up of the SBCs is more complex and the SBCs need more power and more prone to problems.

Raspberry Pi

The Raspberry Pi is a mini computer that can be plugged into the TV and keyboard. The Pi is a bundle of Linux OS and capabilities for audio, video and Internet. This is a well suited platform for projects that require multimedia. The Raspberry Pi Single Board Computer is depicted in Figure 4.

The Raspberry Pi has the ability to interact with the outside world, and has been used in a wide array of digital maker projects, from music machines and parent detectors to weather stations and tweeting birdhouses with infra-red cameras. - Raspberry

BeagleBone

The projects that need a large number of external sensors and high processing power can be designed with the BeagleBone hardware platform. There are lot of pins available in the board to connect with the external devices. BeagleBoard can be used with a USB cable. The BeagleBoard- Single Board Computer is shown in Figure 5.

Figure 4. Raspberry Pi Single Board Computer

Beagle boards are tiny computers with all the capability of today's desktop machines, without the bulk, expense, or noise. Read the step-by-step getting started tutorial below to begin developing with your BeagleBone or BeagleBone Black in minutes. - Beagleboard.org

Dragon Board

The DragonBoard 410c is the first development board that is designed based on a Qualcomm® Snapdragon™ 400 series processor which is shown in Figure 6. Its special features include advanced processing power, Wi-Fi, Bluetooth connectivity, and GPS, all packed into a board the size of a credit card. The DragonBoard 410c is designed to support rapid software development, education and prototyping based on the 64-bit capable Snapdragon 410E processor. This board is well suited for enabling embedded computing and the Internet of Things (IoT) products, including the next generation of robotics, cameras, medical devices, vending machines, smart buildings, digital signage, casino gaming consoles etc.

WyzBee

The WyzBee IoT platform offers a fully inclusive sensing, computing, communicating and cloud support. This board consists of Redpine's Wireless Secure MCU (WiSeMCU™) with multi-protocol wireless module providing Wi-Fi, Bluetooth 4.1, and ZigBee connectivity, six-axis inertial sensors, an infrared receiver, a debug port, push-buttons, LEDs, USB ports, and WyzBee THING™ expansion connector. The WiSeMCU module can be operated on embedded TCP/IP networking stack with SSL/TLS/HTTPS security and also on complete Wi-Fi, Bluetoojh 4.1 and ZigBee stacks. The WyzBee THING expansion header has the provision of adding additional

Figure 5. BeagleBoard-Single Board Computer

Figure 6. Dragon Board 410c

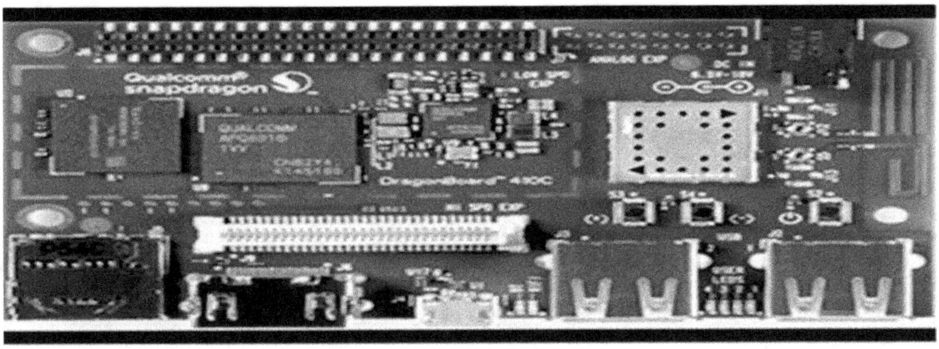

peripherals such as audio, GSM, GPS, capacitive touch display and rechargeable battery provided by Redpine provides additional peripherals which are known as 'Things'. The developers can choose the application development environment which include IAR, Keil and free CoIDE from CooCox based on their requirements. The WyzBee IoT platform is presented in Figure 7.

The Table 1 describes the technical specifications of the prototyping boards.

SENSORS

The Sensing phase of the IoT workflow is established by the *Sensors*. The Sensors are the hardware placed in the fields that are performing the critical work of monitoring processes, taking measurements and collecting data. The deployment of sensors depends upon the application requirements. Precision Agriculture is achieved by

Figure 7. WyzBee IoT Platform

Table 1. Technical Specification of the prototyping boards.

Board	Technical Specification
Arduino Microcontroller Board	• Open source • ATmega328 microcontroller • 14 digital and 6 analogue I/O pins • 32k Flash Memory • Internet connectivity via Shields.
Raspberry Pi	• Open source • Broadcom BCM2835 700MHz processor • GPIO: 8 pins • SD Card socket HDMI • Connectivity via Ethernet.
BeagleBone Black	• Open source • AM335x 1GHz ARM® Cortex-A8 • 2GB of on-board flash • a microSD card reader, HDMI • GPIO 65 pins • Connectivity via Ethernet
Dragon Board	• Qualcomm Snapdragon 410, Quad-core ARM® Cortex® A53 • 1GB LPDDR3 533MHz • 1080p@30fps HD video playback and capture with H.264 (AVC), and 720p playback with H.265 (HEVC) Integrated ISP with support for image sensors up to 13MP • PCM/AAC+/MP3/WMA, ECNS, Audio+ post-processing (optional) • WLAN / Bluetooth / GPS • One USB 2.0 micro B (device mode only) • Two USB 2.0 (host mode only) • One 40-pin Low Speed (LS) expansion connector • One 60-pin High Speed (HS) expansion connector
WyzBee	• ARM Cortex-M4F processor, running at a frequency of up to 160 MHz • Debug options: JTAG and Embedded Trace Macrocells (ETM) • 1MB on-chip flash program memory with flash accelerator and 32KB work flash memory • 128 KB SRAM for code and data use • USB2.0 (Function/Host) Full-Speed with support for up to 6 Endpoints • Up to 36 high speed general purpose I/O ports. • CMSIS-DAP Debug processor: WyzBee™ comes with an on board CMSIS-DAP debug processor

deploying a number of sensors in the field for providing data that helps farmers monitor and optimize crops, as well as adapt to changing environmental factors.

Temperature Sensors

Most of the IoT environments, from industry to agricultural fields use these temperature sensors. The temperature of a machine can be measured by these sensors continuously to ensure it remains within a safe threshold. In the agricultural farm, the temperature of soil, water and plants are measured.

Proximity Sensors

The motion detection is carried out by the Proximity Sensors. A retail shop keeper can use a customer's proximity to a product to send deals and coupons directly to their smart phone. The availability of parking spaces in large venues like airports, malls and stadium is also monitored by the proximity sensors.

Pressure Sensors

The pressure of gas or liquid can be measured by the pressure sensors. This sensors act as a transducer. Pressure sensors can be used to monitor the flow of water through the pipes and make the respective authority alert when something goes wrong. The water wastage due to leaky irrigation system in agriculture can be monitored by these sensors. Smart vehicles and aircraft industry utilize these sensors for determining force and altitude.

Water Quality Sensors

Water Quality sensors are used in the applications like Precision agriculture, water treatment and rainwater quality monitoring.

Chemical/Smoke and Gas Sensors

These devices can be used for air quality control management in smart buildings and throughout smart cities.

Level Sensors

Level sensors detect the level of liquids and other fluids. Smart garbage management and recycling can be done with the help of the level sensors. Tank level measurement, diesel fuel gauging and irrigation control are other applications.

Infra Red (IR) Sensors

There are lots of applications implemented with the Infra Red (IR) sensors. The heat leaks in houses, blood flow in the human body and environmental chemicals can be visualized by these sensors. These sensors can be embedded with wearable electronics.

Humidity Sensors

A humidity sensor senses relative humidity that is both air temperature and moisture. All the control systems in industries and in domestic the humidity sensing is necessary. High volume and cost sensitive applications like office automation, home appliances and industrial process control systems uses these sensors.

Location Sensors

Location Sensors determine whether the latitude, longitude, and altitude of the signals received from the GPS satellites are within feet. Three satellites minimum are required to triangulate a position. Precise positioning is the basis of precision agriculture.

Optical Sensors

The soil properties are measured by the optical sensors using light. The sensors measure different frequencies of light reflectance in near-infrared, mid-infrared, and polarized light spectrums. Sensors can be deployed on vehicles or aerial platforms such as drones or even satellites. Optical sensors have been developed to determine clay, organic matter, and moisture content of the soil.

Electrochemical Sensors

The primary properties needed in the precision agriculture, the soil nutrient levels and pH level of the soil can be measured by the electrochemical sensors. Sensor electrodes work by detecting specific ions in the soil.

Mechanical Sensors

The soil compaction or "mechanical resistance." is measured by the mechanical sensors. The sensors use a probe that go through the soil and trace resistive forces through the use of load cells or strain gauges. Large tractors use the similar form of this technology to predict pulling requirements for ground engaging equipment.

Operating Systems

RIOT OS is a microkernel based operating system suitable for several software requirements for IoT devices and it is designed for energy efficiency, hardware independent development, a high degree of modularity. RIOT has an adaptive network stack, providing support for protocols like 6LoWPAN, RPL, IPv6, TCP and UDP. RIOT's rich set of developer friendly API across all platforms manages a wide range of IoT devices.

FreeRTOS is a small and simple operating system which has a kernel consists of only three or four C files. It offers methods for multiple threads or tasks, mutexes, semaphores and software timers. FreeRTOS has the key features like very small memory footprint, low overhead, and very fast execution.

Contiki is an open source operating system that runs on the tiny low power micro controllers and makes it possible to develop applications that makes efficient use of the hardware while providing standardized low power wireless communication for a range of hardware platform. It is a wireless sensor network operating system and consists of the kernel, libraries, the program loader, a set of processes. Contiki supports a fully standardized IPv6 and IPv4 internet standards. Also supports the recent low power wireless standards 6LoWPN, CoAP and RPL. Contiki applications are written in C programming language, so it is portable to different architectures like TI MSP430. Since it is an open source operating system, it can be freely used in commercial and non commercial systems which can run IoT applications. Contiki provides mechanisms for estimating the system power consumption and to understand where the power was spent. The memory allocation module provides memory block allocation and standard C memory application (Dunkels P.et.al, 2004).

TinyOS is a component-based embedded operating system and an implementation for wireless sensor networks. TinyOS is written in the nesC programming language as a set of cooperating tasks and processes. TinyOS programs are built out of software components, some of which present hardware abstractions. Components are linked to each other using interfaces. TinyOS offers interfaces and components for common abstractions such as packet communication, routing, sensing, actuation and storage.

Embedded Linux is developed using OpenEmbedded, the build framework for embedded Linux. OpenEmbedded offers a best-in-class cross-compile environment. The 32 bit processor embedded systems like ARM, PowerPC, ColdFire etc have enough amount of flash and RAM memory. The response of the kernel is in real time. For example, the linux kernel running in Video Streaming Application is performing the function of converting video format into MPEG4 and sending video stream on network

Mbed is an open source embedded operating system developed by ARM. Mbed is a single threaded OS which will be able to run applications on the smallest and low power devices. Mbed is an operating system platform containing a core, security and key Internet of Things networking and communication technologies and allows the developers to contribute only application code.

Google Brillo is a developer platform for an Android-based embedded OS developed by Google. Brillo operates with a communication protocol called *Weave*.

The smart devices need not have embedded Andriod as their OS. They need to have only the ability to communicate using the *Weave* protocol. This makes the vendors to incorporate Weave in their IoT products to make them Brillo compatible.

Windows 10 For IoT is an embedded operating system released by Microsoft. There are three subset operating systems under Windows 10 for IoT. Windows 10 for IoT Mobile that supports the ARM architecture, Windows for IoT Core that supports Raspberry Pi and Intel Atom and Windows 10 for IoT Enterprise to run a single IoT application are the subsets of Windows 10 for IoT. This OS is suitable for developing in-house applications.

The choice of the hardware and the operating system for developing IoT projects vary widely. The developer can start the project by prototyping using generic off-the-shelf hardware with an iterative design and requirement validation process. Then he can move into deploying the IoT solution.

SMART IRRIGATION SYSTEM

Even though there are different processes involved in smart farming, we consider the Smart Irrigation system as an experimental implementation. As water is the core input for enhancing agricultural productivity, expansion of irrigation has been a key strategy in the development of agriculture in the country. The advanced technologies in the field of IoT can be used improve the agriculture productivity. Smart irrigation is a system which automatically controls and monitors weather conditions, soil conditions, evaporation and plant water and also maintained by some application without direct human interactions. The main aim is to avoid water wastage and improve the quality of crop growth. Various sensors are used to detect the real time conditions of weather and soil. Moisture sensor is used to sense the moisture level of the soil; temperature sensor is used to sense the temperature in the environment and send the information to the microcontroller. Microcontrollers (information gateway) are used for decision making. Soil moisture sensor examines the dielectric content of the soil surface to estimate the volumetric water content in the surface; Temperature sensor use advanced Resistance Temperature Detector Component to detect soil temperature. The cost reduction is the main benefit of the smart irrigation system. The farmer's workload is also minimized. The Table 2 depicts the different hardware needed for implementing the Smart Irrigation System.

Table 2. Technical specifications of Smart Irrigation System

DESCRIPTION	
Moisture sensors- Arduino compatible High Sensitivity Moisture Sensor	The low power device has a working current which is below 20mA, making it useful for battery operated applications. The sensor works bests at 3.3V and has a maximum output signal of 3V as well.
Fire sensors- Arduino compatible IR Flame Sensor Module Detector	The Flame sensor module is a IR sensor module that is particularly sensitive to IR wavelengths emitted from flames. The module has a on-board sensor and a comparetor that can compare the output of the sensor against a threshold voltage and give a digital output. The comparetor threshold can be adjusted using the on-board preset
Arduino Uno R3 with USB Cable	The Arduino Uno is a microcontroller board based on the **ATmega328** (datasheet). It has 14 digital input/output pins (of which 6 can be used as PWM outputs), 6 analog inputs, a 16 MHz crystal oscillator, a **USB** connection, a power jack, an **ICSP**header, and a reset button
Level sensor-Think-Bots Water Level Sensor Liquid Water Droplet Depth Detection	Water sensor is simple and easy to use, and high cost performance, water identification detection sensor, exposed through a series of parallel wire line mark to measure the water droplets, size to determine water level. Easy to complete the water to the conversion of the analog signal, output of the simulation values can be read directly by Arduino board, to achieve the effect of water level alarm
Leakage sensor-Water Sensor Module for Arduino	This sensor monitors the water and identifies water leakage in the storage system
L293D Dual H-Bridge DC Motor Driver module for Arduino	This motor driver is perfect for robotics and mechatronics projects for controlling motors from microcontroller, switches, relays, etc.
INVNT_36 Active Buzzer Alarm	• A module using 9012 transistor drive • 2 Operating voltage 33V-5v • 6 With fixed bolt hole for easy installation • 1 Vcc external 33V-5v voltage (5v microcontroller and can be directly connected to 33V mcu) • 3 I or o external microcontroller io port
ESP8266 ESP-12 Serial WiFi Wireless Transceiver Module for Arduino	• ESP8266 for the Arduino Uno • Highly integrated chip designed for the needs of a new connected world • Self-contained Wi-Fi networking solution

Smart irrigation systems developed with the latest IoT technology can help preservation of water resources by monitoring irrigation through remote sensing technologies. Smart sprinklers connected with smart sensors work automatically to water the agriculture field. The sensors can be deployed in the farm soil to measure the level of moisture. These sensors transmit the data of the moisture level to the smart sprinkler to sprinkle the required amount of water on the soil. The motor driver will control the sprinklers. The figure 8 depicts the block diagram of the Smart Irrigation System.

Figure 8. Smart Irrigation System

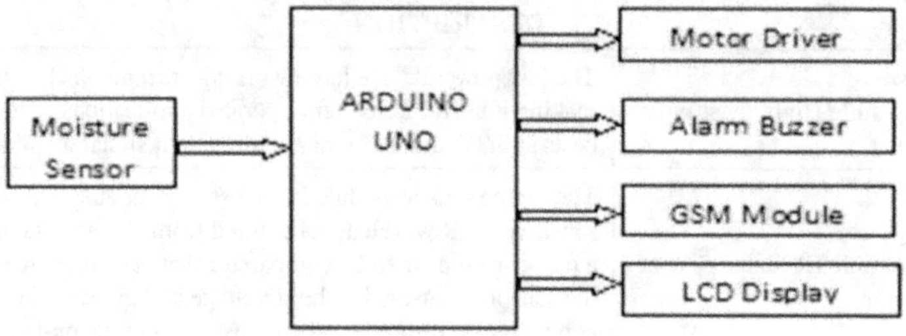

SUMMARY

This chapter presented the Internet of Things (IoT) technologies and communication standards that can be adopted for making a smart environment. We considered the automation of the irrigation in the agriculture field which becomes a Smart Irrigation System. This system would be more beneficial to the farmers at the time of climatic change. The different components of the implementation platform are also discussed.

FUTURE DIRECTIONS

The smart irrigation system can be improved by incorporating the features like uploading the data into the IoT cloud so that the data can be used for future analysis. The data retrieved from the cloud can be analyzed for decision making and predictive analysis.

REFERENCES

Akshay, S., Aishwarya, B. P., & Dr, T. H. (2016). Srinivas (2016), Smart Farm - An Android Based Monitoring System for Agricultural Land. *International Journal of Modern Trends in Engineering and Research*, *03*(04).

Atzori, L., Iera, A., & Morabito, G. (2010). The Internet of Things: A survey. *Journal of Computer Networks*, *54*(15), 2787–2805. doi:10.1016/j.comnet.2010.05.010

Ashton. (2009). Internet of Things. *RFiD Journal*.

Brock. (2002). *Integrating the Electronic Product Code (EPC) and the Global Trade Item Number (GTIN), Updated edition*. MIT Auto-Id center.

Kanjilal, Singh, & Reddy. (2014). Smart Farm: Extending Automation To The Farm Level. *International Journal of Scientific & Technology Research*, *3*(7).

Dunkels, Gronvall, & Voigt. (2004). Contiki - A lightweight and flexible operating system for tiny networked sensors. *Local Computer Networks, 2004. 29th Annual IEEE International Conference on*, 455 – 462.

IEEE 802.15.4. (2011). IEEE Standard for Local and metropolitan area networks--Part 15.4: Low-Rate Wireless Personal Area Networks (LR-WPANs), (pp.314).

Karagiannis, V., & Chatzimisios, P. (2015). A survey on application layer protocols for the internet of things. *Transaction on IoT and Cloud Computing*, *3*(1), 11–17.

Medaglia. (2011). Services, Use Cases and Future Challenges for Near Field Communication. In The Stolpan Project, Deploying Rfid - Challenges, Solutions, and Open Issues. Academic Press.

Tolentino. (2013). *How QR Codes Can Deliver the Internet of Everything*. News.

Gondchawar & Kawitkar. (2016). IoT based Smart Agriculture. *International Journal of Advanced Research in Computer and Communication Engineering*, *5*(6).

Park, M. (2015). IEEE 802.11ah: Sub-1-GHz license-exempt operation for the internet of things. *IEEE Communications Magazine*, *53*(9), 145–151. doi:10.1109/MCOM.2015.7263359

Maharjan. (2010). *RFID and IOT: An Overview*. White Paper. Simula Research Laboratory, University of Oslo.

Series Y: Global Information Infrastructure, Internet Protocol Aspects and Next-Generation Networks,Overview of the Internet of things, ITU-T Y.2060.

Shelby, Z., & Bormann, C. (2010). *6LoWPAN: The Wireless Embedded Internet*. Wiley.

Suma, Samson, Saranya, Shanmugapriya, & Subhashri. (n.d.). IOT Based Smart Agriculture Monitoring System. *International Journal on Recent and Innovation Trends in Computing and Communication*, *5*(2), 177 – 181.

Umamaheswari, S., & Vanitha, K. (2016). *Interaction Model of Service Discovery Using Visa Processing Algorithm and Constrained Application Protocol (CoAP)*. Springer.

Vaibhavraj, Roham, Pawar, Patil, & Rupnar. (2015). Smart Farm using Wireless Sensor Network. *International Journal of Computer Applications, National Conference on Advances in Computing*.

Vilmos, Medaglia, & Moroni. (2011). *NFC Technology and Its Application Scenarios in a Future Of IoT*. Stolpan Project.

Winter. (2012). *IPv6 routing protocol for low power and lossy networks*. Internet Engineering Task Force, RFC 6550.

Winter, T., & Thubert, P.ROLL Team. (2011). *RPL, Internet-Draft*. RPL.

KEY TERMS AND DEFINITIONS

CoAP: Constrained application protocol is a specifically designed application layer protocol for use with constrained devices and constrained networks in the internet of things.

GSM: Global system for mobile used for communicating information between the devices in different locations.

Humidity Sensor: A humidity sensor senses relative humidity that is both air temperature and moisture.

IoT: Internet of things refers to the interconnection of the IP smart objects, such as sensors and actuators.

Microcontroller: A microcontroller is a SoC that offers data processing and storage capabilities. Microcontrollers consists of a processor core (or cores), memory (RAM), and erasable programmable read-only memory (EPROM) for storing the custom programs that run on the microcontroller.

Chapter 5
Internet of Things for Automation in Smart Agriculture:
A Technical Review

Dharm Singh Jat
Namibia University of Science and Technology, Namibia

Anton S. Limbo
Namibia University of Science and Technology, Namibia

Charu Singh
Sat-Com (PTY) Ltd, Namibia

ABSTRACT

By combining the different monitoring and automation techniques available today, we can develop cutting-edge internet of things (IoT) systems that can support sustainable development through smart agriculture. Systems are able to monitor the farming areas and react to the parameters being monitored on their own without the presence of human beings. This automation can result in a more precise way of maintaining the aspects that affect the growth of plants, leading to an increase in the food production on farmlands. This chapter focuses on IOT for automation in smart agriculture and provides a pathway to develop automation system in the smart environment.

DOI: 10.4018/978-1-5225-5909-2.ch005

INTRODUCTION

Internet of Things is the interconnection of physical objects like home appliances, vehicles, and other devices that can be embedded with electronics. This interconnection enables these devices to exchange data with each other as well as other networked devices. IoT devices use sensors to sense the physical world and transmit this sensing in form of an electrical signal. IoT has been applied in many sectors, including agriculture. Applying IoT to agriculture can enable farmers to grow crops or raise animals using the precision of today's modern technology, which results in high outputs of food production. This is achieved by enabling the farmer to monitor aspects that affect production on a farm, these include moisture, humidity, soil nutrients, weather. The monitored data can be analyzed and used to counter the negative effects of these aspects. If used effectively, IoT can simulate an ideal environment for agriculture that is prone to negative effects of parameters that affect the food production in agriculture.

The use of IoT in agriculture has been done using different approaches in the past. Krishna, Silver, Malende & Anuradha (2017) designed and implemented a novel wireless mobile robot equipped with various sensors to monitor different environmental parameters that are suitable for crop yield. In this design, a Raspberry Pi 2 Model B is the main controller. All the sensors such as thermo hygro sensor, soil moisture, humidity, ultraviolet, CO2, ultrasonic and pH sensor are interfaced to Raspberry Pi 2 Model B which is located on the wireless mobile robot. A Camera is also interfaced to the Raspberry Pi 2 Model B to capture the crop field and to observe the live events occurring on crop fields. The novel wireless robot is remotely controlled using necessary commands from the PC section in the receiver side.

Patil and Kale (2016) propose a three modules system for crop monitoring on a farm. The modules are farm side, server side, and client side. The system consists of six methods namely, sensing local agricultural parameters; identification of the location of sensor and data collection; transferring data from crop fields for decision making; decision support and early warming based on data analysis, domain knowledge and history generated; actuation and control based on the decision; crop monitoring via camera module. The paper, further suggests using Ubi-Sense motes for monitoring crop fields. Where data from Ubi-Sense mote will be transferred to Ubi-mote Server side module. And implementing a Decision support system for alerts and crop monitoring. Client-side module will consist of a web application as well as mobile application on Android OS.

Jiber, Harroud, and Karmouch (2011) propose an iFarm Architecture Framework for a more precise way of Agriculture monitoring using Wireless Sensor Networks (WSN). Chang, Zhou, Zhao, Cao, Tan, & Zhang (2014) also proposes an agriculture monitoring system using WSN. In this approach, a layered system consisting of three layers is proposed.

However, most of the work being done in Smart Agriculture is done with little aspects of automation of processes in agriculture, as a result, most of the systems target monitoring. Few systems integrate both monitoring and process automation, and the ones that do, only have a limited automation features. Automation is key to maintaining an ideal environment in which agriculture can be performed with little negative effects of aspects like the weather. This implies that we should be able to monitor parameters that directly affect the form of agriculture we performing, analyze this data, use the analyzed data with the known knowledge of the form of agriculture to perform automation that will counter the negative effects of the aspects we monitoring. Performing this monitoring and automation constantly can result in an environment with little negative effects of the parameters that affect the form of agriculture being performed.

In this chapter, we look at how an IoT system can be designed for both monitoring and automation. This will include analyzing different aspects of an IoT system, which include the Hardware to use; Software to run on the hardware; How data is measured, transmitted, stored and analyzed; and how Automation can be achieved using the results from the analyzed data.

CONSIDERATIONS OF AN IOT SYSTEM DESIGN

Based on the related work done in monitoring and automation in Smart Agriculture, they are considerations that need to be done when designing an IoT system. Here are some of those considerations.

Hardware Considerations

In the field of IoT, new devices are continuously being developed. This continuous development has enabled users to have a wide range of hardware devices to choose from. Because of many devices being released, there is need to consider some characteristics of these devices, when choosing which device to use for an IoT implementation. These characteristics are based on how the device acquires data, processes this data, how it connects to other devices and the power consumption of this device while acquiring and processing this data.

These characteristics are complemented by the ever reduction of both the size and the cost of hardware devices. When considering the type of device to use for smart agriculture, it is key to look at the characteristics of a device in conjunction with the environment where this device will be deployed and the parameters to measure and control in this environment which can be referred to as the "thing".

Microchip produces several Digital Signal Controllers (DSCs) for solar power generation and Motor control for various applications. The Microchip dsPIC33 DSCs provides high-efficiency motor control and other specific purpose applications and the device can operate up to 150°C (Microchip 16-bit MCUs and DSCs, 2018).

Here are some things to consider when deciding on hardware devices to use for monitoring and automating in Smart Agriculture.

Data Acquisition

Data Acquisition is how devices monitor environmental conditions and convert this monitoring to an electrical signal. In IoT most data is acquired using sensors. When considering which hardware (sensors) to use, there is need to understand the kind of data needed to be acquired from the environment. After which we can determine the correct sensor to use in this environment.

Therefore, when deciding on the hardware to use for acquire data from an environment, it is important to consider physical signals needed to measured, which will assist in determining the type of sensors needed to measure that signal. Other things that might be worth considering include the rate at which the signal is measured, as well as the accuracy of the signal.

Data Processing and Storage

IoT devices capture massive amounts of data but unfortunately, they have limited storage capabilities. Therefore, when deciding on the kind of device to use, we have to put the data storage and processing in mind.

Because of the limited storage, some IoT devices simply capture data and transmit it. While others capture data and perform some processing on this data and only transmit the semi-processed data to another device which can perform more processing on this data. The approach to take when considering how data will be stored and processed has to have the environment data in mind. They are conditions which do not require monitoring at a regular frequency. While they are conditions which can be predicted using previous analysis and confirmed with the readings from a device in the field. In such situations, we need a device that only transmits readings when requested without the device storing data.

Hence, when deciding the hardware to use there is need to consider how the data will be processed, and ask ourselves questions like, do we want the data to be processed and stored by the IoT device, or do we rather use the IoT device to just capture data and transmit it to another device dedicated to storing and processing data.

CONNECTIVITY

Devices in IoT system need to communicate, which is worthwhile to consider the way in which these devices communicate with each other as well as other networked devices. IoT devices can communicate in different ways, this includes wireless technologies like Wi-Fi, Bluetooth, Infrared, or Cellular Networks or wired technologies like Ethernet, USB, Serial connections. All these different communications methods have to be selected based on the environmental factors to monitor and automate.

Most IoT devices come with embedded communications modules, the ones that do not can use external modules that can be integrated to the device, as these devices come with many options of adding external peripherals and modules. One such device is a Raspberry Pi Model 3, which comes with Wi-Fi and Bluetooth Connectivity, as well an Ethernet port. It also has the ports for a screen, camera and serial port for other sensor and modules.

Power Source

Because IoT devices are electronic devices, we have to also consider how to these devices will be powered by an IoT system. When deciding on the kind of device to use, it is a good idea to consider the amount of power that this device will require to perform the operations required of it. This can assist in determining the best way to power the device.

For example, if monitoring a few plants in a small garden, devices can be operated using wall sockets as power sources. But other devices might need to be deployed in places where running power cables to power these devices are not ideal. In such cases, there is need to consider other sources of power, like solar, if the area to be measured gets enough sunshine or batteries if the area does not get enough sunshine or a solar system that charges batteries during the day and runs off batteries at night or when there is little sunshine.

Gerber (2017a) suggests that some devices in an IoT system can be hibernated periodically, depending on the need of the device. This can be applied when monitoring sunshine in a field, as an example. In this case, there will be no need to have the light sensor operating during the time that there is no sunshine.

Physicality of the Device

The physicality of the device includes the overall physical design and size of the device (Gerber, 2017a). It is important to put the design of the device in mind when developing an IoT system, especially one for Smart Agriculture, as most devices will be exposed to so many environmental factors like sunlight, moisture, and other harsh weather conditions. For example, when monitoring temperature, the minimum and maximum temperature that the device can operate at should be known, as well a history of the temperature patterns of the area where the device will be placed.

COST

The cost of the devices is also key as it will assist in the estimation of the cost of establishing a fully functioning system as well as the maintaining of the system. The cost of acquiring devices has to be in correlation with what the device will be used for. They are devices that are able to measure multiple parameters (for example, temperature, humidity, air pressure), while some devices are specialized to measure a simple parameter. If the cost of buying one device with multiple sensors is same or less than the one of a specialized device and we need multiple devices to cover a large area, and we not very conscious about accuracy, then the multiple sensor devices will be ideal. This will enable the coverage of a large area at a low cost of buying hardware. But if we want very accurate reading and only need a single device to do the job then, of course, the specialized device will be ideal as we only interested in one parameter.

Software Considerations

After careful consideration and determining suitable hardware that we can use to develop an IoT system. There is need to find suitable software or a platform that will assist in facilitating the instructions to and from IoT devices. Like hardware, there are a number of platforms that can be used in developing an IoT system, some of the hardware manufacturers provide software for their devices, which makes it difficult for these devices to interoperate with other devices.

Other manufacturers' hardware can be implemented with multiple platforms, making these devices more desirable for IoT as there will be a list of libraries that can be used to send and receive data from IoT devices. One such device is an Arduino, which is open source computer hardware. A program for Arduino may be written in any programming language with compilers that produce binary machine code for the target processor (SparkFun, n.d.). Some platforms for IoT devices are not free and there might need to but a license. So when designing an IoT system, they also need to consider the kind of software that will be running on the IoT devices, and if this software will be utilizing the hardware capabilities of the device fully.

Security Considerations

Hess (2014) identified five security considerations after conducting numerous interviews with professionals who work with customers involved in implementing IoT infrastructures. These were Device security, Network security, Server security, Data security, Operating system security.

MONITORING USING IOT

As outlined in the related work, IoT can be used to monitor physical phenomena and relay these reading to other devices as a form of electrical signal, all this can be done without the physical presence of human beings, implying that these ways of monitoring can be done without interfering with how environments naturally operate. In this section, we look at various methods involved in monitoring an environment for a Smart Agriculture approach, while the bearing in mind the considerations mentioned in the previous section.

Sensor and Data Acquisition

The main aim of having an IoT system is to acquire data by means of using sensors. In this section we look and how this data can be acquired and the sensors that can acquire this data. As stated earlier, they are so many devices on the market for IoT systems today, making the choice of which device to use very challenging.

When monitoring an environment for a Smart Agriculture approach we need to determine which environmental factors affect what we are farming. When growing crops, we need to consider the moisture, pH balance and nutrients of the soil, and other factors like temperature, wind, air pressure, and sunrise. All these factors will require a sensor, which is there is also the need to narrow down on the kind of factors directly affect the growth of the plant. Some plants might require only good moisture and temperature in the soil and less affected by wind, as an example. Here are a few parameters we can monitor and type of sensors we can use to monitor them.

Soil Moisture

Soil moisture plays a key role in the life of the plant. Certain plants require constant moisture in the soil, hence the need to monitor this soil parameter. Azzola (2017) used an Arduino MKR1000 and a SparkFun Soil Moisture Sensor. In this approach, the Arduino MKR1000 was getting data from the SparkFun sensor and relaying these data using Wi-Fi to a cloud platform. The cloud platform was then used to get a better (graphical) representation of this data.

Data generated here can be used to analyze the effects soil moisture had on the growth of the plants in a field. This data can also be used to automate a process to maintain the moisture in the soil at an almost constant level.

Soil PH Balance and Nutrients

Soil pH is a measure of the acidity and alkalinity in soils. The desirable pH range for optimum plant growth varies among crops. While some crops grow best in the 6.0 to 7.0 range, others grow well under slightly acidic conditions. Knowledge of the soil and the crop is important in managing soil pH for the best crop performance (Mosaic Crop Nutrition, n.d.). Just as with pH, soil nutrients are also important for adequate plant growth.

Hence, there is need to monitor the pH balance and nutrients in the soil to maintain it at an ideal level for optimum plant growth. Demir (2017) used a Photon microcontroller with a built-in Wi-Fi shield. A SEN0161 pH probe was attached to the microcontroller. SEN0161 is a cheap pH electrode, the disadvantages of this electrode, however, is its' limited lifespan.

Weather (Temperature, Humidity, Rain, Wind, Sunshine)

Other factors that can be ideal to monitor include temperature, to see patterns on how it fluctuates over time and how this fluctuation affects plant growth. Relative humidity levels affect when and how plants open the stomata on the undersides of their leaves (Polygon, n.d.). This makes relative humidity another factor to monitor when growing plants.

Another factor to monitor can include rainfall in the area, and see how this area rainfall can affect the soil moisture, as in how much rainfall will be required to increase the moisture in the soil at an ideal level. Wind and Sunshine affect the temperature and humidity, hence the need to also measure these two parameters to have a clear indication of how they affect plant growth.

Many projects have been done with cost-effective hardware for mini weather stations. The interesting one was done using a Raspberry Pi. This weather station was able to measure Rainfall, Wind speed, Wind gust speed, Wind direction, Ambient temperature, Soil temperature, Barometric pressure, Relative humidity, Air Quality. This project was done for schools, and the idea was to then train learners to develop their own weather stations (Honess, 2015).

Image Monitoring

Image monitoring of crop fields is also an interesting way to get information on what is going on in the farm area. The images generated can be used to analyze the state of plants and identify things pests on plants, or diseases that might be affecting the plants.

Many IoT systems can easily integrate a camera to capture images. A Raspberry Pi comes with a webcam slot on the chip. The only thing to consider when using image monitoring is the size of the images to be captured and send, this might cause delays in the transmission of other readings being monitored in the IoT system.

Data Transmission

After acquiring data using sensors, this data will have to be transmitted for analysis. In this section, we look at ways in which IoT system transmit data. They are many methods in which devices can send and receive data which each other in IoT system. Ray (2015) explains four commonly used protocols for communication in IoT devices.

According to Uddin, Mansour, Jeune, & Aggoune (2017) for agriculture the Internet of Things (IoT) is a fast-growing newer technology where various sensors are connected with hardware for monitoring and controlling the crop parameters to get quality and quantity of agriculture product. For implementing the IoT in agriculture study proposed a dynamic clustering and data gathering mechanism. Unmanned Aerial Vehicle (UAV) is used to identify and assist the operation IoT devices and forming a cluster which is useful in the establishment of reliable uplink communication for crop data transmission.

Prathibha, AHongal, & Jyothi, (2017) developed smart agriculture using automation and IoT monitoring system for temperature and humidity in an agricultural field using various sensors and CC3200 single chip. For image capturing a Camera is connected with CC3200 chip and transmit collected pictures through MMS to farmer's mobile using IEEE 802.11 connectivity.

IEEE 802.15.4: ZIGBEE

A wireless technology currently gaining traction in the LPWAN group, ZigBee is an open global standard and is designed specifically to be used in M2M networks (Ray 2015, para. 3).

IEEE 802.11: WI-FI

Wi-Fi uses radio waves (RF) to allow two network devices to communicate with one another (Ray 2015, para. 6). This protocol is widely used which makes it easy for multiple devices to communicate with other, which makes it ideal for a Smart Agriculture approach as there will need to have multiple devices from different vendors communicating with each other.

Data Storage and Processing

The data is captured and transmitted in enormous, this means that they have to be means of storing this data, and processing it for us to get meaningful information out of it. Devices capturing data normally have limited storage capabilities, hence the need to transmit this data to a central location where it can be analyzed and compared to data captured by other devices. In this section, we look at methods of this data storage and processing.

IoT devices can capture massive amounts of data. The problem with this massive data is that most IoT devices do have large storage capacity, this is further complicated by the fact that IoT devices do not capture, store and process data in a uniform way, meaning that each IoT system will have to have a tailor-made approach to data management.

This then brings a new problem on how to manage this data. In some IoT systems, the sensing devices do not store much data, they capture the data, wait for transmission time, once the data is transmitted, delete data to make space for the next capturing cycle. Some IoT devices can be configured to only capture and transmit data, which can cause congestion in the network.

This approach can be applied in IoT systems which can have a dedicated storage device, with good connectivity between the sensing devices and the storage device. In this approach, the processing can also be done on the storage device.

Because of the volumes of IoT data, it becomes a challenge to analyze this data. This is further complicated if there is need to quickly analyze data and automate a process. Many researchers are working on effective methods of analyzing IoT data. Just as in storing of data, there is no one approach that can work in all aspects of IoT, because of this, when developing an IoT system, we need to have clear guidelines on what it needs to be revealed from the data. Do we just need graphs from the data or do we need more analysis, and what analysis algorithms can work best on this data?

AUTOMATION

Once data from the sensors has been analyzed, we can use this data to automate certain tasks to counter the effects of the measured parameters. In this section, we look at examples of implementing automated systems that react to the monitored data.

A soil moisture IoT system used to measure the levels of moisture in the soil can integrate an irrigation system that is automated. The system will be receiving readings of how much moisture is in the soil and use the irrigators to maintain the moisture at an almost constant level which will best suit the plant being grown.

An IoT system measuring the pH Balance in the soil can be automated. If the pH sensors are measuring high acidity levels in the soils, the system can act on this automatically without the need for human inference. This can be automated by having a system that releases alkaline in the water used for irrigation to reduce the acidity levels in the soil to adequate level best for plant growth. The system can then just notify the farmer on the action that was taken and the reason why.

A more complex system will be to use image processing for pest and disease control in growing plants. This system will use an image being generated by a camera and automatically detect pests or disease on the plants using image processing. This will also include identifying the type of disease or pest and determining the remedy for such a pest or disease, all without the presence of the farmer. The system can then trigger the irrigation system to spray the correct pesticide. The pesticides to be integrated into the system will be based on the type of pests common in the area. This system can also apply some learning algorithm to be able to perform automation of controlling pests.

All these systems can be developed with the already existing devices. It will just be a matter of getting the right devices to perform the tasks that have been previously been done using traditional methods of agriculture.

These days Agriculture farming facing some challenges like water shortage, tough weather conditions, electricity shortage, pest control. The measure challenge is electricity shortage and now solved by Solar Powered Irrigation systems and can be helpful to control, measure, monitor and cultivate (Internet of Things- Irrigation System, 2018). Farmers can easily observe the irrigation systems whether it is working or not. They can also measure temperature and soil moisture by using the wireless connectivity through WiFi enabled a mobile phone or other smart devices. They can also access all the information form remotely using internet connectivity and by knowing weather forecasts information they can take a decision about where and when to fertilize, water, and pesticide the plant. The smart farming system includes electric solar generation system for supply power to the IoT devices, wireless remote monitoring, and controlling system, driving the water pump, interfacing with sensors.

CONCLUSION (COMBINING MONITORING AND AUTOMATION)

By combining the different monitoring and automation techniques available today, we can develop cutting-edge IoT systems that can support sustainable development through Smart Agriculture. This implies combining different monitoring technologies with automaton technologies so that the systems are able to monitor the farming areas and react to the parameters being monitored on their own without having us human being. This automation can result in a more precise way of maintaining the aspects that affect the growth of plants, leading to an increase in the food production on farm lands.

REFERENCES

Ammad Uddin, M., Mansour, A., & Le Jeune, D. (2017). Agriculture internet of things: AG-IoT. In *2017- 27th International Telecommunication Networks and Applications Conference (ITNAC)*. IEEE. DOI: 10.1109/ATNAC.2017.8215399

Azzola, F. (2017). *IoT System To Monitor Soil Moisture With Arduino*. Retrieved from https://create.arduino.cc/projecthub/jfrankie/iot-system-to-monitor-soil-moisture-with-arduino-5e370c

Chang, H., Zhou, N., Zhao, X., Cao, Q., Tan, M., & Zhang, Y. (2014). *A new agriculture monitoring system based on WSNs*. Paper presented at 2014 12th International Conference on Signal Processing (ICSP), Hangzhou, China. 10.1109/ICOSP.2014.7015295

Demir, A. (2017). *Remote Monitoring of the Soil Quality with the Internet of Things*. Retrieved from https://sites.tufts.edu/eeseniordesignhandbook/files/2017/05/Azure_Demir_F1.pdf

dsPIC® Digital Signal Controllers. (n.d.). Access on 13 April 2018, from http://www.mouser.com/ds/2/268/00001032Q-1131382.pdf

Gerber, A. (2017). *Choosing the best hardware for your next IoT project*. Retrieved from https://www.ibm.com/developerworks/library/iot-lp101-best-hardware-devices-iot-project/index.html

Gerber, A. (2017). *Making sense of IoT data*. Retrieved from https://www.ibm.com/developerworks/library/iot-lp301-iot-manage-data/index.html

Hess, K. (2014). *Five security considerations for IoT implementations*. Retrieved from http://www.zdnet.com/article/five-security-considerations-for-iot-implementations/

Honess, D. (2015). Raspberry Pi Weather Station for schools [Blog post]. Retrieved from https://www.raspberrypi.org/blog/school-weather-station-project/

Internet of Things Irrigation System. (2018). Retrieved from http://www.microchip.com/design-centers/internet-of-things/applications-pages/irrigation-system

Jiber, Y., Harroud, H., & Karmouch, A. (2011). *Precision agriculture monitoring framework based on WSN.* Paper presented at 2011 7th International Wireless Communications and Mobile Computing Conference, Istanbul, Turkey. 10.1109/IWCMC.2011.5982844

Krishna, K. L., Silver, O., Malende, W. F., & Anuradha, K. (2017). *Internet of Things application for implementation of smart agriculture system.* Paper presented at 2017 International Conference on I-SMAC (IoT in Social, Mobile, Analytics and Cloud) (I-SMAC), Palladam, India. 10.1109/I-SMAC.2017.8058236

Mosaic Crop Nutrition. (n.d.). *Soil Acidity.* Retrieved from http://www.cropnutrition.com/efu-soil-ph

Patil, K. A., & Kale, R. N. (2016). *A model for smart agriculture using IoT.* Paper presented at 2016 International Conference on Global Trends in Signal Processing, Information Computing and Communication (ICGTSPICC), Jalgaon, India. 10.1109/ICGTSPICC.2016.7955360

Polygon. (n.d.). How Humidity Affects the Growth of Plants [Blog post]. Retrieved from https://www.polygongroup.com/en-US/blog/how-humidity-affects-the-growth-of-plants/

Prathibha, Hongal, & Jyothi. (2017). IOT Based Monitoring System in Smart Agriculture. *2017 International Conference on Recent Advances in Electronics and Communication Technology*. Retrieved from https://ieeexplore.ieee.org/document/8081906/authors

Ray, B. (2015). 5 Types of Wireless Technology For The IoT [Blog post]. Retrieved from https://www.link-labs.com/blog/types-of-wireless-technology

SparkFun. (n.d.). *What is an Arduino?* Retrieved on 13 April 2018 from https://learn.sparkfun.com/tutorials/what-is-an-arduino

Chapter 6
Automation of Marginal Grass Harvesting:
Operational, Economic, and Environmental Analysis

Paulo Peças
Universidade de Lisboa, Portugal

Gonçalo Manuel Fonseca
Universidade de Lisboa, Portugal

Inês Inês Ribeiro
Universidade de Lisboa, Portugal

Claus Grøn Sørensen
Aarhus University, Denmark

ABSTRACT

In the European Union (EU), the use of fossil fuels brings several disadvantages, as they are the main culprits responsible for pollutants and GHG emissions. The increasing demand for sustainable fuels leads to the research of alternative technologies, such as biogas production from lignocellulosic materials. Therefore, the acquisition of biomass from marginal areas under Danish conditions has been evaluated in terms of alternative harvesting equipment: an automated robot (Grassbot) versus a regular tractor for key grass materials used for biogas plants (chopped, unchopped, and baled grass) and compared regarding operational, economical, and environmental performances. The evaluation uses two operations models (IRIS and DRIFT) to consider the field characteristics, machinery characteristics, etc. Selected results show that in terms of fuel consumption, chopping, and mowing are the most demanding operations, and that there is no significant difference between the harvesting equipment regarding CO2 emissions.

DOI: 10.4018/978-1-5225-5909-2.ch006

INTRODUCTION

The increasing demand for sustainable fuels leads to the research of alternative technologies, such as biogas production from lignocellulosic materials. These materials are the most common worldwide and are mainly located in forests or marginal area. However, operational problems have emerged in terms of exploiting them in a cost and environmental efficient way (Christian, 2000; Sørensen and Bochtis, 2010; Pavlou *et al.*, 2016). Difficult terrain topography, difficult access infrastructures and poor soil conditions require different approaches. In particular, efforts are being made to develop smaller, lighter and manoeuvrable machines which are able to perform harvesting operations autonomously (Sørensen and Nielsen, 2005; Sørensen and Bochtis, 2010; Orfanou *et al.*, 2013).

This chapter analyses a marginal grass supply chain in Denmark using alternative harvesting equipment: an automated robot (GrassBot) versus a regular tractor (conventional alternative). In addition, key grass materials accepted by biogas plants (chopped, unchopped and baled grass) are evaluated and compared regarding operational, economical, and environmental performances.

The GrassBot is designed to operate with minimum human intervention but in an efficient and safe way. From the performance point of view, a conventional tractor weighting 2 tons has an output power of around 30kW, the present GrassBot model offers 74 kW with the same weight allowing to process more biomass. This machine is able to cover all the field areas autonomously. To do that, an optimized field path generated by a heuristic algorithm that considers the field boundaries, obstacles, working width, headland passes and driving direction, creates a set of coordinates which represent the parallel fieldwork tracks for traversing the fieldwork area (Sørensen and Nielsen, 2005; Bochtis and Sørensen, 2009). In addition, it is also possible to control the GrassBot using a remote control.

To accomplish the comparison of the harvesting equipment, and also taking into account the supply-chain implications, a process-based model called IRIS is proposed and integrated with a process requirements model (DRIFT) previously developed by Aarhus University (Sørensen and Nielsen, 2005). DRIFT is a program for the modelling and estimating process requirements, namely labor, equipment and time required for each process and also field capacity and efficiency depending on field characteristics and process parameters. The proposed IRIS model receives inputs from the DRIFT model and calculate the resources requirements for a specific production scenario depending on field dimension, harvesting and collection processes. IRIS include materials, energy consumption, labor and equipment required from the equipment manufacturing to the biogas plant gate, where the biomass product is delivered. The parametric characteristics of the IRIS model allow for varying the input parameters such as type of equipment, type of collection or fieldwork conditions. This enables to see how the system is affected by the operational parameters and therefore find the best compromise for each scenario considered. In a second step, IRIS calculate the cost for each supply chain alternative as well as the energy consumed and the green-house gas (GHG) emissions. Through the use of these models, the total supply chain economic and environmental performances of the use of the autonomous robot are compared with conventional harvesting methods for several scenarios of marginal grassland dimensions, as well as for different biomass products, namely unchopped, chopped and bailed biomass. In addition, the transport of the biomass is also considered for the overall economic and environmental analysis, since different biomass density causes different transport requirements.

BACKGROUND

The Biomass in Marginal Grass Lands

The demands to find sustainable alternatives for fossil fuels have never been so high. In the European Union (EU), the use of fossil fuels brings several disadvantages, being the main responsible for pollutants and GHG emissions (The European Commision, 2000). Furthermore, Europe dependency on external fuel sources is expected to increase to 70% until 2030, if no energy alternative strategies are implemented (The European Commision, 2000). In order to mitigate this problem, some mechanisms have been being developed to compel countries reducing their emissions and their dependency on fossil fuels, such as Kyoto t Protocol. For renewable energy sources, the goal is to increase the use of liquid biofuels from 2% in 2005 to 10% in 2020 (Comission, 2007). In general, it is expected that cellulosic biofuels reduce GHG emissions compared to the use of fossil fuels for energy purposes (Demirbas, 2008).

Denmark is one of the leading countries on using renewable energies, namely biomass. To accomplish this, a big research effort has been made to improve the energy efficiency and to reduce the emissions of power plants, having as target achieving a neutral carbon impact in 2050 (Sørensen and Nielsen, 2005; Peeters, 2009; Sørensen and Bochtis, 2010). In fact, according to the current EU polices, biofuels must reduce CO_2 emissions by at least 35% compared to petroleum, which means a limit of 53 kg CO_2 GJ^{-1} (Comission, 2007). However, emission requirements is currently being tightened, since after 2017 this percentage will be increased to 50%, which is translated to a maximum of 42 kg CO_2 GJ^{-1}, (Helgadottir and Hopkins, 2013). The use of biomass fuels instead of fossil fuels can contribute to achieve this target.

However, regarding biomass energy production this increasing effort has been translated into an increasing need for corn and other feedstock commodities (sugarcane or corn for fuel production). If this tendency continues, large quantities of these crops must be harvested every year and the area required per year around supply will not be sustainable. Thus, it is vital to find and explore alternative resources for biomass fuel production (Sharma *et al.*, 2013).

One of the very promising alternatives of biomass for a more sustainable process to produce energy is the use of the biomass from marginal grass lands (MGL). This kind of lands has high phreatic level, difficult access and difficult terrain morphology. Therefore, these lands are not suited for normal agricultural production (Schweier and Becker, 2013).

In the past, MGL were used by farmers to feed their animals by grazing. However, the paradigm has now changed since animals are now fed inside the farms. Therefore, grasses and other plants have the opportunity to grow, accumulate organic matter and attract wild animals, creating problems related to phreatic water contamination. In this sense, measures should be taken to decrease nitrate and phosphate pollution (Peeters, 2009). This represents an opportunity to explore these fields for energy purposes with side benefits for the society and economy. By harvesting these lands, nitrogen is removed, which reduces the leaching. Mowing also prevents scrublands to appear. In addition, visual impact is improved since some vegetation can grow up to several meters if not harvested, which can be avoided applying this process (Mortensen, 2014). These benefits make the use of MGL biomass an attractive process.

However, to achieve a good understanding of its potential the analysis of the harvesting phase is not sufficient. The consideration of the entire supply chain from the MGL to the biogas plant is mandatory in order to have a comprehensive view of the total cost involved as well as of the energetic and emissions balance. The biomass supply chain depends on several factors, namely on the type and quantity of materials, the way they are harvested and aggregated, soil quality sustaining requirements, and finally the transport distance and efficiency (e.g. depending on density of the transported biomass). The amount of biomass extracted from a certain field can be considerably different, depending on various factors. Low production fields only produce 2-3 tdm ha^{-1} (tons of dry matter per hectare), although a higher production of 10-12 tdm ha^{-1} can be obtained by using fertilizers and other inputs (Zhang and Hu, 2013). However, when fields are located in protected landscape, the average yield used in the fields is 4 tdm ha^{-1} since no additional inputs can be applied. The amount of added substances per ha in MGA (i.e. fertilizers, herbicides, irrigation, etc.) induces big differences on the field material output such as the yield or material quality (Peeters, 2009).

Regarding agriculture, this sector was responsible for up to 15.5% of total GHG emissions in 2004 (Caffrey and Veal, 2013). Using biomass fuels such as biogas, the GHG emissions due to transport and electricity in this sector can be reduced (Jungbluth, Büsser and Frischknecht, 2008).

Description of the Supply Chain Stages

A biomass supply chain involves the integration of biomass harvesting, load and collection processes, the storage facilities, transportation networks and processes to convert the raw material into an energetic resource or value added product (Sharma *et al.*, 2013).

There are also several aspects that must be taken into account, for instance the raw material quality. The product should have homogenous density and the geometry should be carefully selected according to each use, in order to provide the highest efficiency energy turnover. Regarding the logistic efficiency, the moisture content (MC) is a key aspect, as it plays an important role on technical assets such as transport and storage capacities. Storage is affected by MC, since higher moisture levels cause higher matter losses. If MC is too high, adding additional pretreatment processes, such as drying, is a common practice.

The biomass supply chain can be divided in five main steps as illustrated in Figure 1.

Agricultural fields can be harvested using a mower, a forage chopper, or a sickle bar cutter. After mowing the material, if the grass is too wet, it must be dried in order to reduce its MC. To perform this task, tedders are used to spread the grass and promote a faster drying process. Once biomass is windrowed, it should be prepared for transportation. The collection processes such as balling, chopping or a self-loading wagon are the main ways to perform this task (Figure 2). Balling can be seen as a collecting and densification process for which grasses and forage are

Figure 1. Biomass Logistic Chain (LSC: local storage center). (adapted from (Mafakheri and Nasiri, 2014)

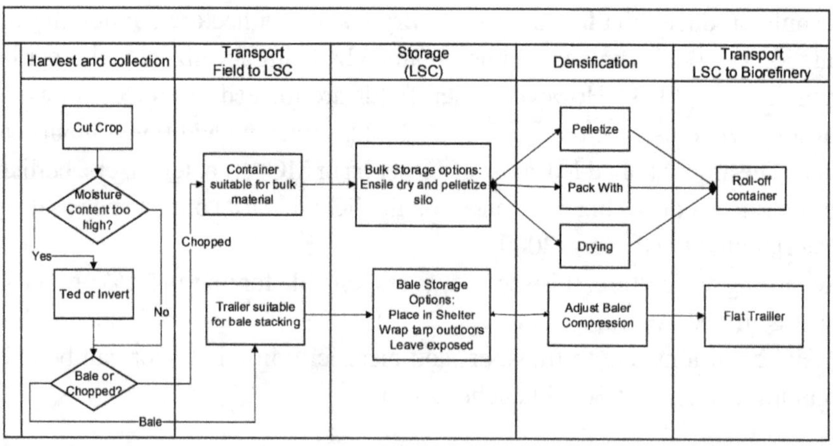

Figure 2. Biomass products

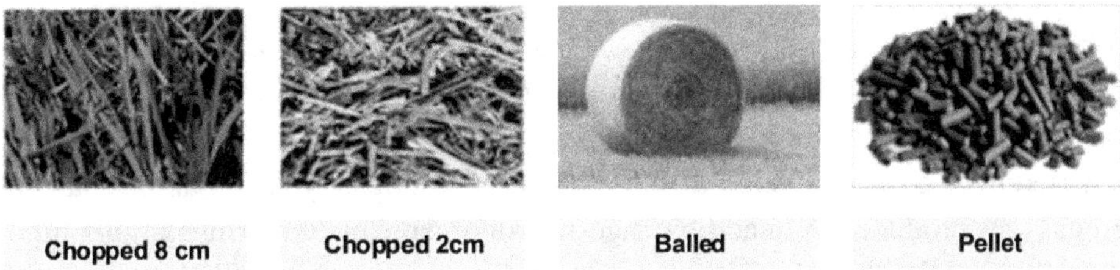

gathered into units called bales. Forage chopper or silage chopper deliver a very high quality product with homogeneous density. There are two types of choppers: the trailed, which is the most affordable and the self-propelled ones. Chopping delivers a very high material processing ratio, being considerably faster than baling operations (Cheng, 2009).

The collection method influences transport costs between the field and local storage center (LSC), higher density biomass products allow lower transport requirements. Transport can be executed by a tractor or a lorry, depending also on the field area, transport distance and road infrastructures. In most of the cases, a tractor is used to perform the task, since usually the LSC is located near the fields.

The storage is an essential step of the biomass supply chain. In this step, the material is stocked and it is ensured that degradation related to microbial activity and climate is minimized. In particularly, in Denmark the biomass is usually stored in covered areas before being transported because of the high wet conditions.

The densification, usually done by grinding and pelletizing, is the primary mean to achieve higher densities of the biomass product. Through this process, biomass is broken down mechanically into small particles which are compressed under high temperature and pressure conditions. Densification is required on baled and unchopped type of collection. This operation can be done in a central collection point or in the biogas plant (Hamelinck, Suurs and Faaij, 2005).

There are three main ways used to transport the processed biomass from the LSC to the biogas plant: road, rail and by sea. Despite these alternatives, biomass is usually transport over small distances. So, trucks are typically used by its flexibility and efficiency in distances up to 100km (Smeets, Lewandowski and Faaij, 2009). The trailer types vary according to the type of transported biomass. The two main trailer types used are roll-off containers and flat trailers.

Producing Energy Using Biomass

Biomass is a renewable source for producing energy (Ekpeni *et al.*, 2014). Energy can be obtained from two types of biomass: manure and plants. The energy of biomass can be converted into high density fuels with high energy density instead of being released in direct combustion reactions. These high density fuels can be charcoal, biogas (gas produced by anaerobic digestion or organic matter fermentation), liquid fuels (biodiesel or bioethanol) and producer gas (Sharma *et al.*, 2013). In this work, only biogas is studied. In Table 1 a comparison between different energy sources is shown based on their densities and specific energies. As it can be noticed the biomass materials are almost a third less energy efficient compared with diesel.

The Drift Model

Aiming to understand the joining effects of field and equipment characteristics Sørensen & Nielsen (2005) developed the DRIFT model: an informatics tool to estimate the process requirements for all field operations and also for harvesting and collecting the biomass in the fields (Figure 3). Based on empirical, operational and physical correlations, this model outputs the required task times used to perform the operation and the quantity of equipment and workers necessary. This tool allows for simulating and understanding the effects on the resources required for different types of harvesting equipment and field characteristics and for different types of biomass collection

Table 1. Biomass typical Bulk and Energy Densities

	Density kg m-3	**Avg. MJ t-1**
	Biomass	
Lose Chops 60mm	70	13,000
Lose Chops 10mm	118	13,000
Bales	150	14,300
Pellets	550	19,100
	Other sources of energy	
Methane (1.013 bar, 15 °C)	790	55,100
Coal	900	32,500
Diesel	840	45,600

In this way, the DRIFT model outputs the process requirements that are mandatory if a subsequent economic or an environmental performance analysis is to be done. Several authors have pointed out that besides the economic performance assessment of the biomass supply chain, it is very important to assess also sustainability related aspects like fuel consumption, energy balance and GHG emissions of these activities (Sokhansanj, Kumar and Turhollow, 2006; Schweier and Becker, 2013; Pavlou *et al.*, 2016).

Nevertheless, no published studies were found that address altogether operational, economic and environmental analysis of MGL supply chains. Therefore, in the present work, the DRIFT model is integrated with a proposed model (IRIS) to estimate those performances on two types of harvesting equipment and for three different collection alternatives. These integrated analyses are applied to MGL giving a contribution to increase the knowledge about biomass supply chain.

MEANS AND METHODS OF THE STUDY

Motivation and Scope

As explained earlier, MGL has characteristics regarding area size, topography and harvesting conditions that create the need to develop equipment more adequate to these areas than the conventional ones. In this work, the performance of an autonomous machine developed at Aarhus University, called GrassBot was compared with a conventional one (a regular tractor).

Figure 3. DRIFT model inputs and outputs

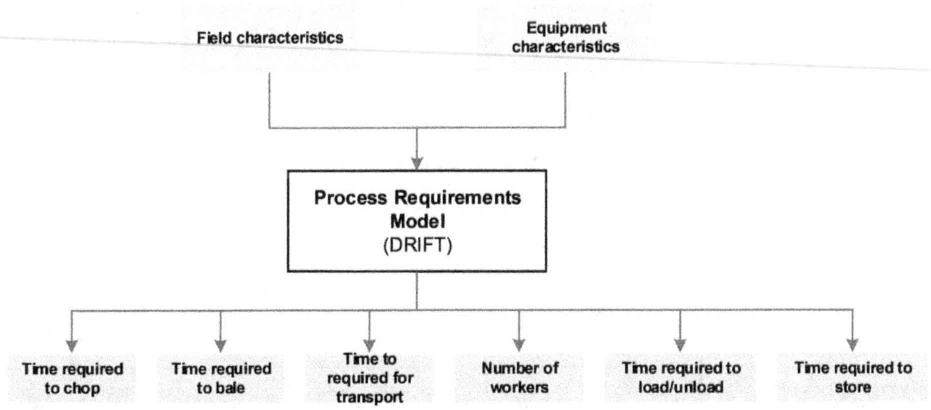

The GrassBot is designed to operate with minimum human intervention but in an efficient and safe way. From the performance point of view, a conventional tractor weighting 2 tons has an output power of around 30kW. The present GrassBot model offers 74 kW having the same weight (Figure 4). A comparison between these two machines is presented in Table 2. The GrassBot has lower acquisition cost and weight and requires lower power to take off, but its belts has a higher unitary cost than the tractor tyres. This are static characteristics that must be analyzed under operational conditions. Therefore, a complete analysis of the implication of these characteristics in the harvesting performance is necessary to understand in which scenarios each of the alternatives performs better.

The main specific aim of this study is to compare an automated solution (the GrassBot) with a conventional solution (regular tractor) performance for the harvesting and collection of MGL for a variety of specific field conditions and operations under Danish conditions. Another aim is to analyse a grass supply chain from harvesting to the biogas plant gate, considering three biomass types of collection (product):

Figure 4. a) GrassBot and b) John Deere 6630 tractor. Courtesy of Lynex A/S and AU

a)

b)

Table 2. GrassBot and John Deere 6630 specifications. The field speed considered assumed that thee fields are not cultivated for several years and are very uneven.

	Units	GrassBot	John Deere 6630
Power to take off PTO	kW	74.00	99.00
Field speed	km h^{-1}	6.00	6.00
Fuel consumption	L h^{-1}	18.10	18.10
Weight	Kg	2000.00	5230.00
Selling price	k€	75.67	99.46
Tyres/Belt cost	€ ha^{-1}	10.00	3.35
Type of operation		Autonomous/Remote	Manual

unchopped, chopped and bales (Figure 5). The six alternatives, two harvesting equipment and three collection methods, are compared in terms of costs, energy consumption and GHG emissions.

Also, these alternatives are compared for three different stages of the supply chain (Figure 5). Stage 1 comprise harvesting, loading and collection operations in the fields. Stage 2 characterizes the transport from the field to the LSC using a tractor. Stage 3 models the storage operations and truck transport from the LSC to the biogas plant and the densification process when required. Meaning that the effect of the different collections alternatives in the biogas plant is out of the scope of the study, nevertheless some considerations are given in the end of the results discussion.

Stage 1 is common for the three types of collection relying on the different performance on the two types of harvesting equipment. These machines are only alternatives in Step 1 being the other steps mainly influenced by the type of collection. In Stage 2, tractor transport to the LSC was assumed for all alternatives. The type of collection has effects on the type of wagon necessary and for the number of transports because of different densities and shapes. A contractor was considered as responsible for the tasks of Stage 3, so the different performances of the collection types are based on general industrial information available.

The final considerations regarding the study scope is the location of the fields studied: Western Denmark; and the type of exploitation: contractor perspective. The contractor main task is to deliver biomass from rented areas and act as producer and distributor of materials, linking farmers and biogas plant.

Figure 5. Biomass supply chain Stages and types of collection

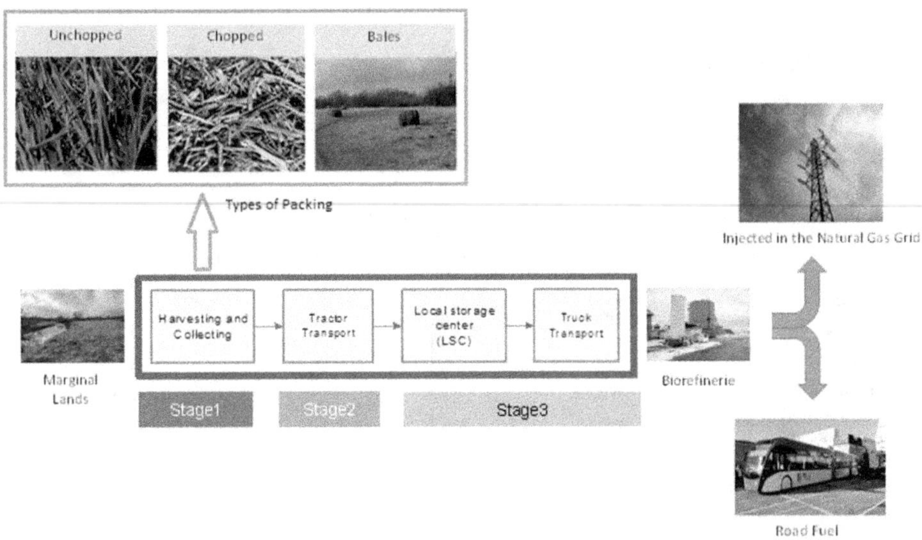

Boundaries and Information Used in Models

The considered boundaries of this analysis include the cost and environmental impact (measured in GHG emissions and energy consumed) of the field works for biomass (MGL) harvesting and collection and also the transport of the biomass to the LSC and to the biogas plant. The biogas plant processes will not be taken into account, but the biomass densification and the biomass energetic content will be considered for energy balanced calculations. In addition, the energy and emissions required to produce, maintain and disposal of the used equipment and consumed material will also be taken into account. Regarding economic assessment, ownership costs are used to account for these aspects and assuring the same scope as for the environmental assessment.

The data related to the MGL, equipment, collection processes and transport is input in the DRIFT model that calculates the process requirements for each alternative and specific scenario (Figure 6). The proposed model, IRIS uses this requirements to calculate the resources required and then it output the economic and environmental performance indicators. For sake of normalization between different biomass products specific values are used: cost factors and emissions and energy per quantity of dry matter, measured in tons of dry matter (tdm). The analysis is done for each supply chain step and also for the complete scope and boundaries of the analysis. Both models are integrated using process-based modelling approach, so allowing sensitivity analysis to the most important design parameters.

Data was collected from different sources. To obtain a concrete idea of the logistics associated with grass supply chains, some interviews were performed. Meetings were conducted with Professors, PhD students and staff from Aarhus University where

Figure 6. Models interdependence and data flowchart

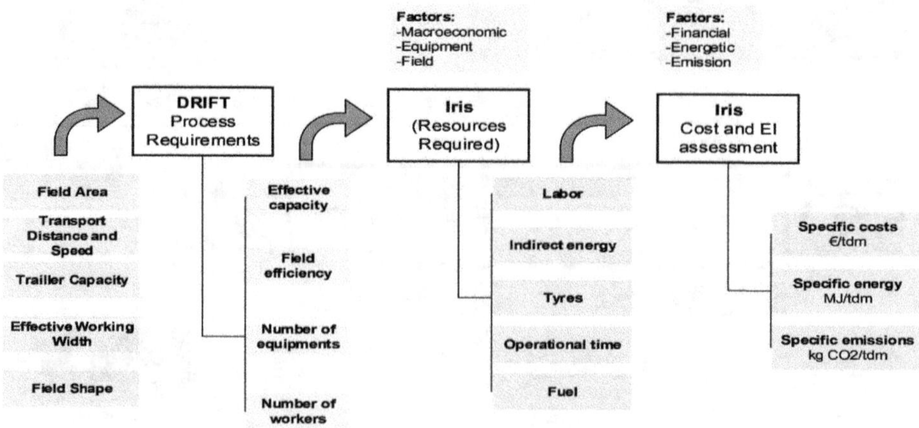

valuable information was gathered. Besides these relevant interviews information, literature regarding harvesting machines and techniques was analyzed in order to achieve coherent assumptions and methods for cost, energy and environmental assessment. In addition, general accepted procedures and information from experts about equipment and harvesting and collection operations were considered to create a reasonable and realistic scenario reflecting Danish conditions. Also, the Foulum biorefinery was visited and an explanation about how the biogas plant works was given. Finally, standard practices were applied to calculate machinery capacity and operations metrics, such as machines life time and maintenance requirements. The standards EP496.3 2006 and D497.5 2006 from the American Society of Agricultural and Biological Engineers (ASABE/ASAE) were used to calculate the usual costs found on the agriculture fleet machines. The considered assumptions are summarized in Table 6 of the Appendix, where the respective sources can be also consulted.

The Use of the DRIFT Model Integrated in the IRIS Model

The calculations of the process requirements were based on the DRIFT model (described in Figure 3) that are connected by IRIS, as represented in Figure 6. The DRIFT model correlations are listed in Table 7 and Table 8 of the Appendix respecting its authors formulations (Sørensen and Nielsen, 2005).

A process-based model based on DRIFT correlations was necessary to build to integrate it with the IRIS model (Figure 7). This process-based model receives inputs about each field area, distance between the fields, LSC, among others. As referred, field and equipment data was obtained from several sources (Table 6 of the Appendix). In addition, complementary information regarding the typically used process parameters and aspects related with field efficiency were obtained from (ASBAE Standards, 2009). In the next section the content of the IRIS model is described and discussed.

THE IRIS MODEL

The IRIS model has two calculation steps: i) the resources requirements and ii) the costs and environmental assessment indicators (Figure 6). IRIS is a process-based model that allows parameters variations for sensitivity analysis through the use of its parametric characteristics. This process-based model communicates with the process-based model based on the DRIFT model.

The first step, related with resources requirements, receives inputs related with process requirements from the DRIFT model and also three types of inputs related with the set of fields, equipment characteristics and energy consumption. The result

Figure 7. Inputs and outputs of the DRIFT model (equations used listed in Table 7 and Table 8 of the Appendix)

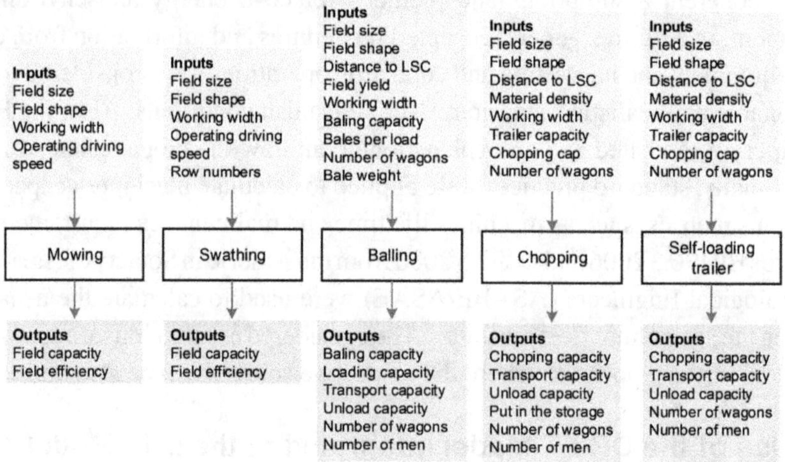

is the consumption of direct energy (fuel for harvesting and other field works) and indirect energy (other equipment for other operations), the consumed tyres, the operational time of the processes and the labor and equipment requirements.

The second step consist of transforming these required resources in cost and in environmental indicators that in this case study are measured in energy and emission per ton of dry matter. Cost factors, financial relations, and energy and emission coefficients are used to perform this calculation.

The First Step of IRIS Model: Resources Required

The calculation of the operational time represents the time of harvesting and is calculated by DRIFT for one field and each specific alternative of collection. But contractors harvest several fields per year, so the IRIS model runs the DRIFT model to calculate the operational time of the several fields under analysis.

The number of operators and the number of equipment required to process each field is also given by DRIFT and then IRIS calculate the accumulated requirements for the several fields considering specific needs as following. The real labor time is calculate by IRIS for each process by using the operational time required and considering an average labor availability per year. An additional factor of 20% is added to the operational time in IRIS calculations as an allowance for unnamed activities related with each process. The amount of equipment needs is also calculated by IRIS and is assumed that the contractor own these equipment despite its use rate is low. The number of equipment required depends on work load and process characteristics as calculated by DRIFT model. So, there is the need to calculate a factor (use factor – UF) that represents the use rate of each equipment for quantifying the fuel consumed and the repair and maintenance requirements (equation 1). To calculate the accumulated use factor of a specific type of equipment, if more than one, equation 2 is applied.

$$Use\,factor = \frac{Working\,time\,per\,year}{Machine\,availability\,time\,per\,year} \tag{1}$$

$$Equipment\,use\,factor_i = \sum Use\,factors\,of\,the\,several\,machines_i \tag{2}$$

The use factor is applied either in cost calculations as in energy consumed and emissions estimation.

The tyres consumed depends on the quantity of necessary harvesting machines, the tyres expected life time and on the working time as represented in equation 3. Their tyres lifetime depends on the type of tyre, soil conditions and the driver behavior. Following the indications of (SEGES, 2015) the tyres life time considered were 2500 hours for the tractors and general agriculture machines, 3000 hours for the GrassBot and 750 hours for the trailers.

$$Tyres\,consumed = \frac{Quantity\,of\,equipments \times Tyres\,per\,machine \times working\,time}{Tyre\,life\,time} \tag{3}$$

The direct energy consumed refers to the fuel consumed during the field works and transport of biomass. Equation 4 and equation 5 were developed by Agricultural Engineering Department of Aarhus University and allow to estimate the direct energy consumed per time:

$$Fuel\,Consumption\,per\,hour = \frac{Engine\,load \times Maximum\,Machine\,power \times Fuel\,efficiency}{Fuel\,density} \tag{4}$$

$$Engine\,load = \frac{Minimum\,power\,required}{Maximum\,power\,required} \tag{5}$$

The real fuel consumption is obtained by the product of the fuel consumption by hour and the working time of each equipment.

The indirect energy includes the energy that was required to produce the equipment used, the one required in its transportation to the farmers and the energy required to the repair and maintenance operations and consumables (e.g. tyres). Since it is not used directly in fieldwork operations it is called as indirect energy. As expressed in equation 6 the total indirect energy was calculated based on the mass of each type of material in each machine and consumable, multiplied by the use factor and by an energy coefficient (the energy required to extract it from earth and have that material available to be used):

$$Indirect\,Energy_i = \sum Material\,use\,weigth_i \times Energy\,coefficient_i \times Use\,Factor_i \tag{6}$$

The energy coefficients used are listed in Table 9 of the Appendix and were based on (Nemecek *et al.*, 2007) and (Sørensen *et al.*, 2014). These references was also used as guide for the equipment production and transportation energy consumption. The GrassBot energy and materials required data was obtained from Agricultural Engineering Department of Aarhus University records. The grinding process energy consumed (as indirect energy) was not estimated using equation 6 because the contractor doesn't own it; this process occurs in the biogas plant. The value is listed in Table 9 of the Appendix and was obtained from (Huisman, Venturi and Molenaar, 1997).

The Second Step of IRIS Model: Specific Cost and Environmental Impact

In the second step of the IRIS model, the required resources computed in step 1 are transformed in economic and environmental performance indicators. Cost factors of materials, labor and equipment are used to obtain the variable running costs and the cost of ownership of the equipment (including its disposal phase). The energy consumed from the equipment and materials used in equipment manufacturing phase to the energy necessary to deliver the biomass to the refinery door is accounted as an environmental impact indicator. It must be clarified that only the energy consumed during the biomass harvesting, collection and transport to LSC is transformed also in cost (fuel related), since the other energy blocks potential cost are included in the cost or price of the acquired equipment, in the products transport and in the grinding process cost. IRIS model also calculate the GHG emissions of the used fuel and materials, consisting in a second environmental performance indicator.

The cost, the consumed energy and the emissions for each alternative of harvesting equipment and collection are presented in specific value, i.e. devised by the quantity of dry matter. Therefore, the obtained values are normalized (or specific) and comparable between alternatives, because each alternative of collection have different levels of MC. In equation 7 is presented the calculation used to obtained total harvested material that is measured in tons of dry matter (tdm), where the yield is the amount of biomass produced by each field:

$$\text{Total Harvested Material} = \sum \text{Yield}_i \times \text{Total Area}_i \times \text{Dry Matter } \%_i \tag{7}$$

The framework of cost calculations is illustrated in Figure 8 where the boundaries of the economic assessment are also indicated (similar for the environmental assessment). Machines maintenance and repair, labor, tyres, fuel, storage, lubricants and grinding are accounted as variable costs while machines ownership which includes equipment depreciation is accounted as a fixed cost. In general, each resource quantity was multiplied by the respective cost factors to obtain the requirement cost, i.e. labor, fuel and tyres. Nevertheless, further explanation is necessary for other requirements since it involves several prior calculations and considerations.

To calculate the cost of ownership a sequence of equations is necessary (ASBAE Standards, 2009) and were included in IRIS. Depreciation cost is related with the continuous decreasing of equipment value over time, and also depends on equipment life spam and residual value (equation 8):

$$Depreciation = \frac{Purchase\,price - Residual Value}{Equipment\,life\,spam\left(EL\right)} \tag{8}$$

The residual value of the equipment (in percentage) was calculated by equation 9 (ASBAE Standards, 2009) and depends on the operating years (n) and working hours (h), expressed as a percentage of the purchased price. According to (ASBAE Standards, 2009) the residual value also depends from three coefficients Ci that depend on the equipment type. The price of the equipment and the Ci coefficients values are listed in Table 10 and Table 11of the Appendix.

$$Residual Value\,after\,n\,years\,of\,use\left(RV\right) = 100\left[C_1 - C_2\left(n^{0,5}\right) - C_3\left(h^{0,5}\right)\right]^2 \tag{9}$$

Figure 8. Framework for cost calculation for each alternative

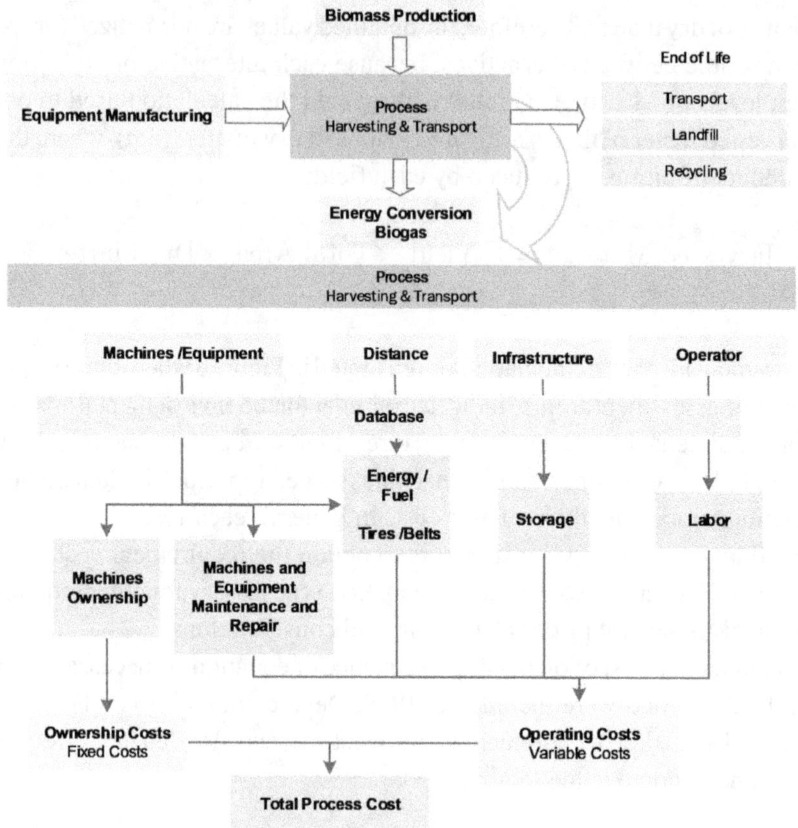

The average annual ownership costs (equation 10), a percentage of the purchase price, is based in three factors: residual value (RV), equipment life spam (EL), interest rate (i) and a taxes related term which accounts the percent of purchase price required for taxes, insurance and housing (ASBAE Standards, 2009).

$$Cost\,of\,ownership\,per\,year\left(\%\right)=100\left[\frac{1-\mathrm{RV}}{\mathrm{EL}}+\frac{1+\mathrm{RV}}{2}\mathrm{i}+\mathrm{Taxes\ \&\ insurances}\right] \tag{10}$$

The variable costs are composed by several parcels.

According to (ASBAE Standards, 2009), the repair and maintenance costs are calculated as percentage of the purchase machine price and they can be estimated by equation 11, that is used by IRIS and then multiplied by the machine price. The repair and maintenance costs depend on the machine characteristics and on the operation hours. The machine characteristics affect the RFMi coefficients (see Table 11 in the Appendix for the values used).

$$Cost\,of\,repair\,and\,maintenance\left(\%\right) = \left(\mathrm{RMF}_1\right)\left[\frac{\mathrm{h}}{1000}\right]^{\left(\mathrm{RMF}_2\right)} \tag{11}$$

To estimate the cost of storage a dried roughage store, cold-air dried was chosen. The investment required to build the infrastructure was found on (Nemecek et al., 2007) and the final cost per ton of dry matter were obtained using a procedure present in (Nemecek et al., 2007; Rotter and Rohrhofer, 2013). The results are in Table 12 of Appendix and take into consideration the fact that the cost of storage increases with the MC, so different types of biomass collection have different storage costs.

The cost of transporting the biomass from the LSC to the biogas plant was based on two studies (Huisman, Venturi and Molenaar, 1997) and (Rotter and Rohrhofer, 2013). The obtained estimations for each type of product is presented in Table 13 of Appendix assuming a distance of 50 km to biogas plant.

The grinding operation is required for the biomass product meet an acceptable quality for biogas production. This quality depends on factors such as MC and particle size. Chopped materials meet these requirements, however, unchopped and baled grass need to be grinded firstly. To perform this task, a grinding machine used in the biogas plant was considered. A variable cost of 7.6 € tdm^{-1} was considered based on (Huisman, Venturi and Molenaar, 1997).

Regarding the environmental impact two main indicators of performance are used: energy consumed and GHG emissions. As referred before the energy consumed considered was accounted from the energy required to produce the materials of the machines and of consumables, to the manufacturing of the machines and consumables, its transport (accounted as indirect energy) and also all the energy for the biomass harvesting, collection, storing and transport to the LSC and to the biorefinery (direct energy), so on a gate to gate basis. Grinding energy consumed can be also named as direct but was considered separately. This total energy consumed is calculated by IRIS in the Step 1, so in Step 2 is only necessary to calculate the energetic balance given by the net energy (equation 12)

$$\text{Net Energy}\left(\text{kWh}\right) = \text{Biomass Energy Output}\left(\text{kWh}\right) - \text{Fossil Energy Input}\left(\text{kWh}\right) \tag{12}$$

The biomass products delivered at the biorefinery door has an energy content (see Table 1). So, the biomass energy output from the field operations is obtained by the product of the total harvested material (calculated in equation 7) by the specific energy content of the specific biomass. The difference of this value with the energy consumed during the all processes, usually from fossil sources, is the net energy.

The emission considered were mainly the ones originated by the electricity consumed and by the fuel used. Electrical energy was considered the one used in the materials and equipment production and also in LSC. The fuel considered for the several transports involved was diesel. These amounts calculated in the Step 1 of IRIS were multiplied by the specific mass of CO_2eq emitted per MJ of energy used coefficients assumed in this study (see Table 6 of the Appendix).

Regarding to the disposal operations path, the assumptions for the several components are present in Table 3. For sake of simplicity only the electricity (and the resultant emissions) necessary for the dismantling the machinery was considered in IRIS. Following a study in the dismantling of agricultural machinery (Nemecek *et al.*, 2007) the value of 0.5 MJ per kg of machinery was used (this energy also accounts the transport to the facility).

RESULTS AND DISCUSSION

The application of IRIS model, integrated with the DRIFT model was applied in the context of MGL existent in Danmark, so the base line used to present the results consider the field characteristics and other aspects listed in Table 6 of the appendix. First the operational and economic performance results for the six alternatives are presented and discussed for the base line. Taking advantage of the parametric characteristics of the process-based models sensitivity analysis are then presented and discussed. Secondly, the energy and emissions obtained results are also presented and analyzed.

Table 3. Components Disposal Path

Material	Final Disposal Place
Used Oil	Incineration Plant
Tyres	50% Cement Plant; 50% Recycled
Metals	100% Recycled
Glass, Papers, Plastic, Rubbers	Municipal Incineration
Transport	Distance 40km to the disposal center by lorry

Operational and Economic Performance

The required fuel and the equipment requirements for the base line analysis are calculated in Table 4. Regarding fuel consumption, chopping and mowing are the most demanding operations since both require the highest engine load. The use factor for Grossbot is higher than for conventional technology, because of the higher total harvested area. When the use factor of self-propelled equipment is null no equipment was used. This is the case for the GrassBot that just need one of these devices. All these aspect together will be accessed by cost analysis.

From the operational point of view, the time spent in one operation can be the limiting factor for the contractor, so their goal is to harvest the maximum area in the shortest time space. Table 5 summarizes all stages and supply chains for automated and conventional technologies where the total time regarding each ton of dry matter of processed biomass is given. In terms of type of harvesting alternative, the regular tractor alternative is the one which requires the least time. On the other hand considering the material type, baling has the fastest process while chopped has the slowest one.

The cost breakdown and the total cost for all the 6 alternatives for the base line conditions is present in Figure 9. The harvesting, loading and collection cost per quantity of dry matter are lower when bales are produced; the use of the regular tractor has a lower cost than the use of the GrassBot in these operations. The higher production cost of chopped and unchopped materials, comparing with balled, derived from the higher number of required machines which increases ownership costs and fuel expenses since more trips are needed. In fact, the chopper needs a continuous storage flow to have optimal efficiency during harvesting, and therefore several wagons are needed. The automated alternatives has a higher cost because the equipment has a higher operation time.

Transport cost to LSC is analogous using both type of harvesting equipment. The lower efficiency in transport of unchopped type of collection, because of its lower density, causes the higher cost is this parcel follow closely the chopped type. The bales present the lower costs of transport for its higher density. This analysis is valid also for the truck transport costs. Since the average distance considered for the LSC to the fields was 3 km, this the transport costs are not an important decision factor in the overall cost. Reversely, the transport to the biogas plant, 50 km were considered, has a relevant influence being the balled collection the one with lower costs for the same reason. Similar conclusions are found in (Huisman, Venturi and Molenaar, 1997) where only for distances higher than 50 km the transport cost of compacted materials can be advantageous.

Table 4. Fuel and equipment required for the base line harvesting and collection of biomass alternatives – Autom. – GrossBot; Conven. – Regular tractor (1 year of activity; biomass type: hay meadow; average field area 16 ha (320 ha per year); number of harvested fields: 20; field shape: pentagon).

	Unchopped Material				Chopped Material				Baled Material			
	Fuel consumption (l h-1)		Use factor % (per unit)		Fuel consumption (l h-1)		Use factor % (per unit)		Fuel consumption (l h-1)		Use factor % (per unit)	
	Autom.	Conven.	Autom.	Conven.	Autom.	Conven.	Autom.	Conven.	Autom.	Conven.	Autom.	Conven.
Trailed Equipment												
Stage 1												
Mower	19.64	20.83	93%	41%	19.64	20.83	93%	41%	19.64	20.83	93%	41%
Swath	6.55	6.55	66%	34%	6.55	6.55	66%	34%	6.55	6.55	66%	34%
Rake	9.82	9.82	66%	34%	9.82	9.82	66%	34%	9.82	9.82	66%	34%
Self loading wagon	13.10	13.10	78%	40%	--/--	--/--	--/--	--/--	--/--	--/--	--/--	--/--
Chopper	--/--	--/--	--/--	--/--	22.02	22.02	62%	33%	--/--	--/--	--/--	--/--
Chopper Wagon	--/--	--/--	--/--	--/--	13.70	10.24	51%	28%	--/--	--/--	--/--	--/--
Baling machine	--/--	--/--	--/--	--/--	--/--	--/--	--/--	--/--	11.90	17.86	95%	50%
Bales trailer	--/--	--/--	--/--	--/--	--/--	--/--	--/--	--/--	0.0	0.0	40%	20%
Front loader	--/--	--/--	--/--	--/--	--/--	--/--	--/--	--/--	6.37	4.76	48%	45%
Stage 2												
Self loading wagon	8.93	8.93	72%	33%	--/--	--/--	--/--	--/--	4.76	4.76	48%	45%
Chopper wagon	--/--	--/--	--/--	--/--	10.24	10.24	105%	49%	14.79	14.79	38%	17%
Self-Propelled Equipment												
GrassBot	- / -	- / -	47%	0%	- / -	- / -	59%	0%	- / -	- / -	60%	0%
Tractor1	- / -	- / -	0%	23%	- / -	- / -	0%	30%	- / -	- / -	0%	29%
Tractor2	- / -	- / -	54%	26%	- / -	- / -	16%*2	8%*3	- / -	- / -	14%*2	7%*2

Table 5. Operation time in hours per ton of dry matter for the several supply-chain stages processes for the base line case.

	Automatic (Grassbot)	Conventional (Regular Tractor)
	Stage 1	
Unchopped	1.75	1.37
Chopped	1.69	1.61
Bales	1.65	1.47
	Stage 2	
Unchopped	0.57	0.51
Chopped	0.97	0.88
Bales	0.37	0.36
	Stage 3	
Unchopped	0.17	0.17
Chopped	0.10	0.10
Bales	0.07	0.07
Unchopped Total Time	2.31	1.88
Chopped Total Time	2.66	2.49
Bales Total time	2.02	1.83

Figure 9. Cost breakdown and total cost calculated by IRIS for de 6 alternatives the base line conditions

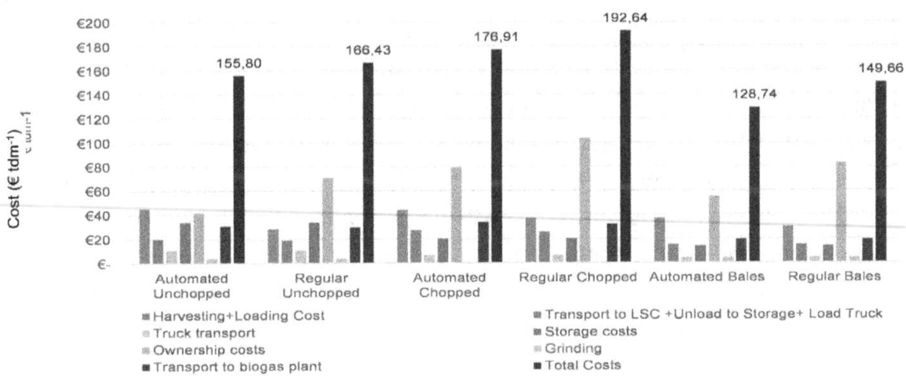

**For a more accurate representation see the electronic version.*

The higher density and lower MC of the bales alternatives also contributes for a lower storage costs, being one of the reasons contributing to this type of collection alternatives to be the one with lower total costs.

The grinding costs are only present in unchopped and balled collection alternatives since the chopped one is already prepared to be used to produce biogas. Their relevance in the total cost is low.

The ownership costs are the ones with more significant difference among alternatives. The higher use factor of the machines and lower quantity required by the autonomy of the GrassBot is decisive for these cost difference (see Table 4). The equipment used in this alternative has a higher use rate, processing more quantities of biomass per year, so its use cost is reduced comparing with the regular tractor harvesting. The chopping collection alternatives are the ones with the higher ownership costs because it requires a higher number of machines and a higher area to operate (see Table 4).

The automated alternative, the Grassbot, allows a lower total cost disregarding the type of biomass collection, mainly because of the higher use rate of the machines. The bale type of collection is the one with lower cost mainly because of lower cost of storage and in harvesting.

Regarding the supply chain stages (not represented in the graph), Stage 1 accounts for 60 to 70% of the overall costs, with Stages 2 and 3 having between 20 to 15% depending on the alternatives.

Sensitivity Analysis on Economic Performance

Using the parametric characteristics of IRIS process-based model integrated with DRIFT sensitivity analysis to several parameters is possible. An analysis of the distance to the LSC and to the biogas plant was done and reveals that balled materials has always the lowest cost in transport and in the supply chain with the difference increasing with the distance (so this analysis is not shown since is an expected result).

From a contractor perspective the scale of the business is very important. If the production capacity is higher, fixed costs per unit are reduced which are translated in bigger operational margins. The best process depends on several parameters, being the total harvest area one of the most important to take into consideration. In Figure 10 the total costs regarding the field area for the all alternatives are presented.

Figure 10. Total cost regarding the field area for the several alternatives

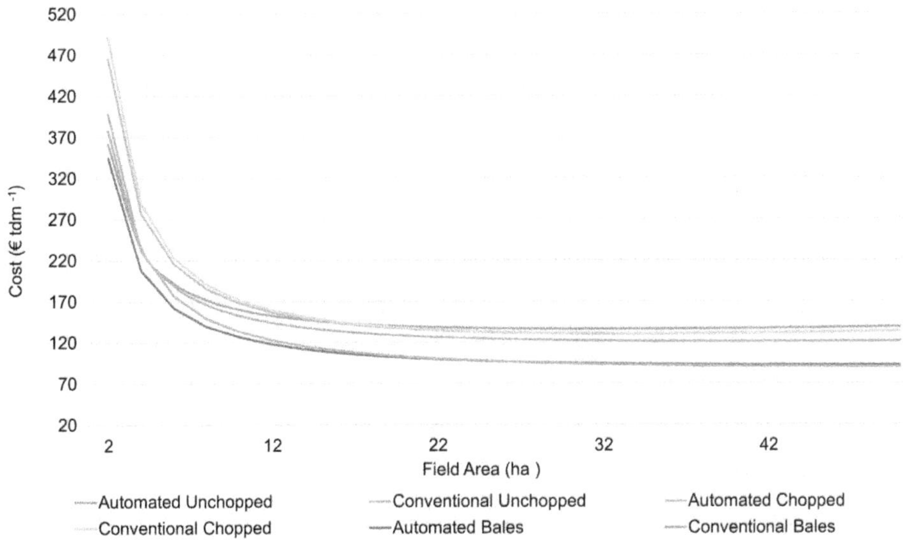

**For a more accurate representation see the electronic version.*

According to Henrik Mortensen from F.R.C. an average contractor harvests per year around 1000 ha. This is translated in 10,000 ha in a 10 year timespan (assumed project duration). In the present case study the base line assumed was 320 ha year^{-1}, due the marginal land limited working conditions, such as the operating speed and field areas. From Figure 10 the bale alternative with conventional tractor became a better choice than the automated one. The increasing of the field size allow the use factor of the machines to increase, reducing the relevance of the fixed cost and increasing the importance of variable costs where the conventional alternatives are better.

The same behavior is not observed for the chopped and unchopped alternatives where for both equipment alternatives the cost begin to rise with the increase of field size. In order to understand it, a variation of each cost component was represented in Figure 12. The fixed costs (or ownership costs) decrease when harvested area increases while specific variable costs increase. The sum of these costs reaches its minimum at 16ha. After that, they increase because of the rise of the variable cost because the repair and maintenance costs were considered variable and depending on the field work time. The validity of this result must be further analyzed in a future study causing eventually the revision of the model used to calculate repair and maintenance associated costs.

Figure 11. Detailed view total supply chain cost regarding area

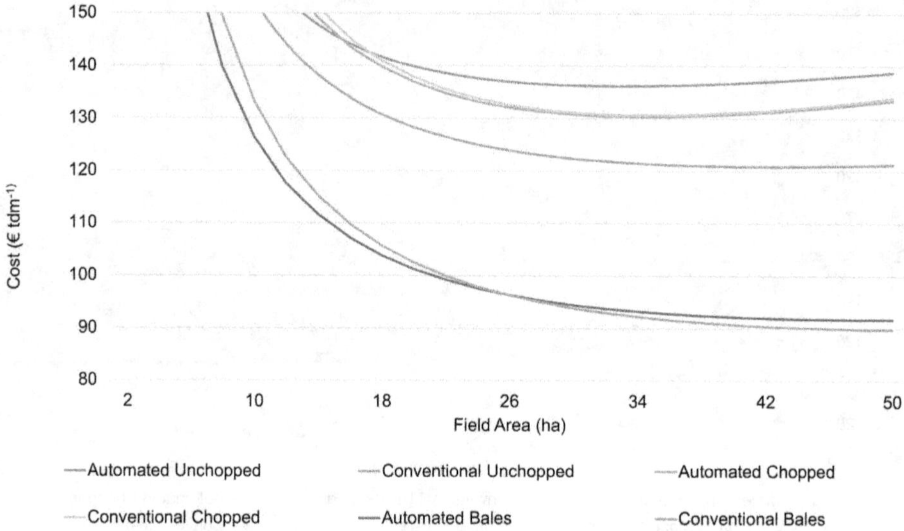

**For a more accurate representation see the electronic version.*

Figure 12. Cost structure automated unchopped

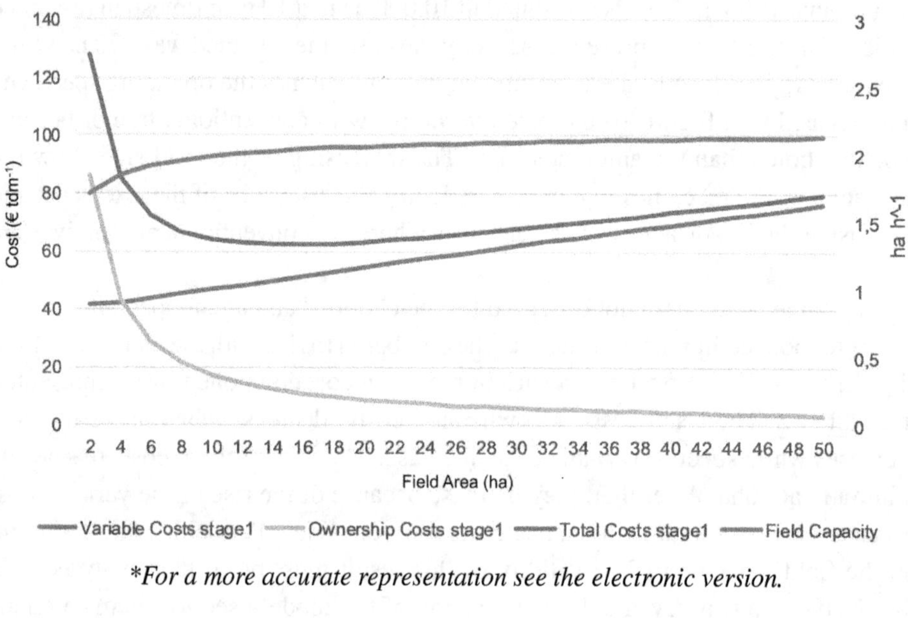

**For a more accurate representation see the electronic version.*

Energy Consumed and Emissions

Energy consumption and also emissions related with GHG for each step were calculated for unchopped, chopped and baled grass and for the two types of harvesting equipment using the IRIS model, as represented in Figure 13 and Figure 14. Since the emissions considered in this study depend largely on the energy consumed, in form of fuel and electricity, the results and analysis are similar.

Unchopped process is the one that requires the most significant amount of energy followed by balling, which was not expectable since chopping operations require more machine time. However, chopped material does not have to be grinded. In fact, grinding contribution for total energy and emissions has significant weight. The considered value in this work derives from a 1997 case study (Huisman, Venturi and Molenaar, 1997). Therefore, it is expectable that newer equipment has lower energetic consumption and lower emissions which influence the final results. This is the reason why this parcel was analysed separately and must be object of analysis in a future study.

Therefore, summing the all contributions, chopped is the most energetic efficient process. This situation was not noticed in the cost assessment because of the low relevance of electrical energy in the cost structure. Without the grinding process the bales are the most energy efficient, followed by unchopped collection.

Regarding the supply-chain stages, the first one is the most demanding regarding direct energy, being responsible for almost 50% of the total energy input. A factor that contributes for this difference is the larger operational time compared with the other stages (Table 5). Regarding indirect energy, used for equipment manufacturing, maintenance and disposal operations, its value is higher using automated technologies. This result is also due the higher use factor of automated process.

The required energy for transport accounts approximately 36% of the total energy. Transport operations are very similar for both processes since the same materials and machines are used. However, it is slightly larger for automated technology because the use factor is higher.

In order to have a clear picture of which are the most sustainable supply chain alternative, the net energy derived from the MGL harvesting was calculated (Figure 15). Chopping technology consumes less energy. Thus it has the highest net energy output, 1547 and 1606 MJ tdm^{-1}, for automated and conventional harvesting technology respectively. If the quality of chopped and unchopped materials were assumed as being constant they would have similar results. By the results' analysis one can be concluded that the energy output from one ton of grass is around 200% higher than the required energy to process that ton of biomass.

Figure 13. Supply chain energy consumed. Stages of the supply chain were defined in Figure 5

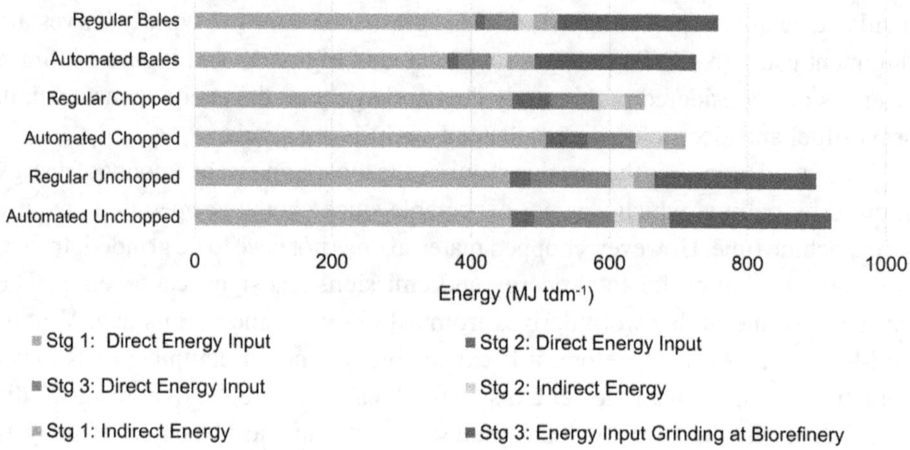

**For a more accurate representation see the electronic version.*

Figure 14 Supply chain GHG emissions. Stages of the supply chain were defined in Figure 5

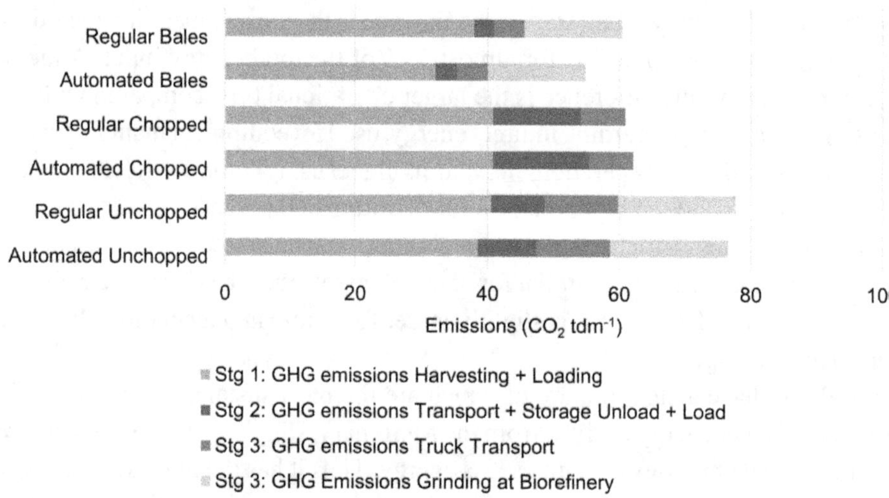

**For a more accurate representation see the electronic version.*

Figure 15. Net energy for all the alternatives of supply chain

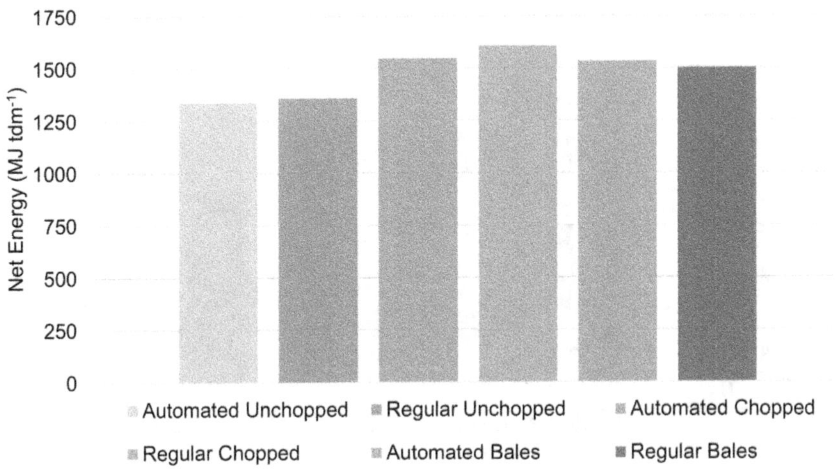

**For a more accurate representation see the electronic version.*

Regarding grass carbon uptake, a supply-chain balance analysis for GHG emissions was conducted, starting in biomass growing, followed by operations and finally the use phase (biogas combustion).

During plants growth, the CO_2 present in air is stored in the plants through photosynthesis. This carbon uptake contributes to a large negative emission in the supply chain analyzed (see Table 6 for carbon capture index consider). Figure 16 sums all the CO_2 taken from the atmosphere and the one released during biomass operations.

Performing these analyses allow concluding that automated and conventional processes have similar CO_2 emissions independently of the biomass material. Bales have the net emissions with 44.8 and 50.4 kg CO_2 tdm^{-1} for automated and conventional respectively. This is in contrast with the net energy, where the chopped collection is the best. The reason rely on the energy used in the grinding process is energy, so the GHG emissions are relatively lower than the ones of transport and equipment (fuel used) where chooped alternative is worst in terms of efficiency.

During harvesting and loading, both harvesting technologies have similar GHG emissions. Transportation GHG emissions depend on the biomass density and collection process. They are higher for the automated processes, due the higher operational time. Tractor transport represent between 5 to 17% of the net GHG. For truck transport it is seen that emissions only depend on the material density. For the unchopped material which has the highest amount of GHG emission, truck transport was found to be responsible for 15% of GHG releases.

Figure 16. Process net emissions

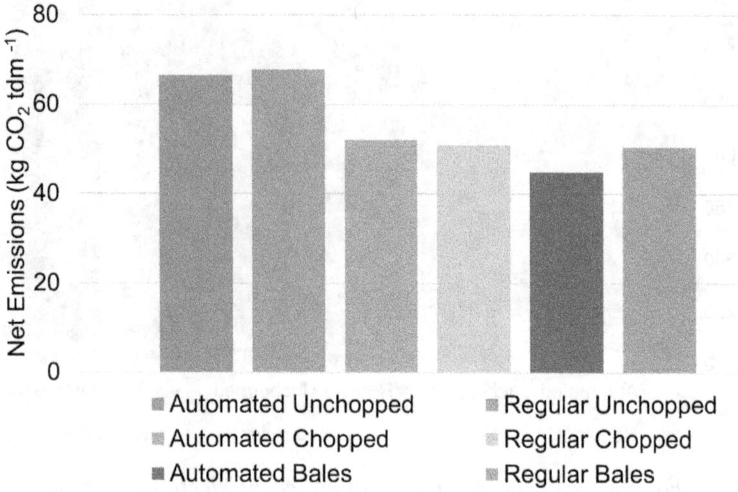

**For a more accurate representation see the electronic version.*

In order to assess the advantages of this energy source, the present work results are compared with other energy sources. To perform this comparison, a common way is to use the emission intensity indicator. This is obtained gathering net GHG emissions per unit of electricity generated gCO_2 kWh^{-1}.

The sustainability of biofuels using ley crops in marginal lands was studied by (Helgadottir and Hopkins, 2013). In their work, it was concluded that net GHG emissions reach acceptable limits for yields above 3 tdm ha^{-1} yr^{-1}. Comparing biogas energy and emissions using grasses with other energy sources, the main emissions are in line with the most efficient process available.

Regarding sustainability criteria for renewable biofuels set by the EU, Figure 17 shows the goals for emission reduction. Bales and chopped materials are suitable to meet these requirements to produce energy for the reasons already explained.

CONCLUSION

The acquisition of biomass from marginal areas under Danish conditions has been evaluated in terms of alternative harvesting equipment: an automated robot (Grassbot) versus a regular tractor (conventional alternative) for key grass materials used for biogas plants (chopped, unchopped and baled grass) and compared regarding operational, economical, and environmental performances. The evaluation uses two operations models (IRIS and DRIFT) to consider the field characteristics, machinery

Figure 17. EU emission reduction goals vs biomass supply chain studied

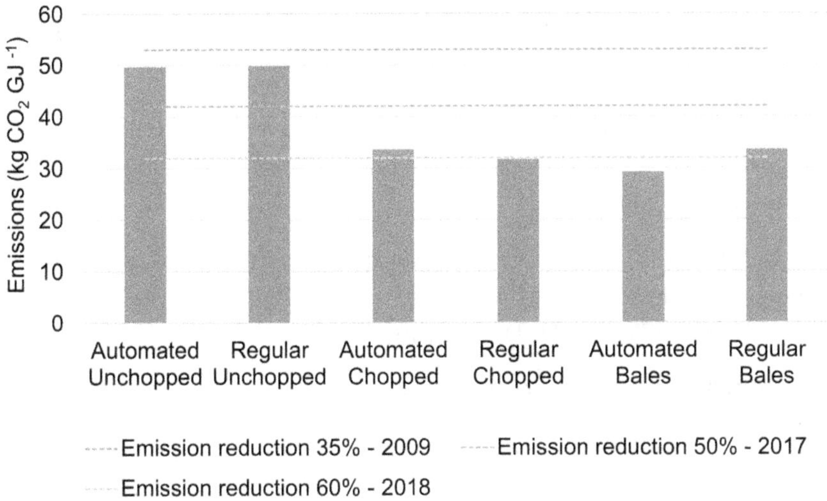

**For a more accurate representation see the electronic version.*

characteristics, etc. The results show that in terms of fuel consumption, chopping and mowing are the most demanding operations since both require the highest engine load. The regular tractor alternative is the one which requires the shortest operational time. Baling has the highest capacity, while chopped has the lowest capacity. The harvesting, loading and collection cost per unit of dry matter are lower when bales are used; the use of the regular tractor has a lower cost than the use of the GrassBot. Higher number of machines are required for chopped and unchopped materials, causing higher production costs (increased ownership costs and fuel expenses) compared with balled materials. The automated alternatives has the highest cost as a consequence of the increased operation time. The harvesting, loading and collection cost per quantity of dry matter are lower when bales are produced; the use of the regular tractor has a lower cost than the use of the GrassBot in these operations. It is also concluded that automated and conventional processes have similar CO_2 emissions independently of the biomass material. Bales have net emissions with 44.8 and 50.4 kg CO_2 tdm^{-1} for automated and conventional respectively. This is in contrast with the net energy, where the chopped method is the best. Both harvesting technologies have similar GHG emissions, whereas transportation GHG emissions depend on the specific biomass density and collection process. Tractor transport represent between 5 to 17% of the net GHG. For truck transport, the emissions solely depend on the material density. For the unchopped material, which has the highest amount of GHG emission, truck transport was found to be responsible for 15% of GHG releases.

REFERENCES

ASBAE Standards, . (2009). *Agricultural Machinery Management Data*. St. Joseph, MI: Author.

Bochtis, D. D., & Sørensen, C. G. (2009). The vehicle routing problem in field logistics part I. *Biosystems Engineering*, *104*(4), 447–457. doi:10.1016/j.biosystemseng.2009.09.003

Caffrey, K., & Veal, M. (2013). Conducting an Agricultural Life Cycle Assessment: Challenges and Perspectives. *The Scientific World Journal*, *2013*, 1–13. doi:10.1155/2013/472431 PMID:24391463

Cheng, J. (2009). *Biomass to Renewable Energy Processes*. CRC Press.

Christian, D. G. (2000). Biomass for Renewable Energy, Fuels, and Chemicals. *Journal of Environmental Quality*, *29*(2), 662. doi:10.2134/jeq2000.00472425002900020040x

Demirbas, A. (2008). Biofuels sources, biofuel policy, biofuel economy and global biofuel projections. *Energy Conversion and Management*, *49*(8), 2106–2116. doi:10.1016/j.enconman.2008.02.020

Ekpeni, L. (2014). Energy Diversity through Renewable Energy Source (RES) – A Case Study of Biomass. *Energy Procedia*, 61, 1740–1747. 10.1016/j.egypro.2014.12.202

European Comission. (2007). Renewable Energy Road Map. Renewable energies in the 21st century: building a more sustainable future. Brussels, Belgium: Author.

Hamelinck, C., Suurs, R., & Faaij, A. (2005). International bioenergy transport costs and energy balance. *Biomass and Bioenergy*, *29*(2), 114–134. doi:10.1016/j.biombioe.2005.04.002

Helgadottir, A., & Hopkins, A. (Eds.). (2013). The role of grasslands in a green future: threats and perspectives in less favoured areas. In *Proceedings of the 17th Symposium of the European Grassland Federation*. Borgarnes: Agricultural University of Iceland (Grassland Science in Europe).

Huisman, W., Venturi, P., & Molenaar, J. (1997). Costs of supply chains of *Miscanthus giganteus*. *Industrial Crops and Products*, *6*(3), 353–366. doi:10.1016/S0926-6690(97)00026-5

Jungbluth, N., Büsser, S., & Frischknecht, R. (2008). *Life Cycle Assessment of Biomass to Liquid Fuels*. ESU-services Ltd.

Mafakheri, F., & Nasiri, F. (2014). Modeling of biomass-to-energy supply chain operations: Applications, challenges and research directions. *Energy Policy*, *67*, 116–126. doi:10.1016/j.enpol.2013.11.071

Mortensen, H. (2014). *Sustainable recovery of biomass on low-lying areas Agromek 2014*. Academic Press.

Nemecek, T. (2007). Life cycle inventories of agricultural production systems. *Data v2. 0, Ecoinvent report*, (15).

Omonode, R., & Vyn, T. (2006). Vertical distribution of soil organic carbon and nitrogen under warm-season native grasses relative to croplands in west-central Indiana, USA. *Agriculture, Ecosystems & Environment*, *117*(2–3), 159–170. doi:10.1016/j.agee.2006.03.031

Orfanou, A., Busato, P., Bochtis, D. D., Edwards, G., Pavlou, D., Sørensen, C. G., & Berruto, R. (2013). Scheduling for machinery fleets in biomass multiple-field operations. *Computers and Electronics in Agriculture*, *94*, 12–19. doi:10.1016/j.compag.2013.03.002

Pavlou, D., Orfanou, A., Busato, P., Berruto, R., Sørensen, C., & Bochtis, D. (2016). Functional modeling for green biomass supply chains. *Computers and Electronics in Agriculture*, *122*, 29–40. doi:10.1016/j.compag.2016.01.014

Peeters, A. (2009). Importance, evolution, environmental impact and future challenges of grasslands and grassland-based systems in Europe. *Grassland Science*, *55*(3), 113–125. doi:10.1111/j.1744-697X.2009.00154.x

Rotter, S., & Rohrhofer, C. (2013). *Biomass based energy intermediates boosting biofuel production*. Academic Press.

Schweier, J., & Becker, G. (2013). Economics of poplar short rotation coppice plantations on marginal land in Germany. *Biomass and Bioenergy*, *59*, 494–502. doi:10.1016/j.biombioe.2013.10.020

SEGES. (2015). *The cost for using belts and tyres*. Author.

Sharma, B., Ingalls, R. G., Jones, C. L., & Khanchi, A. (2013). Biomass supply chain design and analysis: Basis, overview, modeling, challenges, and future. *Renewable & Sustainable Energy Reviews*, *24*, 608–627. doi:10.1016/j.rser.2013.03.049

Smeets, E. M. W., Lewandowski, I. M., & Faaij, A. P. C. (2009). The economical and environmental performance of miscanthus and switchgrass production and supply chains in a European setting. *Renewable & Sustainable Energy Reviews*, *13*(6–7), 1230–1245. doi:10.1016/j.rser.2008.09.006

Sokhansanj, S., Kumar, A., & Turhollow, A. (2006). Development and implementation of integrated biomass supply analysis and logistics model (IBSAL). *Biomass and Bioenergy*, *30*(10), 838–847. doi:10.1016/j.biombioe.2006.04.004

Sørensen, C., & Nielsen, V. (2005). Operational Analyses and Model Comparison of Machinery Systems for Reduced Tillage. *Biosystems Engineering*, *92*(2), 143–155. doi:10.1016/j.biosystemseng.2005.06.014

Sørensen, C. G., & Bochtis, D. D. (2010). Conceptual model of fleet management in agriculture. *Biosystems Engineering*, *105*(1), 41–50. doi:10.1016/j.biosystemseng.2009.09.009

Sørensen, G., Halberg, N., Oudshoorn, F. W., Petersen, B. M., & Dalgaard, R. (2014). Energy inputs and GHG emissions of tillage systems. *Biosystems Engineering*, *120*, 2–14. doi:10.1016/j.biosystemseng.2014.01.004

The European Commission. (2000). Towards a European strategy for the security of energy supply. *EUR-Lex*, pp. 1–111. doi: 92-894-0319-5

Zhang, L., & Hu, G. (2013). Supply chain design and operational planning models for biomass to drop-in fuel production. *Biomass and Bioenergy*, *58*, 238–250. doi:10.1016/j.biombioe.2013.08.016

APPENDIX

Table 6. Project Assumptions

	Unit	Value	Reference
Field Factors:			
Country		Denmark	-
Biomass raw material		Hay meadow	-
Project economic life	years	10.00	
Avg. field area - conventional	ha	8.00	Assumption
Avg. field area - automated	ha	16.00	Assumption
Number of harvested fields		20.00	Assumption
Field shape		pentagonal	Assumption
Biomass final MC	%	15.00	Interview, Foulum (Erik Kristensen)
Yield conventional	tdm ha^{-1}	4.00	(Peeters, 2009)
Yield automated	tdm ha^{-1}	4.00	Assumption
Avg. distance field to storage	Km	2.00	Interview, Foulum (Henrik Mortensen)
Distance from the storage to biogas plant	Km	50.00	Assumption
Distance to the disposal field	Km	40.00	Assumption
Bale weight	kg	280.00	Interview, Foulum (Henrik Mortensen)
Equipment Factors:			
Tractor avg. Speed	Km h^{-1}	25.00	Interview, Foulum (Henrik Mortensen)
Truck avg. Speed	Km h^{-1}	55.00	(Rotter and Rohrhofer, 2013)
GrassBot field speed	Km h^{-1}	6.00	Interview, Foulum
Tractor field speed	Km h^{-1}	6.00	Interview, Foulum
Truck fuel consumption	L h^{-1}	30.00	(Rotter and Rohrhofer, 2013)
Grinding cost (densification)	€ tdm^{-1}	7.6	(Huisman, Venturi and Molenaar, 1997)
Price Factors			
Total hourly wage	€ h^{-1}	23.00	Interview, Foulum (Henrik Mortensen)
Opportunity cost	%	5.39	Calculated
Fuel price tractor	€ l^{-1}	0.84	Eurostat database 2014
Fuel price truck	€ l^{-1}	1.40	(Rotter and Rohrhofer, 2013)
Emissions			
Electric energy (Denmark)	g CO_2MJ^{-1}	147	(Omonode and Vyn, 2006)
Diesel	kg CO_2 l^{-1}	2.65	(Omonode and Vyn, 2006)
Grass absorbed carbon	kg C ha-1 yr-1	2.1	(Omonode and Vyn, 2006)
CO2 content in absorbed carbon conversion factor		3.75	(Omonode and Vyn, 2006)

Table 7. DRIFT model field equations (Sørensen and Nielsen, 2005).

Variable	Term	Process field relations	Unit	Eq.
Process Field Capacity	PFC	$\mathrm{PFC} = \left(\frac{600 \times h}{v \times e} + \frac{p \times b \times n}{e \times (1+a)} + k + s \times h \right) \times \frac{(1+q)}{h}$	$\frac{min}{ha}$	(13)
Process Effective Capacity	PEC	$\mathrm{PEC} = \mathrm{PFC}^{-1}$	$\frac{ha}{min}$	(14)
Process Efficiency	PE	$FE = \frac{10 \times PEC}{e \times v}$	%	(15)
Number of men	men	$men = \left(\frac{sum3 + sum8 + sum9 + sum10 + sum12}{sum1 + sum3} \right) + 2$	-	(16)
Time required for an operation	TR	$\mathrm{TR} = \mathrm{h} \times \mathrm{PFC}$	min	(17)
Transport capacity on field	TCapF	$\mathrm{TCapF} = \frac{\mathrm{t1} \times \mathrm{u} \times 2 \times 60}{\mathrm{r1} \times \mathrm{v2}} \times (1 + \mathrm{q})$	$\left(\frac{min}{ha} \right)$	(18)
Transport on Road	TCapR	$\mathrm{TCapR} = \frac{\mathrm{t2} \times \mathrm{u} \times 2 \times 60}{\mathrm{r1} \times \mathrm{v3}} \times (1 + \mathrm{q})$	$\left(\frac{min}{ha} \right)$	9)
Machine Material capacity	MMC	$\mathrm{MMC} = \frac{\mathrm{Yield}}{PFC}$	$\left(\frac{ton}{min} \right)$	(20)
Baling capacity	BCap	$BCap = \frac{10 \times Net\,Baler\,Cap \times u^{-1}}{e \times v}$	%	(21)
Baling speed	Bspeed	$\mathrm{Bspeed} = \frac{60 \times \mathrm{MMC}}{\mathrm{e} \times \mathrm{u}}$	$\left(\frac{\mathrm{km}}{\mathrm{h}} \right)$	(22)
Time to tie up the bale	TTB	$TTB = \left(\frac{u \times 1000}{Bale\ \ Weight \times ts} \right) \times (1 + q)$	$\left(\frac{min}{bale} \right).$	(23)
Total Baling capacity	TBCap	$TBCap = BCap + TTB$	$\left(\frac{min}{ha} \right)$	(24)

Table 8. Field variables of Table 7 of the DRIFT model.

Variable	Description	Unit
Field process model		
h	Field size	ha
v	Equipment field velocity	km /h
e	Effective working width	m
p	Turning	min/turning
u	Field yield	ton/ha
t	Time for turning	min / turning
b	Field width	m
n	Number of turnings per pass (normally n= 2)	-
a	Dependent on field shape and travel pattern	-
k	Turnings on headland	min /field
s	stochastic crop and soil stops, adjustments, control, tending of machine	min /ha
q	Rest allowance time (5%)	min
ts	Time to tie up a bale with string	min /bale
t1	Transport distance in field	m
t2	Transport distance on road	m
r1	Net load weight	kg /load
v2	Travelling speed in field	Km /h
v3	Traveling speed on road	Km /h
sum 1	Field work	min/ha
sum 3	Changing over wagon	min/ha
sum 8	Transport in the field	min/ha
sum 9	Transport on the road	min/ha
sum 10	Unloading	min/ha
sum 12	Deposit into store	min/ha

*Table 9. Materials and processes energy coefficients (Sørensen et al., 2014) and (Huisman, Venturi and Molenaar, 1997)**

	Energy Coefficient (MJ/kg)
Materials	
Aluminum, alloy 85% recycling	191
Copper, primary 85% recycling	100
Fuel and oil	46.8
Glass, laminated 85% recycling	26.2
Cast iron, un-alloyed 85% recycling	27.6
Plastic	90
Steel, low-alloyed 85% recycling	33
Tyre rubber, synthetic 100% recycling	110
Processes	
Machine manufacturing energy	24.4
Transportation	1.29
Repairs and maintenance	17.07
Grinding*	233.7 MJ/tdm

Table 10. Equipment specifications

Selp Propelled Equipment:					
	GrassBot	**Tractor - John Deere**			
PTO Claimed kW	74	99			
Speed km/h	6	6			
Fuel Consumption L/hr	18,1	18,1			
Weight	2000	5230			
Purchased Price (€)	75667	99456			
Rubber Cost €/ha	10	3,35			
Implements:					
MOWING and SWATH TREATMENT of GRASS			**CHOPPING and HOME RUN of STRAW**		
Mower:	**Kongskilde CM 305 F**	**Kverneland Taarup 4336 CT**	**Chopper:**	**Kongskilde FCT 960 Blower Chopper**	**Kongskilde FCT 960 Blower Chopper**
Working Width (m)	3,1	3,6	Kg/m³ on waggons:	118	118
Transport Width (m)	3,1	3,4	Net chopping cap dry tonne/hour	25	25
Real Speed driving speed (km/h)	6	6	Cutting width, meter	1,8	1,8
Min. Power requirements kW	66	70	Min. power requirement kW	74	74

continued on following page

Table 10. Continued

a kW	0	0	a kW	0	0
b kW/m	10	10	b kW/m	0	0
c kW/ton	0	0	c kW/ton	5,7	5,7
Ppto	31	36	Ppto	44,06	44,06
Purchased Price (€)	16800	26880	Purchased Price (€)	50400,00	50400,00
Estimated Life	2000	2000	Weight (kg)	1600	1600
Weight (kg)	950	2040	Cutting lenght (mm)	10	10
Swath Tedder - Drying	**Kongskilde Z HYDRO Z555H**	**Kongskilde Z HYDRO Z555H**	**Tipper**	**T653/3 double axle tipping trailer**	**T653/3 double axle tipping trailer**
Number	1	1	Cargo Capacity m3	8,2	8,2
Working width (m)	5,4	5,4	Load Capacity kg	6000	6000
Real Speed driving speed (km/h)	6	6	Max Speed km/h	35	35
Min. Power requirements kW	22	22	Speed km/h	30	30
Weight (kg)	505	505	Min Power Required kW	34,4	34,4
Field Speed (Km/h)	6	6	Purchased Price (€)	7200	7200
Estimated Life	2500	2500	Weight (kg)	2380	2380
a kW	0	0			
b kW/m	1,5	1,5			
c kW/ton	0	0			
Purchased Price (€)	7500	7500			
Rake - Windrowing:	**Kongskilde R 460 DS**	**Kongskilde R 460 DS**	**Self loading wagon**	**Pottinger euroboss 330t**	**Pottinger euroboss 330t**
Number	1	1	Cargo Capacity DIN m3	20,5	20,5
Working Width (m)	4,45	4,45	Load Weight (kg)	9000	9000
Real Speed driving speed (km/h)	6	6	Weight (kg)	2800	2800
Min. Power requirements kW	33	33	Net Chopping Capacity (ton/hr)	3,7	3,7
Weight (kg)	526	526	Max Speed km/h	25	25
Field Speed (Km/h)	6	6	Min Power Required kW	44,0	44,0
Estimated Life	3000	3000	Min Power Required kW Transport	30,0	30,0
a kW	0	0	Real Pick-Up Width	1,55	1,55
b kW/m	5,1	5,1	Material Density	70	70
c kW/ton	0	0	Cutting Lenght (mm)	43	43
Ppto	22,695	22,695	Purchased Price (€)	27000,00	27000
Purchased Price (€)	5500	5500			
BALING					
Baller:	**Round Bales Pronar Z500**	**Claas Variant 385**	**Truck**		
Bale diameter (m)	1,2	1,2	Engine Power	315,00	

continued on following page

Table 10. Continued

Chamber width (m)	1,2	1,2	Diesel consumption, empty	20	
Min. power requirement (kW)	40	74	Diesel consumption, full	40,00	
Net press cap (ton/hr)	13,6	15	Fuel Consumption rate l/100km	30	
Cutting width (m)	1,8	2,1	Operating Life years	8,00	
Bale Weight (258 - 296 - 333 kG)	280	280	Investment Costs €	115000	
Number of wagons/units	1	1	Residual Value	30000,00	
Purchased Price (€)	11000	47040			
a kW	2,5	2,5			
b kW/m	0	0	**Truck Trailer**		
c kW/ton	1,8	1,8			
Weight	2250	2845	Roll-off container chopped m3	80,00	
Balle Transport Trailler:	**Pronar T022**	**Pronar T022**	Cargo Space bales (m3)	120	
Balles Load (number)	26	26	Payload (ton)	25,50	
Carrying Capacity (kg)	7300	7301	Fuel Consumption (l/km)	0,325	
Width m; Length m; Height m (m3)	30,723	30,723			
Plataffom lenght m	6,5	6,5			
Maximum Speed (km/h)	35	35			
Min Power Required kW	49,7	49,7			
Purchased Price (€)	8700	8700			
Weight (kg)	2640	2640			
Front Loader:	**John Deere Model 653**	**John Deere Model 653**			
Number of Units	1	1			
Lift capacity (kg)	1950	1950			
Working speed (km/h)	6	6			
Maximum Lift (m)	4,12	4,12			
Min. power requirement (kW)	16	16			
Purchased Price (€)	9000	8064			
Weight (kg)	200	200			
Time to load/unload bale (h/bale)	0,025	0,025			

Table 11. Ownership and Operating Cost Coefficients (ASBAE Standards, 2009)

Equipment Type	C1	C2	C3	RMF1	RMF2
Self Propelled Harvest Equipment					
Cotton picker	1.132	0.165	0.0079	0.11	1.8
Forage harvester	1.132	0.165	0.0079	0.03	2.0
Grain combine	1.132	0.165	0.0079	0.04	2.1
Tractor, <60kW	0.981	0.093	0.0058	0.007	2.0
Tractor, 60-112 kW	0.942	0.100	0.0008	0.007	2.0
Tractor, >112 kW	0.976	0.119	0.0019	0.007	2.0
Harvest Implements					
Baler, rectangular	0.852	0.101	0	0.23	1.8
Baler, large rectangular	0.852	0.101	0	0.10	1.8
Baler, round	0.852	0.101	0	0.43	1.8
Forage harvester	0.791	0.091	0	0.15	1.6
Mower	0.756	0.067	0	0.47	1.85
Mower conditioner	0.756	0.067	0	0.17	1.8
Rake	0.791	0.091	0	0.17	1.4
Forage wagon	0.943	0.111	0	0.16	1.6

Table 12.Storage characteristics (Nemecek et al., 2007; Rotter and Rohrhofer, 2013)

Storage Specifications	Local Storage Center (Closed)		
Investment €	108,930		
Material	Bales	Chopped	Unchopped
Filling Height m	7	7	7
Reduced Storage Capacity %	10	10	10
Capacity m^3	45900		
Density	170	118	70
Capacity tdm yr^{-1}	7803	5416	3213
Cost € tdm^{-1}	13.96	20.11	33.90

Table 13. Truck maximum load (kg) and costs from (Huisman, Venturi and Molenaar, 1997)

	Truck withtTrailer 120m³ (ton)	Truck with Container 80 m³ (ton)	Cost € km⁻¹
Max load ton	25.50	22.50	—
Baled - 153 kg m⁻³	18.36	12.24	1.06
Chopped (2cm) 118 kg m⁻³	14.16	9.44	1.12
Chopped (8cm) 70 kg m⁻³	8.40	5.60	1.12

Chapter 7
Sustainable Smart-Farming Framework:
Smart Farming

Pankaj Agarwal
Amity University Jaipur, India

Vijander Singh
Amity University Jaipur, India

G. L. Saini
Amity University Jaipur, India

Deepak Panwar
Amity University Jaipur, India

ABSTRACT

Agriculture and allied activities play a vital role in a country's economic prosperity. The conventional methods in agricultural practices have become insufficient to cater to the increasing needs. To fulfill the demands, new technologies are to be introduced to raise agricultural standards. Over the past few years, there has been significant interest in designing smart agricultural systems. The manageability of agricultural frameworks has turned into a noteworthy concentration for discusses about future human survival. A significant part of the contention seems to depend on shortsighted understanding of biological models and flops enough to define what maintainability goals are being looked for. To adapt to the undeniably multifaceted nature and between connectedness of current cultivating frameworks with regards to globalization and potential bothers like environmental change, we require a pluralistic way to deal with strategy, which can adapt to the abnormal amounts of vulnerability in these territories and which enables most extreme flexibility of reaction to evolving conditions.

DOI: 10.4018/978-1-5225-5909-2.ch007

INTRODUCTION

Sustainability is improvement that fulfils the requirements of the present without bargaining the limit of who and what is to come, ensuring the harmony between financial development, watch over the earth and social prosperity

Sustainable improvement is an idea that showed up without precedent for 1987 with the distribution of the Brundtland Report(Beckerman, 1994) cautioning of the negative natural outcomes of monetary development and globalization, which endeavoured to discover conceivable answers for the issues caused by industrialization and populace development.

Environmental Sustainability

At the ecological level, sustainability keeps nature from being utilized as an unlimited wellspring of assets and guarantees its insurance and normal utilize.(Islam & Huosong, 2016) From different aspects, for example, natural protection, interest in sustainable power sources, sparing water, supporting feasible versatility also, development in manageable development and engineering, add to accomplishing this ecological supportability on a few fronts.

Social Sustainability

At the social level, sustainability can cultivate the advancement of individuals, groups and societies to help accomplish sensible and genuinely dispersed personal satisfaction, medicinal services and instruction over the globe.(Bornstein, 2007) The battle for sex fairness, particularly in creating nations, is another perspective which in coming years will shape the premise of social sustainability.

Economic Sustainability

Manageability focusses on break even with financial development, that generate money for all, without hurting nature.

Venture and an equivalent conveyance of the financial assets will reinforce alternate mainstays of supportability for an entire improvement.(Unger, 1998)

WHAT ARE THE SUSTAINABLE DEVELOPMENT GOALS?

As a part of a new sustainable development roadmap, the United Nations approved the 2030 Agenda, which contains the Sustainable Development Goals, a call to action to end poverty, protect the planet and guarantee the global well-being of people.

This different point are:

- Eradicate poverty and hunger, guaranteeing a healthy life
- Universalize access to basic services such as water, sanitation and sustainable energy
- Support the generation of development opportunities through inclusive education and decent work
- Foster innovation and resilient infrastructure, creating communities and cities able to produce and consume sustainably
- Reduce inequality in the world, especially that concerning gender
- Care for the environment combating climate change and protecting the oceans and land ecosystems
- Promote collaboration between different social agents to create an environment of peace and sustainable development.

The supportability of farming frameworks has turned into a noteworthy concentration for discusses about future human survival (Kaplan & Kaplan, 2009). A great part of the contention seems to depend on oversimplified understanding of biological models and fail satisfactorily to define what supportability targets are being looked for.

For a considerable length of time, horticulture has been related with the creation of basic nourishment crops. At exhibit, horticulture well beyond cultivating incorporates ranger service, dairy, organic product development, poultry, honey bee keeping, mushroom, subjective, and so forth. Today, handling, advertising, and circulation of harvests and domesticated animals' items and so forth are altogether recognized as a major aspect of current agribusiness. In this manner, horticulture could be alluded to as the creation, preparing, advancement and conveyance rural items.

Farming assumes an imperative part in the whole existence of a given economy. Agribusiness is the foundation of the monetary arrangement of a given nation(Byerlee & Alex, 1998). Notwithstanding giving nourishment and crude material, agriculture additionally gives work chances to extensive level of the population. Below is the importance of agriculture:

- Source of Income
- Role in National Revenue
- Food supply
- Significance to International Market
- Marketable Surplus
- Raw Material
- Significance in Transportation
- Foreign Exchange Resources
- Employment Opportunities
- Food Security

Indian agriculture is plagued by several problems; some of them are natural and some others are man-made. Some of the problems majorly identified are:

- Traditional Farming
- Climatic Condition
- Soil Erosion
- Irrigation
- Lack of mechanisation
- Agricultural Marketing
- Transportation
- Inadequate Storage Facilities
- Scarcity of Capital

Tomorrow's challenges of doubling food supply put sustainability of agriculture at one level with ensuring food security. The global food system needs to be resource efficient and at the same time sustainable. Efficient use of water, reduction of soil erosion and degradation to the minimum, minimization of energy input and maximization of yields under uncertain natural conditions are the goal. They pose highest requirements on the underlying information and knowledge infrastructure and make future farming a knowledge business and a very sophisticated management task(Boisot, 1998).

Agribusiness has seen numerous unrests, regardless of whether the domestication of animals and plants a couple of thousand years prior, the precise utilization of yield turns and different changes in cultivating practice a couple of hundred years back, or the "green revolution" with deliberate reproducing and the across the board utilization of man-made manures and pesticides a couple of decades back. Agriculture is experiencing a fourth upheaval activated by the exponentially expanding utilization of data and correspondence innovation (ICT) in farming(Latchem & Jung, 2009).

The Role of Information and Communication Technology in Agriculture

Today, mechanical improvements for concerning human variables are vital. These incorporate innovations that are sheltered to ranchers, add to enhance agriculturists' personal satisfaction; and utilization of instruments to evaluate cultivate work change estimation, bring down work costs, lessened damage unlucky deficiencies, diminished consumptions for therapeutic care. The field of ICT in the agricultural space is very new to be presented efficiently. Legitimate coordination of ICT in agriculture can help take care of the issue of sustenance emergency and can add to the accomplishment of maintainable agrarian advancement in the area. Convenient and pertinent logical and innovation-based data through ICT is imperative to empower little scale agriculturists to influence successful utilization of their assets, to end up aggressive, and raise their salary. Popularized homesteads may need to depend on privatized augmentation and counselling administrations, since they require their innovative needs as far as ranch administration and promoting procedures.

Wide basing agricultural augmentation exercises; creating cultivating framework research and expansion; having area particular modules of research and augmentation; and advancing business sector augmentation, maintainable agrarian advancement, participatory research, and so forth are a portion of the various territories where ICT can assume an essential part.

A few researches examine led on augmentation associations have uncovered that the conveyance of merchandise is powerful when the grass roots expansion specialist covers a little zone of ward, with various purposes (wide basing). The current arrangement of substantial locales, each with a tight scope of exercises, is less compelling. Be that as it may, expansive basing requires grass attaches laborers to be at the front line of augmentation and ace of numerous exchanges, which isn't generally conceivable. IT can help here, by empowering augmentation specialists to assemble, store, recover and disperse a wide scope of data required by agriculturists, in this way changing them from expansion laborers into information laborers.

The development of such learning laborers will bring about the acknowledgment of the much-discussed base up, request driven innovation age, appraisal, refinement and exchange. Rural augmentation frameworks in most creating nations are under-supported and have had blended impacts. A significant part of the expansion data has been observed to be outdated, immaterial and not material to little agriculturists' needs, leaving such ranchers with next to no data or assets to enhance their efficiency.

ICT helps the augmentation framework in re-arranging itself towards the general horticultural advancement of little creation frameworks. With the suitable learning, little scale makers can even have an aggressive edge over bigger activities. At the point when information is tackled by solid associations of little makers, key arranging can be utilized to give individuals minimum cost inputs, better storerooms, enhanced transportation connections and aggregate transactions with purchasers.

AREAS OF IT CONVERGENCE

Applications of IT in support of agricultural and rural development fall into five main areas, as outlined by Don Richardson. These are:

- Economic improvement of agriculture makers;
- Community improvement;
- Research and training;
- Small and medium ventures improvement; and
- Media systems.

Some agricultural development services that can be provided in the developing world, using ICT, are:

- Online administrations for data, instruction and preparing, checking and discussion, finding and observing, and exchange and handling;
- E-business for coordinate linkages between nearby makers, dealers, retailers and providers;
- The help of communication among scientists, augmentation (information) specialists, and agriculturists;
- Question-and-answer administrations where specialists react to inquiries on specific subjects' ICT administrations to piece and locale level formative authorities for more prominent proficiency in conveying administrations for general farming improvement;
- Up-to-date data, provided to agriculturists as ahead of schedule as could be expected under the circumstances, about subjects, for example, bundles of practices, advertise data, climate estimating, input supplies, credit accessibility, and so on.;
- Creation of databases with points of interest of the assets of nearby towns and villages, site-particular data frameworks, master frameworks, and so on.;
- Provision of early cautioning frameworks about infection/bother issues, data with respect to provincial advancement projects and product protections, postharvest innovation, and so on.;
- Facilitation of land records and online enrollment administrations;
- Improved promoting of drain and drain items;

- Services giving data to ranchers in regards to cultivate business and administration;
- Increased effectiveness and profitability of agreeable social orders through the PC correspondence arrange and the most recent database innovation;
- Tele-instruction for ranchers;
- Websites built up by agrarian research establishments, making the most recent data accessible to expansion (information) specialists and acquiring their input.

Data driven savvy cultivating is a general pattern to be seen in farming. Environmentally and monetarily significant measures to enhance profitability are connected in savvy cultivating. The procedure depends on the standards of Precision Farming, i.e. on the utilization of GPS-direction to apply site-specific rural measures. Be that as it may, while the focal point of Precision Farming was essentially on cultivating innovation to for instance take into account auto-guiding of tractors and reapers, the focal point of smart cultivating shifts towards a more adjusted, all-encompassing methodology—going from "most astounding spatial accuracy" to "smartest treatment".

Independent, automated vehicles have been produced for cultivating purposes, for example, mechanical weeding, use of compost, or reaping of organic products. The improvement of unmanned ethereal vehicles with self-governing flight control, together with the advancement of lightweight and effective hyperspectral depiction cameras that can be utilized to ascertain biomass advancement and treatment status of harvests, opens the field for modern homestead administration exhortation. Additionally, choice tree models are accessible now that enable agriculturists to separate between plant infections in view of optical data. Virtual fence advances permit dairy cattle group administration in view of remote-detecting signs and sensors or actuators appended to the domesticated animals.

Taken together, these specialized enhancements constitute a specialized transformation that will produce problematic changes in farming practices. This pattern holds for cultivating in created nations as well as in creating nations, where organizations in ICT (e.g., utilization of cell phones, access to the Internet) are being embraced at a quick pace and could turn into the distinct advantages later on (e.g., as occasional dry season estimates, atmosphere brilliant farming). Such significant changes by and by come with open doors as well as large difficulties.

It is essential to call attention to out at an almost phase of this insurgency to evade "bolt ins": supporters and doubters of innovation need to take part in an open exchange on the future improvement of cultivating in the advanced period. Just if parts of innovation, decent variety of harvest and domesticated animals frameworks, and systems administration and establishments (i.e. markets and arrangements), are thought about together in the discourse, should cultivating in the advanced period be named "smart farming"

Ample Opportunities

Smart cultivating lessens the environmental impression of cultivating. Limited or site-particular use of data sources, for example, composts and pesticides, in exactness horticulture frameworks will relieve filtering issues and also the discharge of ozone depleting substances. With current ICT, it is conceivable to make a sensor organize taking into consideration relatively ceaseless observing of the ranch. Correspondingly, hypothetical and pragmatic systems to interface the conditions of plants, creatures, and soils with the requirements for generation inputs, for example, water, manure, and drugs, are in reach with momentum ICT all inclusive.

Savvy cultivating can make horticulture more productive for the rancher. Diminishing asset sources of info will spare the rancher cash and work and expanded unwavering quality of spatially unequivocal information will lessen dangers. Ideal, site-particular climate gauges, yield projections, and likelihood maps for illnesses and catastrophes in view of a thick system of climate and atmosphere information will permit development of products in an ideal way. Site-particular data likewise empowers new protection and business open doors for the whole esteem chain, from innovation and information providers to ranchers, processors, and the retail area in creating and created social orders alike. In the event that all cultivating related information are recorded via mechanized sensors, the time required for organizing the use of assets and for authoritative reconnaissance is diminished.

Smart cultivating likewise can possibly support customer acknowledgment. On a basic level, upgrading administration additionally allows expanded item quality (e.g., higher measures of cell reinforcements and other auxiliary metabolites in view of ideal fruiting densities in plantations; or physiologically more agreeable drain items in view of individualized sustaining apportions of domesticated animals). These items are more advantageous as well as offer at higher costs, a key procedure in utilizing land all the more productively.

Moreover, the straightforwardness of creation and handling will increment along esteem chains on the grounds that ICT permits enlistment as to which cultivate delivered a specific item under which conditions. This offers the potential for new, more straightforward types of collaboration among agriculturists and shoppers. All things considered, various obstacles must be overcome. Among the significant inquiries to be tended to: Who claims the information? ICT that records the contribution of assets and the yield of items raises inquiries of property rights and utilization of information. Plans of action may make included an incentive by changing over spatially unequivocal enormous information into data and counsel for ranchers as well as for administrative specialists who may utilize the information for reconnaissance and control. Governments must build up an administrative design that ensures superb information while in the meantime cultivating trust among all performers included. The potential abuse of information makes extra lawful and moral difficulties for control and checking.

What's more, ICT will escalate the difficulties of obligation and responsibility of new advancements. There must be responsibility for fumble or blunders prompting financial and ecological results. For instance, who is in charge of hints of fungicides abandoned on gathered natural products when that fungicide has been connected past the point of no return? Is it the agriculturist, the supplier of the product, or the maker of the sensor? Such situations are a long way from being clear, as of late exemplified by mischances with self-driving autos. High expenses to embrace innovation for singular ranches and restricted learning and abilities can be noteworthy appropriation obstacles, particularly in creating nations. Hence, the entrance to the most recent innovation may stay confined to huge and industrialized homesteads.

The advantages of ICT may be restricted to industrialized nations and concentrated on the creation of understood and broadly developed yields, for example, wheat, maize, and rice. This additionally builds the danger of unsustainable increase rehearses. In an industrialized setting, malady episodes might be postponed by fungicides, yet this accompanies an expanded danger of creating safe contagious strains that would then be able to act considerably more devastatingly once they have beaten anticipation measures.

The change to industrialized social orders has prompted fast declines in cultivating inhabitancy rates to estimations of 2% and less of the populace in Europe and North America. Digitalization of horticulture may impact business openings and occupation profiles of ranchers and cultivating related experts much further. Will this advancement spur or demoralize capable people from going into the field of farming? Will the potential loss of agriculturists' duty to information overseeing robots and ICT frameworks increment or diminishing the acknowledgment and the valuation for this calling? Another developing test will join the agriculturists' information and encounters with these new innovations.

Savvy cultivating can give a deliberate way out of secured advancements and practices described by solid polarization and market division. It offers a way toward economical farming by enhancement of innovations, harvest and animals creation frameworks, and systems over all performers of the agri-nourishment division. There is no single strategy approach that can accomplish this vision, which underpins and encourages the suitable utilization of ICT innovation. Or maybe, the thought is to distinguish the predominant components that compel or debilitate a supportable utilization of the innovation and to choose the most proper activities in created and creating nations. This may bring about better access to capital now and again and to particular help of interests in others. In addition, the help of helpfully utilized ranch observing innovation (e.g., together possessed unmanned aeronautical vehicles checking fields of whole towns) or interests in instruction and preparing may likewise bolster the maintainable utilization of these advancements. In all cases, be that as it may, the arrangement condition ought to give a reasonable, legitimate setting that takes into account successful possession and client rights. The conceivable outcomes of the computerized time may prompt new types of enhancement on ranches. Like the level-headed discussion and conceptualization of "brilliant urban communities," the potential outcomes of ICT will probably not prompt one all around uniform and quickly acknowledged cultivating framework however to an assorted variety of cultivating frameworks. Specialized advancements adding to expansion are encouraged by administration counsel if given with high unwavering quality and lucidity, regardless of whether ranchers have not delivered a specific product previously. Current issues with protection, for instance to anti-infection agents and pesticides, could be kept away from with a higher assorted variety of creation frameworks. Nonetheless, in spite of the fact that the "Web of Things," including agrarian hardware, can be utilized to oversee standard cultivating circumstances, the rancher still needs to fill in as both researcher and guard dog, watching out for unanticipated circumstances. Ranchers can contribute the time authorized by digitalization in treatment of illnesses or in observing and treating domesticated animals in a more individualized manner. Product vermin and illnesses require just be handled when certain limits—decided with new ICT applications—are come to. All things considered, such a purposeful increment of decent variety requires that purchasers, ranchers, and chiefs are persuaded of the advantages of these mechanical focal points. Also, it requires another arrangement of information exchange with differentially managed straightforwardness works: the authoritative and generation information being exchanged to providers and the administration must be straightforward to agriculturists. Also, the shopper must be able to acquire understanding into the whole sustenance creation chain. ICT empowers agriculturists to trade data, set up participation and associate audit, and possibly create casual data frameworks that can supplement the formal data arrangement of controlling experts. Such a stream of data among ranchers and amongst agriculturists and shoppers would be scale free and would not be limited by state outskirts. Clear signs for a reception of such frameworks can be seen as of now in created and creating nations, with online networking stages and activities. Institutional developments would be conceivable, prompting systems of agriculturists who are more self-sorted out and adaptable than today.

ICT and information administration can give novel routes into a beneficial, socially acknowledged horticulture that advantages nature (e.g., soil, water, atmosphere), species decent variety, and agriculturists in creating and created nations. Be that as it may, this can just occur with the proactive improvement of arrangements supporting the important legitimate and market design for brilliant cultivating, with a discourse among cultivating innovation supporters and doubters, and with watchful thought of rising moral inquiries.

Brilliant Farming speaks to the use of present day Information and Communication Technologies (ICT) into horticulture, prompting what can be known as a Third Green Revolution.

Following the plant reproducing and hereditary qualities upheavals, this Third Green Revolution is assuming control over the farming scene in view of the joined use of ICT arrangements, for example, exactness gear, the Internet of Things (IoT), sensors and actuators, geo-situating frameworks, Big Data, Unmanned Aerial Vehicles (UAVs, rambles), mechanical technology, and so on.

Smart Farming has a genuine potential to convey a more beneficial and maintainable agrarian creation, in light of a more exact and asset productive approach. In any case, while in the USA conceivably up to 80% of ranchers utilize some sort of Smart Farming Technologies(SFT), in Europe it is close to 24%.

From the rancher's perspective, Smart Farming ought to give the agriculturist included an incentive as better basic leadership or more proficient misuse activities and administration. In this sense, smart cultivating is unequivocally related, to three interconnected innovation fields tended to by Smart Farming Network:

Smart Farming Technology Types

Smart cultivating innovations are divided into three principle classifications that, as expressed above, cover the cyclic arrangement of PA(Bongiovanni & Lowenberg-DeBoer, 2004):

- Data obtaining innovations: this classification contains all looking over, mapping, route and detecting advancements.
- Data examination and assessment advancements: these innovations run from straightforward PC based choice models to complex ranch administration and data frameworks including a wide range of factors.
- Precision application innovations: this classification contains all application advances, concentrating on factor rate application and direction advances.

Data Acquisition Technologies

The Smart Farming Technologies are used for recording and mapping field and crop characteristics are divided into the categories below:

- Worldwide route satellite frameworks innovations (in certainty these advancements record the real position which can be utilized for various purposes, for example, direction, mapping and so on.)
- Mapping technologies
- Data acquisition of environmental properties (Camera based imaging, NDVI measurements, soil moisture sensors)
- Machines and their properties Global navigation satellite systems (GNSS) technologies

Mapping Technologies

Elevation Maps

Elevation is a basic layer in PA since it is extremely valuable to enable ranchers to comprehend yield reaction. It impacts soil arrangement, water development and editing perspectives. It can decide waterlogged regions, disintegration hazard, waste limitations, and frequently is identified with soil write. Utilizing information from GNSS recipients, it is conceivable to deliver a computerized rise demonstrate (DEM) (Austin, DiSera, & Brooks, 2015) of a field or a homestead that can be utilized to group territory attributes, for example, slant, perspective, bend, sunlight-based radiation capture attempt, scene water stream headings and topographic wetness records. Height maps can recognize how geography can influence agronomic outcomes in a field and obviously to level the field. Utilizing this data, it is conceivable to create (I) forms and geography maps, (ii) 3-D displaying of ponding danger, overflow and speed maps, (iii) cultivate format outlines, (iv) shape bank outline, seepage designs and on-ground usage and (v) cut and fill arrive levelling plans.

Soil Mapping

Soil examining is fundamental to gather data about soil surface (sand, residue, earth substance), accessibility of supplements for harvests to develop (N, P, K, Ca, Mg, pH, lime) and other soil synthetic properties (natural issue, saltiness, nitrate, sulphate, substantial metals). What's more, it can be utilized to distinguish soil compaction, dampness content and other mechanical and physical soil properties. Soil examining can be executed utilizing the arbitrary, versatile or matrix system. In irregular examining, soil centres are acquired from arbitrary areas inside the field. In versatile examining, chose test areas rely upon earlier data. Framework inspecting includes efficiently gathering tests from foreordained focuses in the field. None of the current soil testing hones has been perceived as the best. Another technique to delineate field's dirt properties is the utilization of in some the-go sensors that can possibly give profits by the expanded thickness of estimations at a generally ease. These sensors can be either joined with a GNSS collector and create maps of soil properties or they can be utilized as continuous sensors where the yield of the sensor is utilized quickly for variable-rate utilization of composts, lime and fertilizer.

Yield Mapping

Yield mapping or yield observing is a strategy in agribusiness of utilizing GNSS information to examine factors, for example, edit yield and dampness content in a given field. The segments of a grain yield mapping framework incorporate a grain stream sensor that measures grain volume, a grain dampness sensor that evaluates dampness variety, a grain lift speed sensor that measures grain speed to figure grain mass, a GNSS reception apparatus that geo-references grain estimations, a header position sensor that starts grain estimation when the header is brought down and a movement speed sensor that gives the separation that the reaper has secured amid a specific logging interim. There are numerous kinds of grain sensors that are industrially accessible, for example, an oar wheel volume stream sensor, force plate sensor, gamma beam sensor, strain check based effect sensors, infrared sensor. Other yield sensors are likewise found in writing that are not economically accessible, for example, rotated wood screw, piezo-film strips, capacitive sensor, ultrasonic sensor, lift based stream sensor, X-beam systems.

Recording of Environmental Parameters

- Camera Based Imaging
  - **RGB Cameras:** Red, Green and Blue (RGB) cameras consolidate the hues red, green and blue to delineate the scope of hues that exist in the earth and in the farming fields.
  - **LiDAR Sensors:** LiDAR sensors (Light Detection and Ranging) are instruments that measure the separation from the objective by laser. This innovation has been utilized to contemplate the phenotypic variety by making three-dimensional models of plants.
  - **ToF (IR) Cameras:** Time of Flight (ToF) cameras can create moulded and disjointed infrared light in the space. Keen sensors at pixels of the camera record the reflected light and figure an opportunity to return.

- **Light Curtains:** Light shades are another framework that is utilized to think about the phenotypic qualities. The framework comprises of a few bars which are put in parallel. One bar discharges light shafts that end up at the other parallel bar. Along these lines the framework records if the light shafts are obstructed by a protest.
- **Multi- or Hyper-Spectral Cameras:** Multispectral cameras are cameras that can photograph the environment with a limited number of spectra in the visible and infrared spectrum.
- **Thermal Cameras:** Warm cameras can produce pictures identified with the surrounding temperature. This is on account of they work in the long wavelength infrared (to 14,000 nm) bringing about seeing the radiation transmitted by the objective as a result of its warmth. (Calderón, Navas-Cortés, Lucena, & Zarco-Tejada, 2013) Warm cameras have been utilized to think about the phenotypic difference for foreseeing water worry of plants, to distinguish illnesses and pathogens and for the aging of natural products.

Normalized Difference Vegetation Index (NDVI)

The NDVI is a numerical file in view of the obvious and close infrared groups of the electromagnetic range that shows if an objective being watched contains live green vegetation or not; it takes esteems in the vicinity of 0 and 1. There are numerous uses of NDVI for either agribusiness or natural arrangements. It can be utilized to evaluate edit yields, rate ground cover, photosynthetic movement of the plant, surface water, leaf territory file, the measure of biomass, field execution, rangeland conveying limits, and so on.

In a hurry NDVI ground sensors can be joined with either a GNSS collector to deliver maps of NDVI or they can be utilized as ongoing sensors where the yield of the sensor is utilized instantly for variable-rate manure or splashing applications.

Spectral Sensors

Spectral sensors are instruments that can detect the measure of light reflecting from objects, which they change over to an electrical flag. They measure light in the unmistakable (400– 700 nm) and infrared range (700– 2500 nm). Phantom sensors are utilized broadly in farming since it has been discovered that these estimations are identified with a plant's physiology and improvement.

Fluorescence Sensors

Fluorescence incited by bright radiation has been utilized as a non-dangerous technique for evaluating plants status. In particular, the fluorescence of plants caused by UV radiation has been utilized for the distinguishing proof of types of plants, for plant development, for absence of supplements in plants, for absence of water, for temperature impacts on plants and for identifying assaults by pathogens of plants.

Soil Moisture Sensors

Data on the spatial and transient advancement of soil dampness is of awesome significance for the utilization of soils and for vegetation, specifically where the water assets are rare.(Adamchuk & Rossel, 2011) There are a few dependable approaches to quantify soil dampness.

Frequency Domain Reflectometry (FDR) Sensors

At the point when a capacitor utilizes the dirt as a dielectric, its electrical capacitance relies upon the dirt water content. Such capacitors can be made of metal plates or bars. In the event that this capacitor compose is associated with an oscillator to shape an electrical circuit, any adjustment in the circuit's working recurrence shows changes in soil dampness. This is the working guideline of capacitance and recurrence space reflectometry (FDR) sensors.

Time-Domain Reflectometry (TDR) Sensors

These sensors depend on estimating the time it takes for an electromagnetic heartbeat (wave) to engender along a transmission line encompassed by the dirt. In this way, TDR sensors deliver a progression of exactly coordinated electrical heartbeats with an extensive variety of high frequencies, that movement along a transmission line that is worked with a coaxial link and a test.

Amplitude Domain Reflectometry

The working standard of these sensors depends on the impression of a piece of the vitality transmitted (electromagnetic wave going along a transmission line) back to the transmitter when the wave achieves an area with various impedance. A voltage standing wave along the transmission line is delivered when the reflected wave interfaces with the occurrence wave. These sensors limit the impact of soil electrical conductivity by picking a flag recurrence with the goal that dirt water substance can be evaluated from the dirt or test impedance.

Phase Transmission

The standard of these sensors depends on the stage move that an electromagnetic wave at a settled recurrence will express in connection to its stage at the cause in the wake of voyaging a settled separation. The properties that deliver this stage move are the length of movement along the transmission line, the recurrence and the speed of proliferation. In this way, realizing speed of spread is identified with soil dampness content, when a settled recurrence is utilized and the length of movement is steady, soil water substance can be dictated by this stage move. The test of these sensors comprises of two open concentric metal rings to apply stage estimating hardware toward the start and end of the wave guides

Time Domain Transmission

These sensors measure the time that an electromagnetic heartbeat requires to spread along a transmission line (one-way). They are like TDR sensors, yet for this situation an electrical association toward the start and end of the transmission line is required. The test comprises of bowed metal poles to accomplish the addition toward the start and end of the transmission line in the electronic piece.

Tensiometers

Tensiometers depend on the rule of water harmony between the dirt arrangement and the water substance of a fixed water-filled tube introduced in the dirt through a penetrable and soaked permeable material. This harmony comes about because of accomplishing a similar weight potential for both the water in the tube and the water held in the dirt grid. Henceforth, the dirt water matric potential is proportional to the vacuum or suction made inside the tube. These sensors comprise of a fixed water-filled plastic tube with a clay container toward one side and a negative weight check at the other

Gypsum Blocks

These sensors decide soil dampness by estimating the protection between two anodes inside the gypsum pieces, which is corresponding to water substance of the square (low protection when water content gets littler). Gypsum squares are permeable, so their water content is identified with the dampness of the dirt that in which it is being introduced. The condition for solid estimations is ideal contact amongst sensor and soil. The gypsum squares are covered for all time in the dirt at the coveted profundity with a future of 3 to 5 years (contingent upon the sort of soil).

Granular Matrix Sensors

Water conditions inside granular grid sensors change with comparing variety in water conditions in the soil. These progressions inside the sensor are reflected by contrasts in electrical protection between two terminals imbedded in the sensor. Protection between the anodes diminishes with expanding soil water. These sensors have a permeable fired outer shell with an interior grid structure that contains two anodes. An interior tube-shaped gypsum tablet cushions against soil saltiness levels that happen in a few sorts of flooded soil. An engineered permeable layer is encompassed by a stainless-steel packaging or sleeve with gaps.

Heat Dissipation Sensors

These sensors depend on the way that become materials warm scarce speedier than wet materials in light of warmth dissemination delivered by the warm conductivity of water. Subsequently, expanded water content in a permeable material increments in extent to warm stream. A warm warmth test has a permeable piece joined with a warmth source and an exact temperature sensor. The warmth source labors for a couple of moments and the temperature sensor measures the temperature when warming to figure the distinction. These sensors are sold with the adjusted connection between the deliberate change in temperature and soil water potential.

Machines and Their Properties

Travel Speed Sensor

This sensor decides the separation the tractor or consolidate reaper ventures. At times travel speed is estimated with a GNSS recipient or a radar or ultrasonic sensor.

Tractor Sensing Systems With ISOBUS

Data to-activity basic leadership forms, and additionally accuracy farming applications, require sensors for in a hurry information gathering of product and soil variety (e.g. soil dampness content, NDVI, edit thickness, et cetera). (Backman, Oksanen, & Visala, 2013) The ISOBUS convention can assume a critical part in the improvement of exactness agribusiness and encourages data to be traded and put

Figure 1.

away more proficiently between sensors, processors, controllers and programming bundles from various producers inside a similar tractor or vehicle(Abidine, Heidman, Upadhyaya, & Hills, 2002)

Unmanned Aerial Vehicles (UAVs)

An unmanned airborne vehicle (UAV), generally known as an "automaton" is a flying machine without a human pilot on board. The flight of UAVs might be controlled either self-rulingly by on-load up PCs or by the remote control of a pilot on the ground or in another vehicle. There are two fundamental stages for UAVs: settled wing and multi-rotor.

Comparison with manned aircraft

- UAVs can be flown in unsafe circumstances (no pilot or researcher on board).
- UAVs can fly for long terms, on dull missions, for example, mapping or for diurnal estimations without burdening pilot or group.
- UAVs with long perseverance can stay as yet amid a crisis, empowering long-haul consciousness of a circumstance.
- UAVs with a long-range ability can be propelled from or travelled to a remote area.

- UAVs with high height capacity can fly securely above the weather or more air movement. Examination with satellites:
- UAV pictures are not bothered by mists on the grounds that their flying stature is low.
- UAVs can travel to exactly chose areas at absolutely chose circumstances.
- UAVs can be entrusted to stay over discretionary focuses for long lengths
- UAVs can convey an assortment of tradable high determination imaging instruments.
- UAVs are recoverable for upkeep and updates of sensor and correspondence frameworks.

Unmanned Ground Vehicles (UGVs)

Light weight implies that the vehicle requires less vitality and instigates less soil compaction, and they should be little for security reasons, to accomplish more prominent accuracy on their errands and to have greater mobility. Mechanical outline of the models relies upon its primary assignments or designers' objectives. The UGVs can keep running on tracks or wheels. Despite the fact that tracks have numerous favourable circumstances contrasted with wheels with zero turn span, better buoyancy, smoother ride on unpleasant surfaces, more prominent effectiveness over a more extensive scope of soil conditions and greater security on slopes, their fundamental disservices for use on robots are the movement control and the posture estimation.

Data Analysis and Evaluation Technologies

These SFTs are utilized for investigation of the information got from the information securing SFTs and are arranged as takes after:

Management Zone Delineation

All information gathered must be investigated and deciphered if a significance is to be drawn from them. There are by and large an excessive number of information and proper strategies that exist or must be created for the examination should be connected. Basic exploratory measurements can give an early introduction of the qualities, their spread, the range and the dispersion. Geo-measurements, in light of what is called 'the hypothesis of regionalised factors', is essentially a probabilistic strategy for spatial introduction. Last development of the guide at the nearby level is made conceivable from evaluated esteems in light of the variogram by kriging; the variogram depicts the structure of the spatial variety of the examined information.

This sort of data, which can be acquired for various properties and for progressive years, opens new and fascinating conceivable outcomes in agronomic product investigation and administration. Variograms are utilized to survey the spatial variety of the deliberate esteems. For every property semi changes are plotted against the separation (slack) between the focuses. A model is fitted to the trial variogram, which is the hypothetical variogram. Maps covering the entire field can be created and show the variety in the properties. There are a few techniques for information investigation, despite the fact that that there isn't an unmistakable method to think about the maps created. This is as yet in light of an optical impression for examination of the maps. Connections between parts of the field with various properties can be completed to survey their relations.

The investigation of the information plans to characterize parts of the field with basic qualities that can be overseen independently. These are the administration zones. Depiction of administration zones ought to make homogeneous parts of the field where inputs or different practices can be connected similarly. The administration zones ought to be sufficiently extensive to allow variable-rate use of sources of info, yet sufficiently little to be homogeneous.

- *Decision-Support Systems*

A choice emotionally supportive network (DSS) is a PC based framework(Bui & Lee, 1999) that backings business choices. In horticulture it alludes to the choices taken by the rancher for cultivate administration. Accuracy agribusiness is associated specifically to basic leadership by the rancher. It can be depicted for instance of the change of information into choices.

Farm Management Information Systems (FMIS)

Agribusiness has turned out to be extremely perplexing and ranchers utilizing Smart Farming Technologies (SFTs) gain a huge measure of information that need to dissect and determine the best choices for their product administration. The way to progress is access to convenient data and expounded choice making. Decision making is a critical perspective in cultivate administration and has been considered by various creators and with various application. The essential parts of FMIS incorporate particular agriculturist situated outlines, devoted UIs, computerized information preparing capacities, master learning and institutionalized information correspondence and versatility. To enhance usefulness, different administration frameworks, database organize structures and programming models have been proposed, where FMIS have expanded in refinement through the combination of new advancements, for example, electronic applications and applications for PDAs and tablets.

As agribusiness is a mind-boggling framework it consolidates various connections between agriculturists, guides, brokers, merchants, legislative bodies, cultivate hardware, natural controls, financial estimations and others. FMIS can cover a substantial number of capacities, for example, stock, logbook, coordinate deals, site-particular administration capacities. The dispersion of data administration as business advancement in the cultivating group could profit by the extensive research created in the most recent decades on the selection of ICT and internet business among the two shoppers and private companies.

The FMIS could be characterized by the application. There is programming accessible for edit checking, for entire homestead administration, for exactness farming just and particular programming for particular applications (planning water system, showering guess, precise climate estimates)

Table 1. Farm management Information systems functions

Function Title	Function Description
Field operations management	Recording of homestead exercises to enable rancher to upgrade trim creation by arranging exercises and watching the genuine execution of arranged errands. Preventive measures might be started in view of the observed information.
Finance	Estimation of the cost of each ranch movement, input– yields counts, gear charge-outs, work prerequisites per unit territory. Anticipated and genuine expenses are additionally thought about and contribution to the last assessment of the ranch's financial reasonability.
Inventory	Observing and administration of all creation materials, hardware, chemicals, composts, and seeding and planting materials. The amounts are balanced by the customer designs and client orders.
Traceability	Harvest utilized, an ID naming framework to control the deliver of every creation area, including utilization of data sources, representatives, and hardware, which can be effectively filed for quick review.
Reporting	Formation of cultivating reports, for example, arranging and administration, work advance, work sheets and directions, orders buys cost announcing, and plant data.
Site Specific	Mapping the highlights of the field, investigation of the gathered information, age of variable rate contributions to streamline info and increment yield. This is the SFT part. It could be a different programming or could be coordinated.
Sales	Administration of requests, charges for administrations, online deals.
Machinery management	Incorporates the subtle elements of hardware use, the normal cost per work-hour or per unit territory. It additionally incorporates armada administration and coordination.
Human resource management	Representative administration, accessibility of representatives in time and space, taking care of work times, installment, capabilities, preparing, execution, and skill.

(S Fountas et al., 2005)

Table 2. Smart farming technologies economic impact

Smart Farming Technologies	Economic Impact
Global navigation satellite systems (GNSS)	GNSS Technologies don't have coordinate monetary effect, yet it is a prerequisite for most PA applications and in this way the advances may have a backhanded advantage.
Differential GNSS	
Real time kinematic (RTK)	
Network RTK (NRTK)	
Wide area RTK (WARTK)	
Undifferenced GNSS	
Precise point positioning (PPP)	
Fast PPP (FPPP)	
Elevation maps	Rise, soil parameters (ECa, pH, dampness substance) and yield mapping does not offer direct monetary effect in the event that it isn't deciphered from Crop Consultants together with the rancher to apply site-particular product administration.
Soil mapping	
Yield mapping	
Yield monitor display	
RGB cameras	Imaging does not offer direct financial effect, but rather in the event that it is utilized for VR application, the diminishment in sources of info will think about the farmer income
LiDAR cameras	
ToF (IR) cameras	
Light curtains	
Multi/Hyper-spectral Cameras	
Thermal cameras	
Spectral sensors	NDVI estimation does not offer direct monetary effect, but rather on the off chance that it is utilized for VR application, the diminishment in data sources will think about the farmer income
Fluorescence sensors	
Frequency domain reflectometry	Soil dampness sensors estimations don't offer direct financial effect, yet in the event that it is utilized together with water system administrations they can be utilized for the decrease of water system water which ponders the income for the farmer.
Time domain reflectometry	
Amplitude domain reflectometry	
Phase transmission	
Time domain transmission	
Tensiometers	
Gipsum blocks	
Granular matrix sensors	
Heat dissipation sensors	
Travel speed sensor	These advancements don't offer particular direct financial increases for the rancher, yet they may have a circuitous advantage when connected together with VR applications
Tractor sensing systems using ISOBUS	
Unmanned aerial vehicles	Unmanned vehicles can give benefit to the homestead, mostly because of restricted/truant work cost and less fuel costs contrasted and tractor mounted frameworks. Little low weight vehicles may diminish costs in connection to soil compaction and harm.
Unmanned ground vehicles	
Management zone delineation	Depiction of zones does not offer direct financial effect, with the exception of on the off chance that they are thought about for cultivate administration enhancement.
Decision support system	FMIS gives to agriculturists/cultivate directors itemized planning methodology, field arranging, accounting for appropriation applications and for open experts' reviews
Farm management information System (FMIS	

(Spyros Fountas, Wulfsohn, Blackmore, Jacobsen, & Pedersen, 2006)

Table 3. Other Smart Farming Techniques

Techniques	Description
Big Data	Utilized for gathering of information and aides for basic leadership in the way like water administration.
Evapotranspiration	Water request of any yield can be dictated by utilizing reference edit Evapotranspiration and product coefficient.
Wireless Sensor Network	It gives a financially savvy answer for screen and control the air weight, temperature, moistness and soil pH.
Cyber Physical System	Framework which coordinates computational and physical parts and communicates between them to detect the event of progress.
Cloud Computing	IOT with distributed computing is useful to detect the topographical necessities by following area and charge pay per utilization can be actualized which diminishes cost.

(Balafoutis et al., 2017)

Smart Farming Framework

The fundamental segments of the Smart-Farming system are portrayed beneath.

- **Data Wrapper:** It offers a nonexclusive method to depict qualities of sensors utilizing tactile metadata, containing general data about the information stream. A semantic explanation module empowers to comment on the parsed tactile information.
- **Device Administrator:** It naturally oversees IoT gadgets, expelling the requirement for human administrators, giving the important devices to autonomic administration procedures to uphold choices at a later stage. Oversees gadget character and approval, considers unwavering quality of information streams (e.g. constant registering if values fall with specific points of confinement) and blame recuperation.
- **Discovery Module:** It guarantees versatile enrolment and disclosure of IoT gadgets and administrations progressively, in an attachment and play way. These gadgets can be either situated at the same physical space (e.g. inside the homestead) or remotely, got to through the web/web
- **Data Total:** It manages vast volumes of information utilizing time arrangement examination and information pressure methods to lessen the span of crude tangible perceptions conveyed by the information wrappers. (Sasikala, 2013)
- **Data League:** It answers clients' inquiries, e.g. the measure of compost expected to apply over some region. This part first finds pertinent streams as indicated by the prerequisites specified in the demand. At that point, it makes

an interpretation of clients' solicitations into RDF Stream Processing (RSP) questions and assesses the inquiries to acquire comes about. (Kainrath, 2009)

- Event location (source: CityPulse) gives instruments to handling commented on and totaled information streams to get cultivate occasions, for example, requirement for water system, wiped out creatures or vermin identification in crops.
- Real-time versatile thinking (source: CityPulse) considers rancher's inclinations and dynamic logical homestead related data (spoke to by continuous occasions), with a specific end goal to give ideal choice help progressively.
- External specialist (source: Agri-IoT in-house created) addresses interoperability, gadget heterogeneity, information taking care of and convention adjustment. Assumes an essential part to virtualising objects, administrations, techniques and procedures, thinking about client's character and approval.
- Dashboard (source: ThingSpeak, freeboard) gives prompt and natural visual access to the aftereffects of handling and examination of information and occasions.
- Mobile Apps (source: MapYourMeal, FoodLoop) are based over alternate parts, comparably to the dashboard, and utilize their APIs to offer different administrations to their portable clients, either to the agriculturists for constant data and quick basic leadership, or to the shoppers and transport specialists at the business focuses for more straightforwardness (Csikszentmihalyi, 2004).

Figure 2. Smart Farming

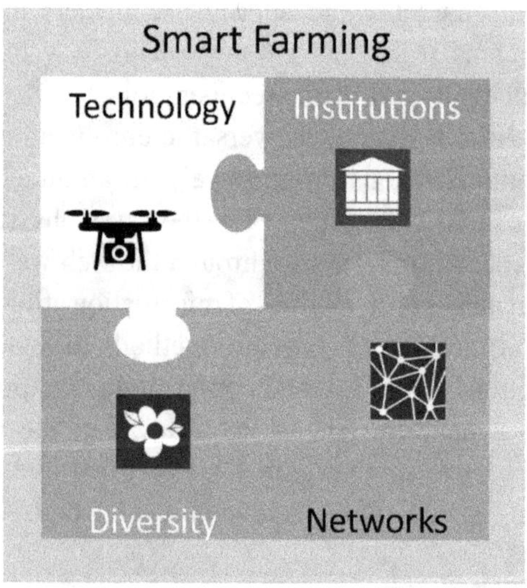

- Knowledge base (source: OpenIoT) gives benefit metadata to sensor/information stream revelation.

Smart Farming based structure applying ongoing stream preparing, examination and thinking in the area of horticulture(Charan, 2017), in light of semantic web advancements, encouraging more educated and exact basic leadership by agriculturists and occasion identification. We have explored the presentation of IoT in brilliant cultivating and its chances, through the consistent blend of heterogeneous advances, and also the semantic joining of data from different sources (sensors, online networking, associated ranches, legislative cautions, controls and so on.), guaranteeing increment of generation and profitability, better items' quality, security of nature, less utilization of assets (e.g. water, composts), speedier response to flighty occasions and more straightforwardness to the customer. Savvy cultivating accomplishes this by offering interoperability between sensors, forms, information streams, cultivates as substances and online administrations, abusing open information, making utilization of semantic innovations and connected web information. Our assessment endeavours concentrating on two sensible and requesting cultivating situations show the great execution of the proposed system in medium-to-expansive homesteads(Charan, 2017), while our talk uncovers the extensive open doors emerging in cultivating by presenting open guidelines and semantics in light of IoT.

REFERENCES

Abidine, A. Z., Heidman, B. C., Upadhyaya, S. K., & Hills, D. J. (2002). Application of RTK GPS based auto-guidance system in agricultural production. The American Society of Agricultural and Biological Engineers Paper, (21152).

Adamchuk, V. I., & Rossel, R. A. V. (2011). Precision agriculture: proximal soil sensing. In *Encyclopedia of Agrophysics* (pp. 650–656). Springer. doi:10.1007/978-90-481-3585-1_126

Austin, R. F., DiSera, D. P., & Brooks, T. J. (2015). *GIS for Critical Infrastructure Protection*. CRC Press. doi:10.1201/b18839

Backman, J., Oksanen, T., & Visala, A. (2013). Applicability of the ISO 11783 network in a distributed combined guidance system for agricultural machines. *Biosystems Engineering*, *114*(3), 306–317. doi:10.1016/j.biosystemseng.2012.12.017

Balafoutis, A. T., Beck, B., Fountas, S., Tsiropoulos, Z., Vangeyte, J., van der Wal, T., ... Pedersen, S. M. (2017). Smart Farming Technologies–Description, Taxonomy and Economic Impact. In Precision Agriculture: Technology and Economic Perspectives (pp. 21–77). Springer.

Beckerman, W. (1994). 'Sustainable development': Is it a useful concept? *Environmental Values*, *3*(3), 191–209. doi:10.3197/096327194776679700

Boisot, M. H. (1998). *Knowledge assets: Securing competitive advantage in the information economy*. OUP Oxford.

Bongiovanni, R., & Lowenberg-DeBoer, J. (2004). Precision agriculture and sustainability. *Precision Agriculture*, *5*(4), 359–387. doi:10.1023/B:PRAG.0000040806.39604.aa

Bornstein, D. (2007). *How to change the world: Social entrepreneurs and the power of new ideas*. Oxford University Press.

Bui, T., & Lee, J. (1999). An agent-based framework for building decision support systems. *Decision Support Systems*, *25*(3), 225–237. doi:10.1016/S0167-9236(99)00008-1

Byerlee, D., & Alex, G. E. (1998). *Strengthening national agricultural research systems: policy issues and good practice* (Vol. 24). World Bank Publications. doi:10.1596/0-8213-4173-1

Calderón, R., Navas-Cortés, J. A., Lucena, C., & Zarco-Tejada, P. J. (2013). High-resolution airborne hyperspectral and thermal imagery for early detection of Verticillium wilt of olive using fluorescence, temperature and narrow-band spectral indices. *Remote Sensing of Environment*, *139*, 231–245. doi:10.1016/j.rse.2013.07.031

Charan, S. (2017). *Enabling Innovation in Agriculture and Improving Access to Key Service by IOT (Internet of Things)*. Gateway.

Csikszentmihalyi, M. (2004). *Good business: Leadership, flow, and the making of meaning*. Penguin.

Fountas, S., Blackmore, S., Ess, D., Hawkins, S., Blumhoff, G., Lowenberg-Deboer, J., & Sorensen, C. G. (2005). Farmer experience with precision agriculture in Denmark and the US Eastern Corn Belt. *Precision Agriculture*, *6*(2), 121–141. doi:10.100711119-004-1030-z

Fountas, S., Wulfsohn, D., Blackmore, B. S., Jacobsen, H. L., & Pedersen, S. M. (2006). A model of decision-making and information flows for information-intensive agriculture. *Agricultural Systems*, *87*(2), 192–210. doi:10.1016/j.agsy.2004.12.003

Islam, M. A., & Huosong, X. (2016). Can Corporate Social Responsibility (CSR) Ensure the Sustainable Growth of Ready Made Garments (RMG) Sector in Bangladesh. *World Journal of Social Science Research*, *3*(4), 615. doi:10.22158/wjssr.v3n4p615

Kainrath, D. (2009). *Ecopreneurship in theory and practice: A proposed emerging framework for ecopreneurship*. Academic Press.

Kaplan, S., & Kaplan, R. (2009). Creating a larger role for environmental psychology: The Reasonable Person Model as an integrative framework. *Journal of Environmental Psychology*, *29*(3), 329–339. doi:10.1016/j.jenvp.2008.10.005

Latchem, C., & Jung, I. (2009). *Distance and blended learning in Asia*. Routledge.

Sasikala, C. (2013). Cloud Computing as Value Addition and Cost Effective Solution to Library and Information Centres. *Electronic Resources Management in Libraries*, *1*, 177.

Unger, R. M. (1998). *Democracy realized: The progressive alternative*. Verso.

Chapter 8
Peregrinating Gardens From Traditional to Most Advanced Handy Approach for Avoiding the Unnecessary Utilization of Resources

Santosh R. Durugkar
Amity University Jaipur, India

Ramesh C. Poonia
Amity University Jaipur, India

ABSTRACT

We always say agriculture is a backbone of Indian economy. In this chapter, the authors have proposed a novel approach based on priority-driven scheduling-based irrigation model (for home garden) which supplies optimum and good quality water to the crops. Technology used in this proposed system is wireless sensor network. Proposed system will be very useful as it immediately irrigates the plants if the moisture level decreases to avoid the future losses. In this process, soil moisture values will be sensed and compared to find out the lowest value. Such systems will start a new era in agriculture and will prove this could be the major requirement in future due to many critical factors such as irregularity of monsoon, less availability of water, etc. Subtopics enlisted in this chapter such as literature survey, design and modeling, technical specification, sampling, results, and analysis will elaborate depth of proposed system.

DOI: 10.4018/978-1-5225-5909-2.ch008

INTRODUCTION

The Scope of this proposed system is limited to smart home garden but it is applicable to a large field with few extensions i.e. nothing but expanding the hardware layer. Climate change is crucial and everyone is responsible for preserving our nature and needs to adopt the modification strategies in day-to-day life, which disturbs the climate. Reducing the utilization cost of natural resources is the important challenge and everyone should contribute to climate-resilient pathways for sustainable development. In day-to-day life, adaptation means to adapt current state of being adapted and moving onwards to the dynamic evolutionary process, which changes periodically that leads to the adaptation. Natural resources oil, water, trees etc. must be utilized in an effective way. Therefore, considering a longer-term perspective, in the view of sustainable development increases the likelihood more people to enhance future options and preparedness must adopt actions. WSN can be implemented where 'n' sensors equipped with various sensors will fetch real time values from the environment and pass them to the controller i.e. coordinator. Upon receiving these values coordinator takes place appropriate action. Systematically algorithm needs to take appropriate action. In the proposed system framework designed with X-Bee approach and all the nodes communicate with each other (Chen, Y, 2015; Waltenegus & Poellabauer, 2010; Karl & Willig, 2007; Sohraby et al., 2007). In future, there can be a requirement for developing a system which will decide how to irrigate, when to irrigate and where to irrigate. These are the upcoming trends in agriculture (Havlak, R. D. 2012).

"Water" Use It Wisely as It Is Important Natural Resource

"Water" is stimulant of life to survive and all developmental activities require water, whether it is agriculture sector or industrial purpose. The population is increasing which in turn results in the increase in domestic demand of water including all the resources. The water resources are available either in presence of surface water, which is less due to the irregularity of monsoon, and groundwater, which is reducing too. It is very important to differentiate between crop water requirement and irrigation, which is nothing but avoiding unnecessary utilization of the water. We have to make sure excess as well as less irrigation is harmful to the plants (Dukes et al., 2012; Aguilar et al., 2015). Strategy for sustainable development and management becomes a challenging task and research organizations, governments are continuously working on applying new strategies. We have proposed a scheme to tie up technology with the agriculture field. Indian economy depends upon the agriculture sector and it depends upon the monsoon. If the monsoon is irregular then it affects economy directly. We can observe region-wise analysis to conclude statistical data will definitely vary. Few regions, where we can found significant increase in rainfall and in other very less rainfall (Gadgil, S. 2013).

WSN Technology

A Wireless Sensor Network consists of multiple detection stations called as sensor nodes, each of which is small, lightweight and portable (vary with the application). Every sensor node can be equipped with a transducer, microcomputer, transceiver and power source (Hac, A. 2003). The use of transducer is to generate the electrical signals based on sensed physical effects. The microcomputer i.e. master node processes and stores the sensor output. The transceiver receives commands from a central computer i.e. either control panel or master node that is taking care of all the sensor nodes and all the tasks initiated by them will be coordinated in an effective manner. Information transmission of wireless sensor networks has various different requirements in performance, such as real-time event detection and responding with it, the reliability of the network should be maintained to minimize the delay (Chen et al., 2008).

Related Work

The reliable lifetime of the network can achieve by assigning 'n' tasks to sensors using the protocols against optimal task assignment and against energy-unaware protocols. All the sensor nodes will pass their tasks to the central coordinator node (Abdel Salam & Olariu, 2011).

Figure 1. WSN in Agriculture

In the field of agriculture, use of proper methods of irrigation is important due to irregular monsoon and less availability of water for agriculture. The main advantage of using this method is to reduce human intervention with technology to ensure proper irrigation. Normally in multi-hop communication where the scalability of the network is maximum, nodes which are near to a sink tend to become congested as they are responsible for forwarding data from nodes that are farther away, i.e. heavy load or traffic can be observed nearby sink node (Salarian et al., 2014). In the proposed approach, we have used star topology (the way nodes communicate with each other) so that sensors nodes communicate with the coordinator node by means of passing instructions (Wightman & Labrador, 2009). Coordinator node should be in the safe and reliable environment if the network consists of 'n' nodes. Normally assumption about the channel with communication takes place is symmetric when the signal noise ratio is certain. The energy consumption for transmitting a message from Node 'A' to Node 'B' equals to that from the Node 'B' to Node 'A' (Chen et al., 2013). Sending the traffic generated by each sensor node to the sink node through multiple paths, instead of a single path, allows significant energy conservation because if already the desired path is selected and in some cases if it won't be available due to any reason then continuously checking whether that path is available or not consumes a lot of energy (Bouabdallah et al., 2009). Cost-effective solution for deployment and implementation of the nodes must be considered, is highly scalable network mobility of 'n' nodes can be considered to provide dynamic event coverage, which is quite difficult to maintain due to periodic updates (Alam et al., 2011). In highly scalable network data aggregation and data packets, issues are higher and directly affects timely and reliable communication and finding the best path among the multipath will be a critical task if the network is highly dynamic (Nandhagopal & Sivanandam,, 2013; Mao et al., 2013). Recent advancements in wireless sensor networks have changed the era of traditional to modern. Many fields can be incorporated into wireless sensor network especially agriculture sector and other potential applications such as banking, weather, health management and many more (Keshtgari & Deljoo, 2011).

PROPOSED APPROACH

Priority-driven approach, in wireless irrigation system is applicable to a Smart Home Garden or for a large field. It means priority must be assign to a task among multiple tasks. Hardware design of the network also has its impact on the performance i.e. placing of the sensor nodes and communication in between them (Hsieh et al., 2014; Liu & Zhao, 2009; Goh et al., 2008). As we are discussing the priority-driven strategy in which using soil moisture sensed values will be collected and if a particular point moisture is less than the set threshold then immediate water will be made available, this is a kind of polling w.r.t. priority (Goh et al., 2008). Hence, we are suggesting in this way 'water" which is a valuable resource can be preserved and utilize in an effective and efficient way (India, Government, 2015, 2014). Recent trends in monthly and seasonal cumulative rainfall depth, number of rainy days and maximum daily rainfall must be analyzed for precision agriculture. This approach is helpful for accuracy and yield of the crops can be increased (Lacombe & McCartney, 2014; P Guharkata, M., 2006). Indian economy is based on the agricultural and agricultural sector totally depends upon the rainfall. Therefore, rainfall directly affects the agricultural sector and recent technological advancements have made the deployment of small, inexpensive, low-power, distributed WSN devices to process the data and communication among each other. Such nodes are called as sensor nodes. Each sensor node is capable of only a limited amount of processing. However, when coordinated with the information from a large number of other nodes, they have the ability to measure a given physical environment in detail. Thus, we can describe a sensor network as a collection of sensor nodes to co-ordinate data for some specific action. Unlike traditional networks, sensor networks depend on dense deployment and co-ordination to carry out their tasks.

Moisture Sensing Sub System

All the values will be stored temporarily to the data center where the comparison will take place and immediately the plant which moisture level will be less then, irrigation will starts for that plant. This method will have advantages such that, only plant at which water is required that plant will be irrigated. Actually, idea is to observe soil moisture daily at the root zone is essential. Irrigation systems assist end users from planting through harvest but utilization of the water should be in an effective manner. Soil monitoring plays an important role in precision agriculture (Daniele Zaccaria, S. O., 2014; Wright et al., 2012; Paige & Keefer, 2008; Werner, H., 1992).

Water Availability at the Root Zone

The following figure represents the water availability at the root zone of plants and the schematic representation of the root zone to categorize into few regions such as excess water region, less water region etc. Applying sensors to get the moisture values from root zone is very important. rated soil water available this is also known as Volumetric Water Contents and rating can be assigned. (Pereira et al., 2013).

We can note the moisture in the soil and can manage it to obtain desired production growth without getting losses (Service, N. R., 2012).

Figure 2. Moisture Sensing Sub System

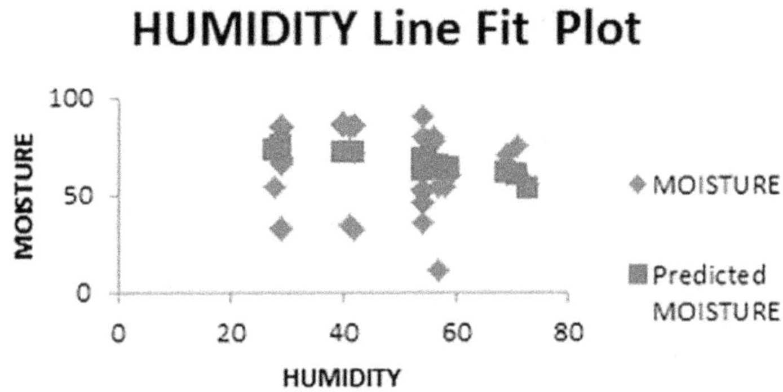

Figure 3. Water Availability at the root zone

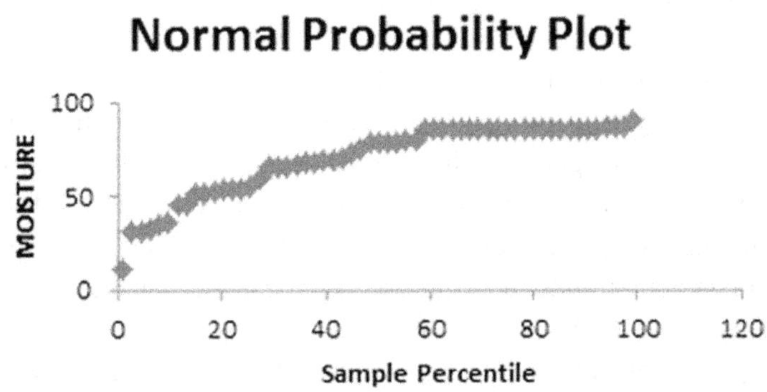

Best Water Utilization

It is assumed that any water can be given to the irrigation but it is not correct and good. As water have its TDS, pH, Salinity values etc. by which we can decide whether water is good for the production yield of the crops. However, it directly influences the growth of plant. Therefore, in the best water utilization diagram we have shown pH, Salinity and TDS of the water must be check at first. Then, if the retrieved values are permissible then only irrigation starts to the plants. It is the ultimate one of the goal of the proposed system. We are expecting that system will immediately display the message on the desktop application to the user if the values are acceptable. In our proposed system, this will be the advantage that we are checking these parameters of the water before utilizing the water. Up till now, existing systems are just focusing on the irrigation but we are focusing on the optimal utilization of best water for the crops. Ultimately, these will affect the crop growth and production yield.

Figure 4. Best Water utilization

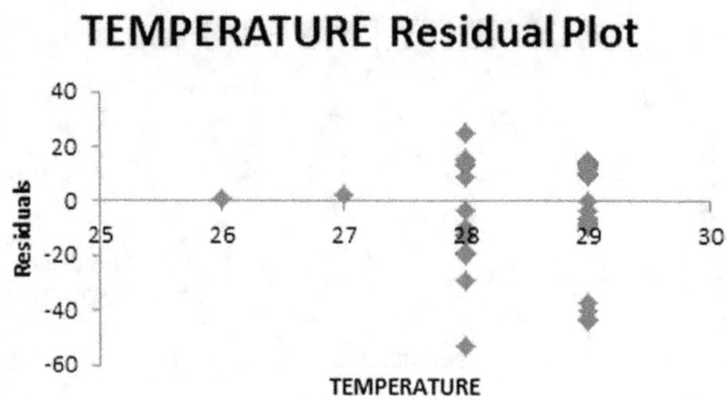

Block Diagram of the System

The topology i.e. structure in which all the sensor nodes communicate with each other. WSNs can vary from a simple star network to an advanced multi-hop wireless mesh network. The propagation technique between the hops of the network can be routing or flooding. A wireless sensor network is a collection of small components, which are randomly dispersed devices to provide functions such as, ability to sense real-time values of environmental conditions for parameters such as temperature and humidity. The ability of WSN components is to operate devices such as switches, motors which control those conditions. Such components also provide efficient, reliable communications in wireless network i.e. WSN.

Figure 5. Block diagram of the system

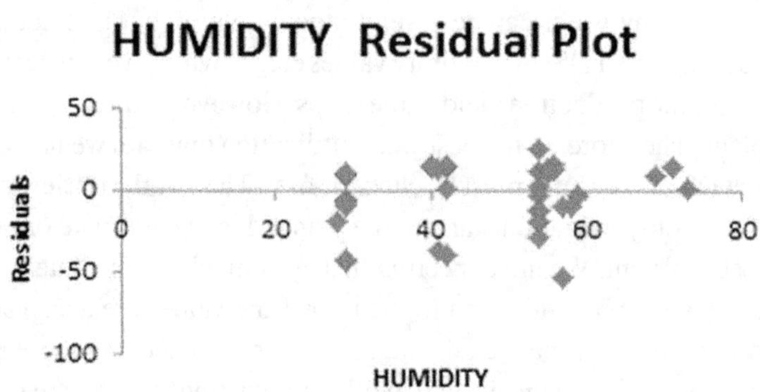

An application is necessary to communicate with the nodes involved in the proposed system. We are storing 'n' values, which have retrieved in the database, e.g. we are using MySQL. Therefore, MySQL query calls will be executed during storing and retrieving the values to and from the database. As we are focusing on the calculation of the weekly, monthly, bi-monthly utilization of the water, these sensed and calculated values must be stored somewhere. We will store these values in the database for future analysis purpose.

Storing and Retrieving Values From Database

The system will execute the priority-driven strategy in which focus is on getting moisture values from the plants and comparing those values with the set threshold. If during this comparison retrieve value is less than the threshold then immediately irrigation will starts. This scenario, we have described with the help of above-mentioned diagram.

- If moisture < threshold
- Then
- Return moisture

This loop will be executed and irrigation with the help of solenoid valve will start.

X-Bee data packets can be transmitted as either unicast or broadcast transmissions. Unicast transmissions route data from one source device to one destination device, whereas broadcast transmissions are sent to many or all devices in the network.

EXPERIMENTAL OBSERVATION

We have given soil samples agriculture University so that at initial stage we can conclude which soil sample is good as compared to other soil samples. Then, we will correlate these values after applying priority-driven Strategy. Our intention

Figure 6. Storing and Retrieving Values from Database

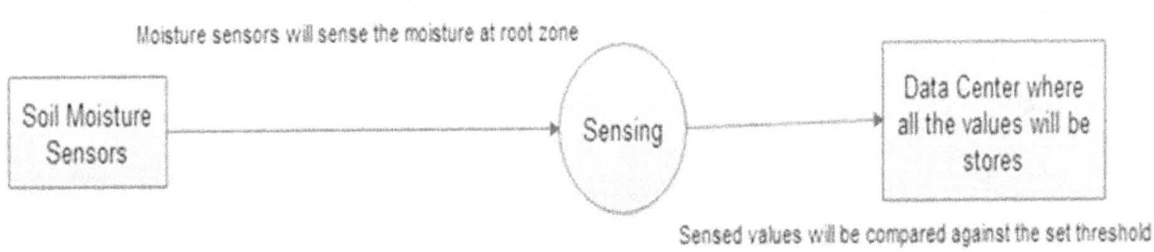

Figure 7. Transmission and Reception of data

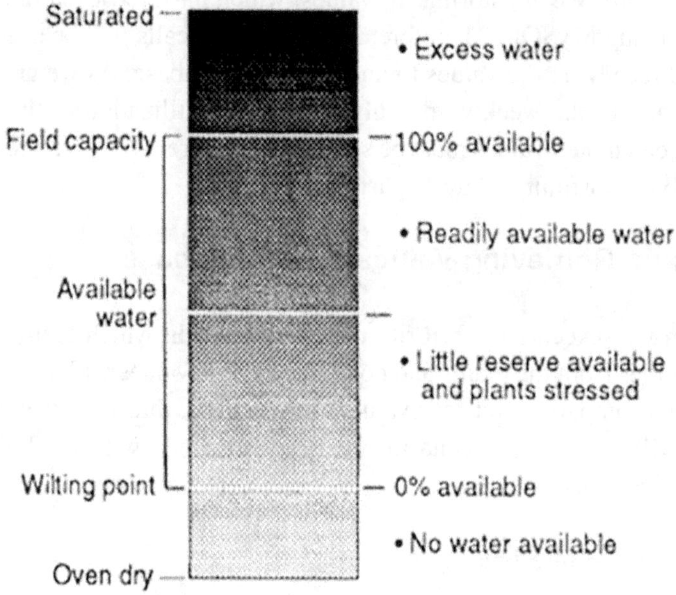

Available water in the soil.

is to show how priority-driven approach will save the unnecessary wastage of the water to maintain the soil moisture. We will show the importance of priority-driven approach in recent days where monsoon is irregular and due to this, it is hard for us to maintain the ground water level.

Readings Fetched

See Table 1.

Covariance Analysis for Temperature and Moisture

See Tables 2-4.

Covariance Analysis for Humidity and Moisture

See Tables 5-7.

Figure 8. Fetching values from nodes and water tank

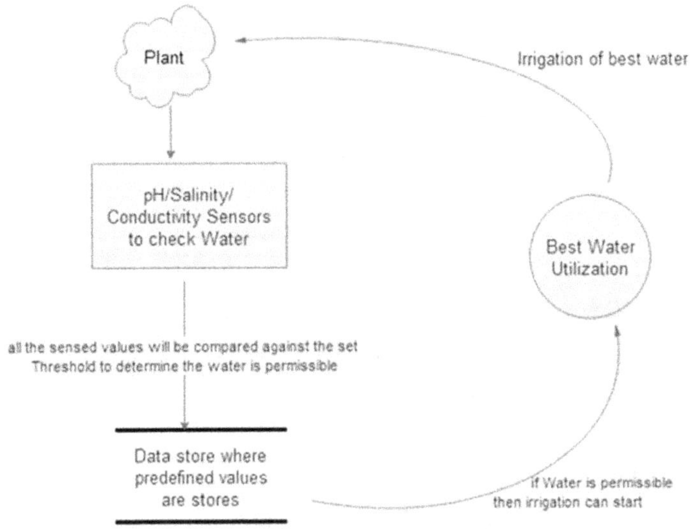

Table 1. Reading Received from nodes

NODE	TEMPERATURE	HUMIDITY	MOISTURE	pH	DATE	TIME
NODE1	27	54	65	10	17/07/2017	16:20:11
NODE2	28	59	60	9	17/07/2017	16:20:34
NODE3	26	73	55	8	17/07/2017	16:20:57
NODE1	28	28	54	6	21/07/2017	14:56:09
NODE2	29	55	78	4	21/07/2017	14:56:30
NODE3	28	71	75	6	21/07/2017	14:56:52
NODE1	29	54	65	10	21/07/2017	15:01:32
NODE3	28	69	70	7	21/07/2017	15:01:52
NODE2	29	54	80	6	21/07/2017	15:02:57

Regression Analysis for 'n' Days

See Table 8.

Regression Analysis Will Be as Follows

Refer Table 8 Node values

Table 2. Covariance Analysis

TEMPERATURE	MOISTURE
27	65
28	60
26	55
28	54
29	78
28	75
29	65
28	70
29	80

Table 3. Covariance Measured

	TEMPERATURE	MOISTURE
TEMPERATURE	0.888888889	
MOISTURE	**5.333333333**	81.432099

Table 4. Correlation Measured

correlation	**0.626869551**

Table 5. Covariance in between humidity and moisture

HUMIDITY	MOISTURE
54	65
59	60
73	55
28	54
55	78
71	75
54	65
69	70
54	80

Table 6. Covariance Measured Humidity and Moisture

covariance		HUMIDITY	MOISTURE
	HUMIDITY	163.3580247	
	MOISTURE	30.04938272	81.4320988

Table 7. Correlation Measured - Humidity and moisture

correlation	0.260535858

Table 8. Regression Analysis for 'n' days

NODE	TEMPERATURE	HUMIDITY	MOISTURE	pH	DATE
NODE1	27	54	65	10	17/07/2017
NODE2	28	59	60	9	17/07/2017
NODE3	26	73	55	8	17/07/2017
NODE1	28	28	54	6	21/07/2017
NODE2	29	55	78	4	21/07/2017
NODE3	28	71	75	6	21/07/2017
NODE1	29	54	65	10	21/07/2017
NODE3	28	69	70	7	21/07/2017
NODE2	29	54	80	6	21/07/2017
NODE1	29	29	69	9	10/08/2017
NODE2	29	42	86	6	10/08/2017
NODE3	28	58	54	8	10/08/2017
NODE1	29	29	69	10	10/08/2017
NODE2	29	42	86	7	10/08/2017
NODE3	28	57	54	7	10/08/2017
NODE1	29	29	68	9	10/08/2017
NODE2	29	42	86	8	10/08/2017
NODE3	28	57	11	7	10/08/2017
NODE1	29	29	68	9	10/08/2017
NODE2	29	42	86	9	10/08/2017
NODE1	29	29	67	9	10/08/2017
NODE2	29	42	32	10	10/08/2017
NODE3	28	56	80	7	10/08/2017

continued on following page

Table 8. Continued

NODE	TEMPERATURE	HUMIDITY	MOISTURE	pH	DATE
NODE1	29	29	66	9	10/08/2017
NODE1	29	29	32	9	10/08/2017
NODE2	29	42	72	5	10/08/2017
NODE3	28	56	78	7	10/08/2017
NODE1	29	29	0	10	17/08/2017
NODE2	29	41	85	4	17/08/2017
NODE3	28	56	78	6	17/08/2017
NODE1	29	29	86	10	17/08/2017
NODE2	29	41	0	6	17/08/2017
NODE3	28	55	78	6	17/08/2017
NODE1	29	29	86	10	17/08/2017
NODE1	29	29	85	10	17/08/2017
NODE2	29	40	87	8	17/08/2017
NODE3	28	54	53	6	17/08/2017
NODE1	29	29	85	10	17/08/2017
NODE2	29	40	87	8	17/08/2017
NODE3	28	54	52	7	17/08/2017
NODE1	29	29	85	10	17/08/2017
NODE2	29	40	87	9	17/08/2017
NODE3	28	54	90	6	17/08/2017
NODE1	29	29	85	10	17/08/2017
NODE2	29	40	86	8	17/08/2017
NODE3	28	54	52	7	17/08/2017
NODE1	29	29	85	10	17/08/2017
NODE2	29	40	86	7	17/08/2017
NODE3	28	54	46	7	17/08/2017
NODE1	29	29	85	10	17/08/2017
NODE2	29	40	86	8	17/08/2017
NODE3	28	54	46	6	17/08/2017
NODE1	29	29	85	10	17/08/2017
NODE2	29	40	86	7	17/08/2017
NODE3	28	54	36	7	17/08/2017
NODE1	29	29	85	10	17/08/2017
NODE2	29	40	86	6	17/08/2017

Table 9. Regression Measured

SUMMARY OUTPUT					
Regression Statistics					
Multiple R	**0.969523**				
R Square	**0.939974**				
Adjusted R Square	0.920701				
Standard Error	18.08553				
Observations	57				
ANOVA					
	df	*SS*	*MS*	*F*	*Significance F*
Regression	2	281709.3	140854.6	430.6346	6.5E-34
Residual	55	17989.74	327.0862		
Total	57	299699			

- **The R Square:** This is how linear the relationship between the variables exists. The closer to 1, the more linear the relationship will be and in our example for sample data collected in Table 1 **R Square value is 93.99%** (Table 1 contains data sample for 04 days only). The more variation in Y variable i.e. moisture values (dependent variable) that is explained by X independent variable), in our calculation X variable is temperature and humidity.
- The F is a measure of the strength of the relationship between variables and it is 430.6346
- Multiple R is 96.95% which is the correlation coefficient. It tells you how strong the linear relationship is.

Descriptive Statistics

See Table 10.

Figure 9. Temperature line fit plot v/s moisture

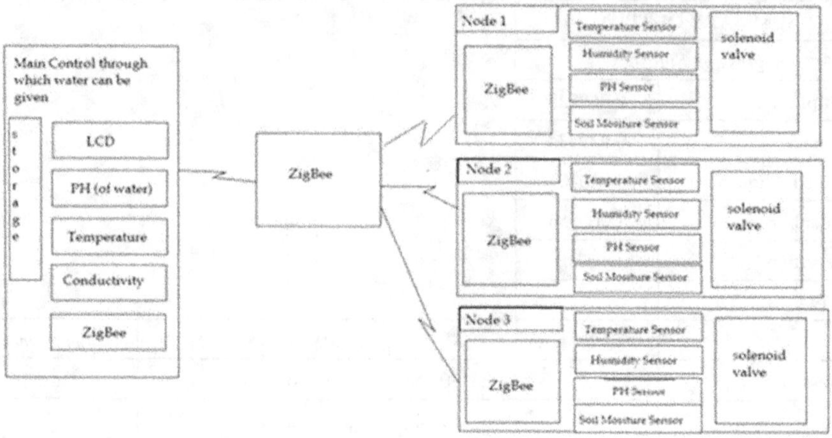

Figure 10. Humidity Line fit plot v/s moisture

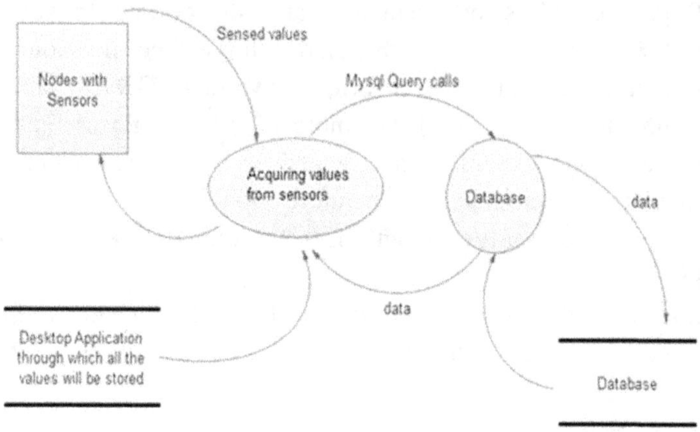

CONCLUSION

Developing such, system requires to consider soil type as pH, EC, moisture varies with the soil sample. Careful placing of soil moisture sensors is important. We have used 03 soil samples (black, brownish and red) in the proposed system. Before using directly in the experimentation we have been advised to test these soil samples.

Figure 11. Normal probability plot of moisture

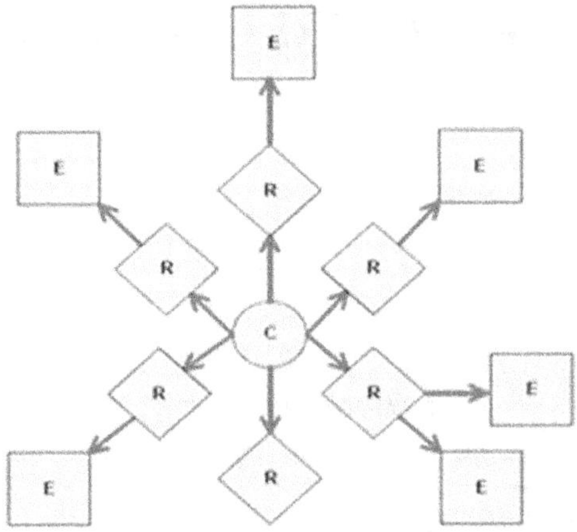

Figure 12. Temperature residual plot

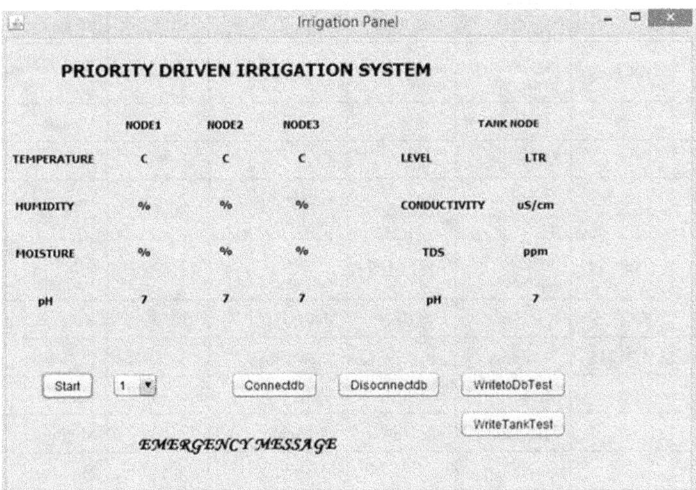

Therefore, samples tested at the Department of Soil Testing of Mahatma Phule Krishi Vidyapeeth Rahuri, Maharashtra. (very popular agricultural University) We have collected the initial test results of these samples and realized among these samples black cotton soil is quite good and will be productive in terms of yield. As the scope of our system is smart home garden, we are using probe designed specifically to sense the moisture at the surface zone of the plant. We will store these values in the

Figure 13. Humidity residual plot

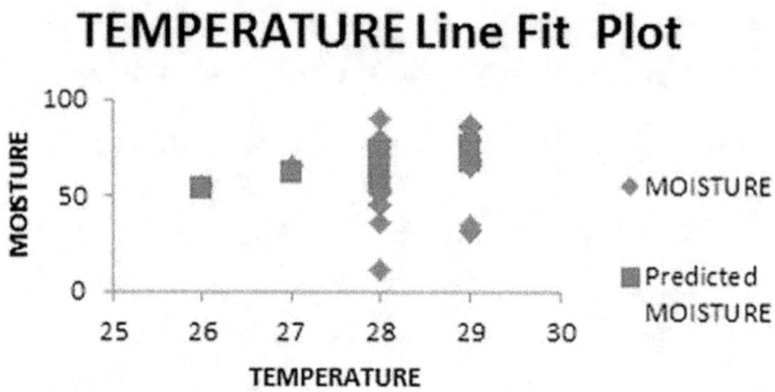

Table 10. Descriptive Statistics

Descriptive Statistics							
TEMPERATURE		***HUMIDITY***		***MOISTURE***		***pH***	
Mean	28.59649	Mean	43.26316	Mean	70.05263	Mean	7.894737
Standard Error	0.082488	Standard Error	1.685952	Standard Error	2.501806	Standard Error	0.227612
Median	29	Median	41	Median	78	Median	8
Mode	29	Mode	29	Mode	86	Mode	10
Standard Deviation	0.622772	Standard Deviation	12.72866	Standard Deviation	18.88822	Standard Deviation	1.718432
Sample Variance	0.387845	Sample Variance	162.0188	Sample Variance	356.765	Sample Variance	2.953008
Kurtosis	4.12756	Kurtosis	-0.86396	Kurtosis	0.568615	Kurtosis	-0.91567
Skewness	-1.75943	Skewness	0.382957	Skewness	-1.12601	Skewness	-0.26882
Range	3	Range	45	Range	79	Range	6
Minimum	26	Minimum	28	Minimum	11	Minimum	4
Maximum	29	Maximum	73	Maximum	90	Maximum	10
Sum	1630	Sum	2466	Sum	3993	Sum	450
Count	57	Count	57	Count	57	Count	57
Largest(1)	29	Largest(1)	73	Largest(1)	90	Largest(1)	10
Smallest(1)	26	Smallest(1)	28	Smallest(1)	11	Smallest(1)	4
Confidence Level (95.0%)	0.165244	Confidence Level (95.0%)	3.377368	Confidence Level (95.0%)	5.011721	Confidence Level (95.0%)	0.455961

database and then immediately system invokes a priority driven strategy to compare the moisture values received. The effectiveness of the monitoring system is dependent upon proper placement and installation. To get accurate results and better irrigation care should be taken to place sensors by avoiding variations in placement. Since there is significant variation across fields, we can recommend using several sensor locations for large fields.

REFERENCES

Abdel Salam, H. S., & Olariu, S. (2011). Toward efficient task management in wireless sensor networks. *IEEE Transactions on Computers*, *60*(11), 1638–1651. doi:10.1109/TC.2010.264

Agency, I. A. (2008). Field Estimation of Soil. Vienna: Academic Press.

Aguilar, J., Rogers, D., & Kisekka, I. (2015). Irrigation Scheduling Based on Soil Moisture Sensors and Evapotranspiration. *Kansas Agricultural Experiment Station Research Reports*, *1*(5), 20. doi:10.4148/2378-5977.1087

Alam, K. M., Kamruzzaman, J., Karmakar, G., & Murshed, M. (2011, August). Dynamic event coverage in hybrid wireless sensor networks. In *Network Computing and Applications (NCA), 2011 10th IEEE International Symposium on* (pp. 348-353). IEEE. 10.1109/NCA.2011.62

Bouabdallah, F., Bouabdallah, N., & Boutaba, R. (2009). On balancing energy consumption in wireless sensor networks. *IEEE Transactions on Vehicular Technology*, *58*(6), 2909–2924. doi:10.1109/TVT.2008.2008715

Chatterjee, A., & Mukherjee, D. (2014, January). Reliable multipath wireless sensor network routing protocol scheme for network lifetime maximization. In *Business and Information Management (ICBIM), 2014 2nd International Conference on* (pp. 64-68). IEEE. 10.1109/ICBIM.2014.6970935

Chen, J., Du, R., Wang, Q., & Yao, S. (2013, July). Secure routing based on network coding in wireless sensor networks. In *Trust, Security and Privacy in Computing and Communications (TrustCom), 2013 12th IEEE International Conference on* (pp. 58-64). IEEE. 10.1109/TrustCom.2013.11

Chen, J., Zhou, M., Li, D., & Sun, T. (2008, November). A priority based dynamic adaptive routing protocol for wireless sensor networks. In *Intelligent Networks and Intelligent Systems, 2008. ICINIS'08. First International Conference on* (pp. 160-164). IEEE. 10.1109/ICINIS.2008.171

Chen, Y. (2015). *Routing algorithm dedicated to environmental data collection: precision agriculture* (Doctoral dissertation). University Blaise Pascal-Clermont-Ferrand II.

Daniele Zaccaria, S. O. (2014). *Advantages and Constraints of Soil Moisture Monitoring*. Academic Press.

Dargie, W., & Poellabauer, C. (2010). *Fundamentals of Wireless Sensor Networks*. Chichester, UK: John Wiley & Sons Ltd. doi:10.1002/9780470666388

Dukes, M. D., Zotarelli, L., Liu, G. D., & Simonne, E. H. (2012). *Principles and practices of irrigation management for vegetables*. Academic Press.

Fritz Westover, K. B. (2014). *Using Soil Moisture Sensors for Vineyard Irrigation Management*. Atascadero, CA: Vineyard Team.

Gadgil, S. (2013). *The Indian monsoon and its variability Clouds, climate and Tropical Meteorology. ICTS*. CAOS.

Goh, H. G., Kwong, K. H., Michie, C., & Andonovic, I. (2008, August). Performance evaluation of priority packet for wireless sensor network. In *Sensor Technologies and Applications, 2008. SENSORCOMM'08. Second International Conference on* (pp. 494-499). IEEE. 10.1109/SENSORCOMM.2008.90

Goh, H. G., Kwong, K. H., Michie, C., & Andonovic, I. (2008, August). Performance evaluation of priority packet for wireless sensor network. In *Sensor Technologies and Applications, 2008. SENSORCOMM'08. Second International Conference on* (pp. 494-499). IEEE. 10.1109/SENSORCOMM.2008.90

Goldsmith, A. (2005). *Wireless communications*. Cambridge University Press. doi:10.1017/CBO9780511841224

Guharkata, P. M. (2006). Trends in the Rainfall Pattern. Pune: National Climate Centre.

Gupta, H. P., Rao, S. V., & Venkatesh, T. (2013, January). Sleep scheduling for partial coverage in heterogeneous wireless sensor networks. In *Communication Systems and Networks (COMSNETS), 2013 Fifth International Conference on* (pp. 1-10). IEEE. 10.1109/COMSNETS.2013.6465580

Hac, A. (2003). *Wireless sensor network designs*. John Wiley & Sons Ltd. doi:10.1002/0470867388

Havlak, R. D. (2012). *Lawn Irrigation: How, When and Where to Water!* Texas AgriLife Extension.

Hsieh, C. M., Samie, F., Srouji, M. S., Wang, M., Wang, Z., & Henkel, J. (2014, October). Hardware/software co-design for a wireless sensor network platform. In *Hardware/Software Codesign and System Synthesis (CODES+ ISSS), 2014 International Conference on* (pp. 1-10). IEEE.

India, Government. (2014). *Agricultural Statistics*. New Delhi: Oxford University Press.

India, Government. (2015). *Statistics Related to Climate Change India - 2015*. New Delhi: Ministry of Statistics and Programme Implementation.

Izadi, D., Abawajy, J., & Ghanavati, S. (2015). An alternative node deployment scheme for WSNs. *IEEE Sensors Journal, 15*(2), 667–675. doi:10.1109/JSEN.2014.2351405

Jose, L., & Chavez, S. R. (2012). using soil water sensors to improve to improve irrigation management. *Proceedings of the 24th Annual Central Plains Irrigation Conference*.

Karl, H., & Willig, A. (2007). *Protocols and architectures for wireless sensor networks*. John Wiley & Sons.

Keshtgari, M., & Deljoo, A. (2011). *A wireless sensor network solution for precision agriculture based on zigbee technology*. Academic Press.

Lacombe, G., & McCartney, M. (2014). Uncovering consistencies in Indian rainfall trends observed over the last half century. *Climatic Change, 123*(2), 287–299. doi:10.100710584-013-1036-5

Liu, Q., & Zhao, D. (2009, December). Priority polling strategy for wireless sensor networks. In *Information Science and Engineering (ICISE), 2009 1st International Conference on* (pp. 4062-4065). IEEE. 10.1109/ICISE.2009.852

Mao, X., Liu, Y., Tang, S., Liu, H., Han, J., & Li, X. Y. (2013). Finding best and worst k-coverage paths in multihop wireless sensor networks. *IEEE Transactions on Parallel and Distributed Systems, 24*(12), 2396–2406. doi:10.1109/TPDS.2012.329

Maughan, T., Allen, L. N., & Drost, D. (2015). *Soil Moisture Measurement and Sensors for Irrigation Management*. Academic Press.

Mishra, D. P., & Kumar, R. (2015). Qualitative Analysis of Wireless Sensor Network Simulator. *International Journal of Computer Applications*.

Nandhagopal, S. M., & Sivanandam, S. N. (2013). Reliable Data Delivery in Mobile Adhoc Networks Using Light Weight Verification Algorithm with High Node Mobility. *IJITEE*, 2278-3075.

Paige, G. B., & Keefer, T. O. (2008). Comparison of Field Performance of Multiple Soil Moisture Sensors in a Semi-Arid Rangeland. *JAWRA Journal of the American Water Resources Association*, *44*(1), 121–135. doi:10.1111/j.1752-1688.2007.00142.x

Pereira, A., Shock, C. A., Wang, F. X., Feibert, E. B. G., Shock, C. C., & Flock, R. (2013). *Irrigation monitoring using soil water tension*. Academic Press.

Peters, R. T., Desta, K. G., & Nelson, L. (2013). *Practical use of soil moisture sensors and their data for irrigation scheduling*. Academic Press.

Rahuri, M. P. (2017). *Soil testing Analysis*. Rahuri: MPKU Rahuri.

Salarian, H., Chin, K. W., & Naghdy, F. (2014). An energy-efficient mobile-sink path selection strategy for wireless sensor networks. *IEEE Transactions on Vehicular Technology*, *63*(5), 2407–2419. doi:10.1109/TVT.2013.2291811

Schroder, J. H. (2006). *Soil moisture-based drip irrigation for efficient use of water, nutrients, and sustainability of vegetables cropped on coarse soils* (Doctoral dissertation). University of Florida.

Service, N. R. (2012). *Irrigation Water Management*. NRCS Montana.

Service, N. R. (2014). *Irrigation Water Management*. National Resource Conservation Service, 01-04.

Short, M. (2010, January). The case for non-preemptive, deadline-driven scheduling in real-time embedded systems. *Lecture notes in engineering and computer science: Proceedings of the World Congress on Engineering*.

Sohraby, K., Minoli, D., & Znati, T. (2007). *Wireless sensor networks: technology, protocols, and applications*. John Wiley & Sons. doi:10.1002/047011276X

Stojmenovic, I. (Ed.). (2003). *Handbook of wireless networks and mobile computing* (Vol. 27). John Wiley & Sons.

Swami, A., Zhao, Q., Hong, Y. W., & Tong, L. (Eds.). (2007). *Wireless sensor networks: signal processing and communications perspectives*. John Wiley & Sons. doi:10.1002/9780470061794

Werner, H. (1992). *Measuring soil moisture for irrigation water management.* Cooperative Extension Service, South Dakota State University, US Department of Agriculture.

Wightman, P. M., & Labrador, M. A. (2009). *Topology Control in Wireless Sensor Networks*. Academic Press.

Wright, J., Wildung, D., & Nennich, T. (2012). *Irrigation Considerations and Soil Moisture Monitoring Tools*. University of Minnesota.

Wu, Y., Li, X. Y., Li, Y., & Lou, W. (2010). Energy-efficient wake-up scheduling for data collection and aggregation. *IEEE Transactions on Parallel and Distributed Systems*, *21*(2), 275–287. doi:10.1109/TPDS.2009.45

Chapter 9
Smart Irrigation Techniques for Water Resource Management

Smita Chaudhry
Kurukshetra University, India

Shivani Garg
Kurukshetra University, India

ABSTRACT

Rising temperatures and increased frequency of extreme events will have direct and negative impacts on natural resources. Water resources are limited on earth; hence, there is a need to manage the utilization techniques of water. The irrigation system improvement using the wireless network is a solution to accomplish water conservation goal as well as improvement in irrigation practices. Smart farming enhances the capacity of the agricultural systems to support food security. The need for adaptation and the potential for mitigation into sustainable agriculture development strategies can be incorporated into such system. The smart farming system includes different techniques of agricultural practices to conserve different resources including water. Solar powered smart irrigation systems are a part of the smart irrigation system. Smart irrigation system includes temperature, moisture, and humidity sensors system. Different smart irrigation systems which are used all over the world will be discussed in this chapter.

DOI: 10.4018/978-1-5225-5909-2.ch009

BACKGROUND

The first canal built appears to have been named the Exefer Canal in post Roman Britain, opened in 1563. In Russia, the Volga-Baltic Water ways was opened in 1718. The archaeological evidence of irrigation in farming first found about 6th millennium B.C. in the Middle East's Jordan Valley (Hillel, 1994). In Peru, archaeologists found remains of three irrigation canals and after radiocarbon dating it was found that these are dated the 4th millennium BCE. These canals are the earliest record of irrigation in the New World (Dillehay, 2005). In the middle of 20th century, the invention and introduction of diesel & electric motors led for the first time to irrigation system that could pump groundwater out of major aquifers faster than it was recharged. Sophisticated irrigation & storage systems were developed, including the reservoirs built at Girnar in 3000 BCE (Rodda, J.C 2004). In America and Oceania, the area equipped for irrigation was 15% and 1% in 2003 & 18% and 0.9% in 1980 respectively (Sieberta et al., 2006). At present, India has the highest irrigated area in the world today with almost 57 Mha ($1/5^{th}$ of world's net irrigated area (Singh, 1997).

INTRODUCTION

Water scarcity already affects every continent. Around $1/5^{th}$ of the world's population live in areas of physical scarcity, and almost one-quarter of the world's population, face economic water shortage (CWAC, 2014). Agriculture is the backbone of the economic base of each country of the world. The daily need for food shows the importance of agricultural development. Food and agriculture are the largest consumers of water, requiring one hundred times more than we use for personal needs. One of the main challenges in agricultural activities is irrigation. As the global climate decreases the source of water throughout the world, it is necessary to take steps for preserving it. The extent of irrigated land has more than doubled, increasing from 139 to 301 million hectares between 1961 and 2009 (FAO, 2011a). However traditional irrigation management is done by the people itself. It requires the presence and continuous monitoring of irrigation by the farmers in the field area. An early canal irrigation system had also found in The Indus Valley civilization in Pakistan & North India (from 2600 BCE). Large-scale agriculture was used for the purpose of irrigation. Sophisticated irrigation & storage systems were developed, including the reservoirs built at Girnar in 3000 BCE (Rodda and Ubertini, 2004). Soil moisture is the primary required information in achieving optimum water requirements for the crops (Schroder, 2006). In 2001, agricultural uses accounted for about 5%, 10%, and 17% of the internal renewable water resources of Africa, the Caribbean, and Asia, respectively. Asia has the largest proportion of global freshwater withdrawal for agriculture about 73%, (Hillel et al., 2008). The concept of this chapter came due to the global climate shifts because of various reasons such as global warming; excessive air pollutants etc. causing lack of seasonal rains and depletion of groundwater table levels. Currently, most of the Indian farmers cultivate their land depending on seasonal rains and water through bore wells. To combat these issues intelligent/ smart irrigation is thought by using scientific and technical methods. Farmer controls the electric water motor using mobile phone without going in paddy field. If the water level reaches at required level, automatically motor will be off without confirmation of farmer (Uddin, 2012). The objective of smart irrigation technologies are to provide water delivery in a programmed manner to the crops to ensure all the crops have enough water for their healthy growth, to reduce water wastage in irrigation, and to minimize the economic cost for the users.

IRRIGATION SYSTEM

Definition-"Irrigation is the technique in which water is supplied to plants at regular intervals according to their need for agriculture" (Snyder and de Melo-Abreu, 2005). Indian agriculture is dependent on the monsoons which are not sufficient source of water. The automatic irrigation system can provide water to farms according to their need of moisture level and soil types. Suitable water supply is required because most of the farmers are dependent upon the monsoon. The farmer follows a schedule for watering in the conventional system which is different for different crops.

The majority of water is used for irrigation of crops and pastures while only a small amount is used for other agricultural purposes such as livestock drinking water, and dairy and piggery cleaning. Water is also needed to raise livestock for raising the animals and also in the production of meats, poultry, eggs, and milk. Irrigation can reduce water availability for other purposes. A reduction in the water supply can cause great distress for farmers who grow crops such as rice and cotton as the decrease in the irrigated area for these crops, due to dry conditions, results in no growth. Without an efficient water supply for irrigation, farmers could not grow crops and provide nourishment for the large world population. Farmers often construct watering ponds (dams) for livestock to fulfil the watering needs of animals.

Factors Affecting Irrigation Systems

The plant needs; hydro zoning, irrigation system design, and components, as well as irrigation scheduling, must all be considered when doing irrigation. Some other factors which affect the irrigation systems are:

1. **Soil Moisture:** Soil moisture content is an essential factor in agriculture. Highly humid soil needs lesser water than low humid ones.
2. **Topography:** Topography is the slope of the ground and how much uneven or levelled it is. The irrigation method is selected accordingly.
3. **Climate:** Speed of wind, humidity, and dryness affect the methods of irrigation.
4. **Means of Irrigation:** The sources of irrigation and the chemical composition such as pH, hardness or softness, mineral level of the water also affect the irrigation method.
5. **Crops:** Different crops require different quantities of water. The irrigation method is selected according to water requirement by the crops and pattern of its sowing like rice crop need a large amount of water as compared to other.
6. **Texture of the Soils:** The method of irrigation is decide by knowing the water storage capacity of the potential root zone, surface drainage; the nature and extent of salts in the surface soil and the sub-soil, also decide. The particles texture is also deciding the irrigation methods to be used in fields, like water holding capacity, penetration etc.

TYPES OF IRRIGATION SYSTEM

Surface Irrigation: Surface irrigation includes a large group of irrigation methods in which, over the surface of the field. This technique needs more labour than other irrigation methods. So that the cost of operation increased during the irrigation process. Surface irrigation methods contain two basic categories: ponding and moving water. Proper design of surface irrigation systems takes into account the

Figure 1. Different types of irrigation methods used all over the world

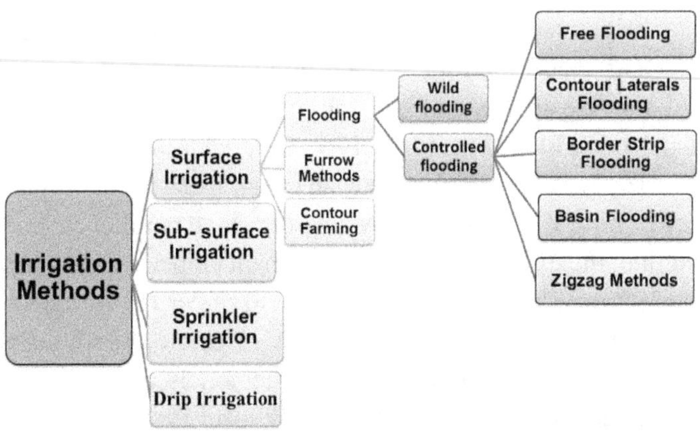

soil type (texture and intake rate), slope, and levelness of the field, stream size. It is very difficult to obtain water distribution with high uniformity in large area on coarse textured soils such as gravel and sands than on fine-textured soils such as loams to clay (Hill, 2008).

This method includes three categories:

1. **Flooding Method:** Flood irrigation is an effective method of irrigation but not as efficient as compared with other options. From total irrigated water, only half of the water irrigates the crops and the other half is lost due to evaporation, runoff, uncultivated areas, and transpiration through the leaves of weeds.
    a. **Wild Flooding:** Wild flooding is when water is allowed to flow freely across farmland. It is a cheap irrigation system but uses water inefficiently and some areas get more water than others.
    b. **Controlled Flooding:** Controlled flooding refers to all methods used to reduce or prevent the negative effects of flood waters such as soil erosion, water logging etc. It is further categorized on the basis of methods of irrigation as:
        i. **Free Flooding:** Water is applied from water source without any levee to guide its flow. Movement of water is not restricted. This technique is beneficial for newly developed fields where the surface is irregular having up and down surface.
        ii. **Contour lateral:** lateral lines or channels are made in the field and free flooding is done in these lines.
        iii. **Border Strip:** in this method, the area is divided into a series of long narrow strips known as border strips. Water is allowed to flow in each strip. Strips are separated with the help of low embankment so that water flows towards lower ends providing moisture to soil.
        iv. **Basin:** it is used for cultivation of crops that are able to grow short-time flooding and used for charge watering, soil salinity control, rice irrigation. This method is most commonly used to irrigate orchards.
        v. **Zig- Zag Flooding:** Area is divided into small plots by low bunds that are arranged in zig- zag manner.
        vi. **Check Flooding:** In this type, the area is divided into checks strips. The water is supplied in each check freely as in ordinary flooding.
2. **Furrow Method**: It is conducted by creating small parallel channels along the field length in the direction of the predominant slope. Water is supplied to the top of each furrow and flows down the field under the influence of gravity. This type of irrigation method is suitable for tree crops and row crops.
3. **Contour Method:** The contour-furrow method is that in which irrigation water is applied by furrows, but the nearly level furrows carry water across

a sloping field rather than downslope. The contour furrows are curved to fit the land surface. They have just enough grades to carry the irrigation stream. Pipelines are run across the slope to feed the individual furrows.

1. **Sub- Surface Irrigation:** Subsurface irrigation is a highly-efficient watering technique that reduces outdoor water use by 30 to 40% (SNWA, 2017). The system consists of drip irrigation tubing planted about five inches below the surface. The water goes straight to plant roots and it does not blow away or run down the sidewalk. Water is applied through underground distribution system consisting of a properly designed main field ditches, laterals, laid 15 to 30 m apart. Water is introduced into different layers of soil through open ditches, mole drains or tile drains.
2. **Sprinkler Irrigation:** Irrigation water is pumped from the source through pipes to the sprinklers and sprayed into the air. Two types of sprinkler systems are used to irrigate 'farm crops. In one, rotating sprinkler heads are spaced equally along the lateral lines. In the other, the lateral lines are a perforated pipe that remains in one place until the required amount of water has been applied and are then moved the same distance for each successive setting.
3. **Drip Irrigation:** Drip irrigation is sometimes also called trickle irrigation and involves dripping water onto the soil at very low rates from small diameter plastic pipes fitted with outlets called drippers. Water is given close to plants or in the roots of plants so that only part of the soil in which the roots grow

Figure 2. (a) Surface irrigation system commonly used by farmers. (b) Flooding method, (c) Furrow method (d) Contour method

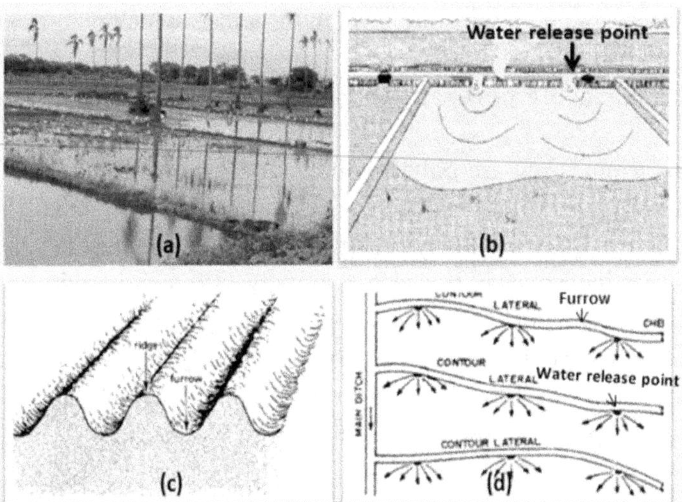

Figure 3. Subsurface irrigation system

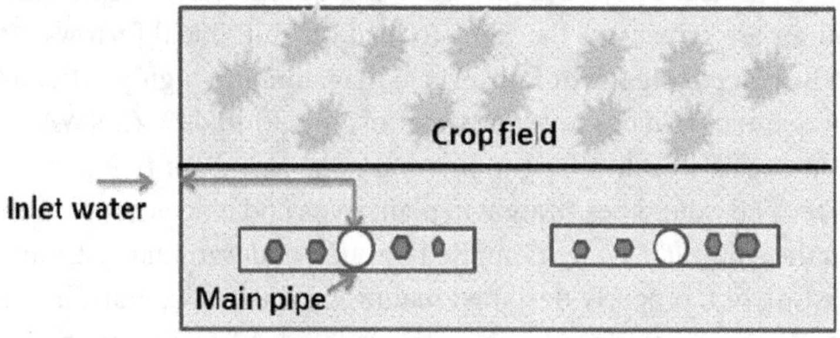

Figure 4. Sprinkler Method of irrigation used for farm crops

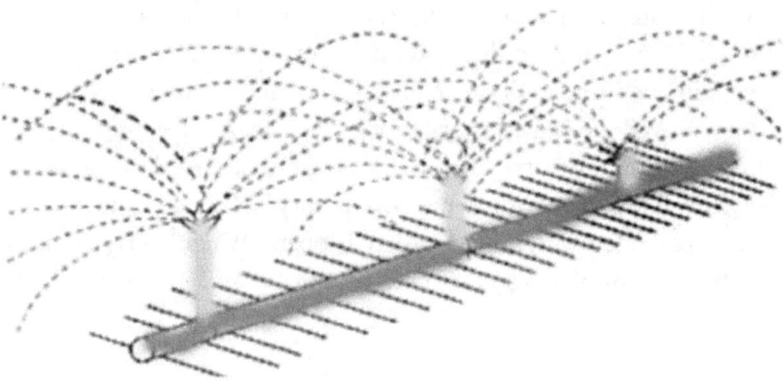

is wet. With drip irrigation water, applications are more frequent and give a very favourable high moisture level in the soil in which plants can flourish.

There are four main types of emitters and also known as types of drip irrigation:

- **Porous Pipe:** They are either manufactured from very porous material or consist of hose which has been drilled with extremely small holes – both which allow water to slowly seep through. Their greatest advantage is cost, being very inexpensive. It is also extremely easy to work with.
- **Pre-Installed Emitter Lines:** This type of irrigation system is made of a number of hoses throughout the garden. A number of evenly spaced emitters are present in each hose. They end up releasing small drops of water into the soil and are particularly good for landscapes with shrubs.

Figure 5. Drip irrigation method used for efficient irrigation

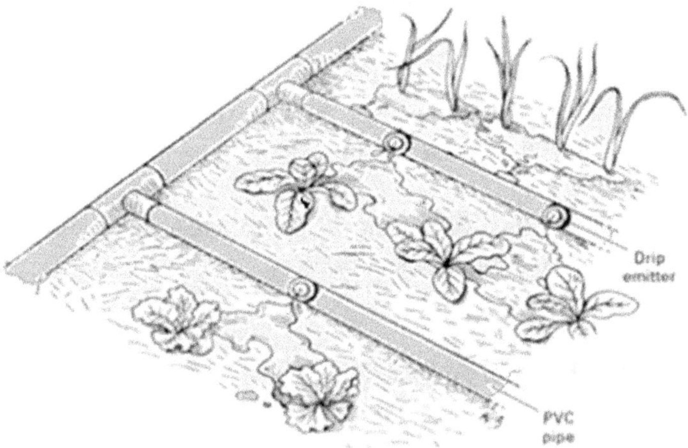

- **Watermatic Drip System:** This system helps in water conservation by reducing evaporation. It makes use of devices such as micro-spray heads, which is used for trees and flower beds. This irrigation method can be used in areas where water conservation is necessary, because there recycled water is used for irrigation purposes.
- **Micro-Sprinkler Heads:** These sprinklers are mainly built for vineyards and orchards, but are now starting to be used in backyards as well. This type of system provides water to the roots evenly, saving water and helping trees to increase their yields.

NEEDS OF SMART IRRIGATION SYSTEM

The essential requirements for adoption of any irrigation method are:

- Stream flow is adequate so that the quality of irrigation is such that the depth of soil wetting is sufficient. The irrigator achieves high productivity so that during a day's work he can irrigate large hectares of land.
- Allows a uniform water distribution in root zone of a crop with as small as 6 cm applications for light irrigation,
- Allows heavy uniform application of water.
- Allows use of large concentrated water flows for reduction of conveyance losses, field channel network, and labour cost,
- Facilitates mechanized farming,

- Occupies minimum land under bunds, etc.,
- Is inexpensive and economically justifiable,
- High efficiency of water application i.e., the ratio of water stored in the root zone to that delivered to the field should be maximum,
- Minimum wastage of water either through surface runoff or through deep percolation below the root zone of a crop.

The selection of an irrigation technique appropriate for a specific situation is an important aspect of water management of crops.

SMART IRRIGATION SYSTEMS

A little advancement in the traditional system involves establishing centralized control of irrigated land using the wired architecture which also leads to more cost and maintenance manual. (Damas et al., 2001). Smart irrigation controllers estimate soil moisture or specific water needs of plants so that excess water use is minimized. Automatic irrigation farming involves the use of a control system to control and monitor the irrigation process. Smart technologies consist of climate-based controllers, soil moisture sensor- based controllers, and rain sensors. If properly installed and maintained, these controllers save water.

COMPONENTS OF SMART IRRIGATION SYSTEM

The improvement of farm irrigation efficiency is important not only to enhance the overall irrigation efficiency of the irrigation system but also to enhance the crop water productivity. Smart irrigation system increases the water use efficiency and saves water from wastage.

Sensing Technologies: Smart irrigation system is dependent on effective measures by using different sensors such as temperature sensor, humidity sensor, soil moisture sensor. Remote monitoring is an effective method to avoid disturbance in environment and improve efficiency. Ethernet network, RF module and ZigBee wireless network are some examples of wireless network used to transmit data in remote Monitoring System. Wireless Sensor Networks (WSN) is a system having components such as of sensors, radio frequency (RF) transceivers, microcontrollers and power sources. These technologies have led to the development of low cost, low power, multifunctional sensor nodes. Sensor nodes allow environment sensing with data processing and instrumented with a variety of sensors, such as temperature, humidity and soil moisture detection that allow monitoring of different environments

(Chavan and Karande, 2014). They are able to network with other sensor systems and exchange data with external users. Sensor networks are used for a variety of applications such as wireless data acquisition, environmental monitoring, smart buildings and highways, machine monitoring and maintenance, site security, safety management, automated on-site tracking of expensive materials, and in many other areas. Currently, there are two standard technologies are available for WSN: ZigBee and Bluetooth. A general WSN process consists of

1. Application layer
2. Transport layer,
3. Network layer,
4. Data link layer,
5. Physical layer,
6. Power management plane,
7. Mobility management plane and
8. The task management plane.

The three sensors used mainly in the system.

1. **Temperature Sensor:** If the temperature of the area is high, the water usage capacity is also high i.e. plants transpire more water in hot weather and absorb more water from the soil. Temperature sensor records the temperature change in the area and that data is utilized during irrigation. Temperature affects several processes in soil and soil ecosystem so that the soil temperature measurement is necessary (Valente et al. 2006). The ZigBee, LM35 series are precision integrated-circuit temperature sensors, whose output voltage is

Figure 6. Different components of smart irrigation techniques

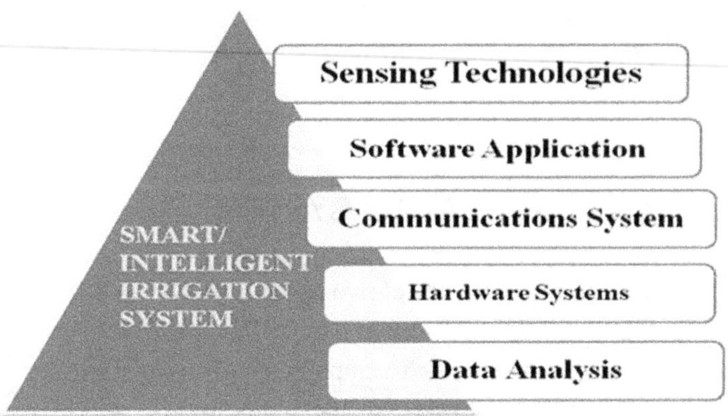

linearly proportional to the Celsius (Centigrade) temperature. These are low cost and small size sensor. The temperature range is -55° to +150°C.

2. **Soil Moisture Sensor:** Soil water is an essential component in relation to plant growth. If the moisture content of a soil is optimum for plant growth, plants can readily absorb soil water. This type of sensors can monitor the moisture content in the soil and according to that data is user supplies water to the soil. The soil moisture sensor used is capacitive type. The sensor gives analog output of zero (0 V) volt when there is 100% moisture and 5V for 0% moisture. There are principally two types of sensors, one is water potential types such as tensiometers and various forms of granular matrix sensors and another type is soil moisture sensors that give a percentage or relative content of soil moisture. Water potential probes are used to measure how hard it is to remove water from the soil, providing the best indication of available water for plants.
3. **Humidity Sensor:** Humidity can also affect plant turgor pressure, which is an indicator of the amount of water in plant cells. When humidity is low, moisture evaporates from plants very quickly. Plants need more water. Humidity sensor detects the moisture requirement of plant and sends information to a user. Humidity detecting instruments usually rely on measurements of some other quantity such as temperature, pressure, mass or a mechanical or electrical change in a substance as moisture is absorbed. By calibration and calculation of data collected by the device, these detecting quantities can lead to a measurement of humidity.
4. **Software Application:** It uses inputs like moisture data, water consumption percentage, and water demand data to calculate and for transmitting the information to the user.
5. **Communications Systems and Hardware Systems:** Commonly mobile phones, wireless networks, computer systems are used to communicate information/ data from sensors to the user and vice versa. Global System for Mobile communications (GSM) /GPRS module consists of a GSM/GPRS modem assembled together with power supply circuit and communication interfaces.

TYPES OF SMART IRRIGATION SYSTEM

The smart irrigation system consisted of two components one is wireless sensor units (WSUs) and another is wireless information unit (WIU), linked by radio transceivers (mobile or monitor) that allowed the transfer of soil moisture and temperature data, implementing a WSN that uses ZigBee technology. A wireless sensor unit is comprised of a RF transceiver, sensors (temperature, humidity, soil moisture etc.), a microcontroller, and power sources (battery or solar panels for energy collector). Several WSUs can be installed in-field to constitute a distributed sensor network for the automated irrigation system. The soil moisture and temperature data from each WSU are received, identified, recorded, and analysed in the WIU. The WIU consists of a master microcontroller, radio modem, a GPRS module, two electronic relays, two pumps for driving the water from the tanks, and a deep cycle rechargeable battery (recharged by solar panels or any other source).

The smart irrigation system is categorized into following classes according to the type of sensor it has. In the process of developing an automated wirelessly monitored irrigation system, several methods were followed.

- Remote Monitoring using GSM.
- Monitoring using fuzzy controller.
- Monitoring using humidity sensors.
- Monitoring in drip irrigation.
- Remote Monitoring using sensors.
- Solar powered smart irrigation system.

REMOTE MONITORING USING GSM

In this system, the user sends an activation command to the system through SMS. The system will check the moisture level and if it is less than the prescribed level, the system will start the motor. If the moisture level reaches the sufficient level, the motor will be turned off. The corresponding event of the operation of the motor is notified to the user using SMS. This motor is controlled by a starter which is indirectly activated by a relay circuit. Using GSM technique, an automated remote monitored irrigation system is provided (Shen et al., 2007). The system sets the time period depending on the temperature and humidity of the soil for irrigating the land. The humidity and temperature level of the soil and the crop varies for various types of crops.

MONITORING USING FUZZY CONTROLLER

The system consists of two units, Wireless Sensor Networks (WSNs) and a monitoring centre. Nodes in the monitoring area collect information of the soil moisture and growth information of different crops in different periods using solar power. WSN contains many sensor nodes and controller nodes. The data sensed from the WSN is

Figure 7. Flow diagram of working of Remote monitoring GSM smart irrigation method

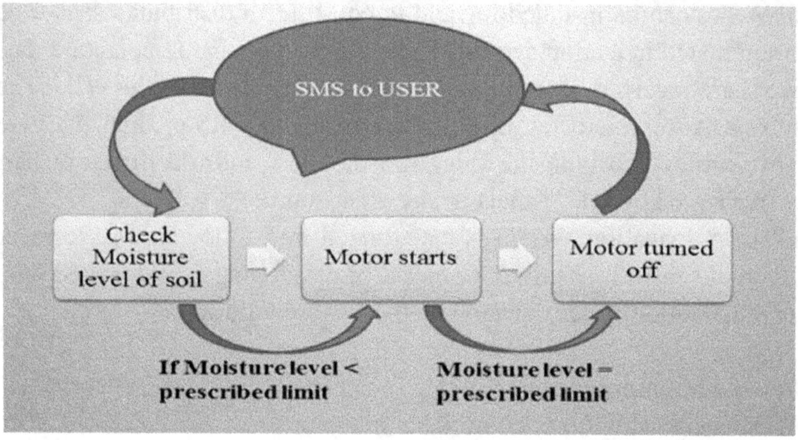

given as inputs to the monitoring centre which in turn gives the information about the irrigation, demand of the water level and control over the opening or closing of the valves. Some of the nodes are assigned as routers and coordinators. Sensor nodes sense the temperature and humidity whereas the routing nodes route the communication information and forwarding the data to the relevant nodes and the coordinator node receives data from the routing nodes and sends it to host computer monitoring centre. The monitoring centre record the real-time moisture value form all the nodes and calculate crop irrigation water requirement and output the result to WSN and control to the valves. The soil moisture sensor is used for measuring the moisture value.

MONITORING USING HUMIDITY SENSORS

For measuring the humidity value of a region, the humidity sensor is used. It works on the operating temperature range of 0 – 60°C. The temperature value is sensed by the sensor in which the output voltage is linearly proportional to the Celsius temperature. A low-cost soil moisture sensor is used to control water supply in water-deficient areas. The sensor communicates the information wireless communication module to a centralized server. The server provides the control to the water supply and displays the moisture data. Two galvanized metallic electrodes are kept at a distance of 30mm inside the soil area in an acrylic sheet. The change of impedance between two galvanized metallic electrodes due to the varying moisture content in

Figure 8. Flowchart of working of Monitoring sing fuzzy controller based smart irrigation system

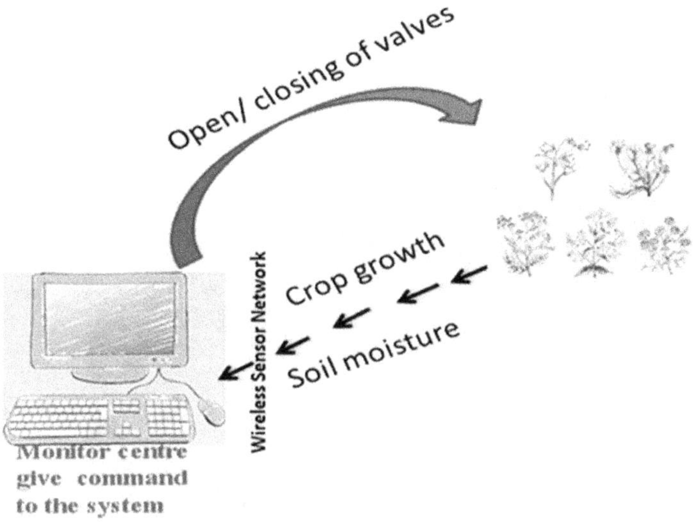

the surrounding medium is measured. Based on all these measurements, the data is transferred to the server for processing.

MONITORING IN DRIP IRRIGATION

In order to achieve the effective use of water in drip irrigation, sensors are scattered throughout the field region. Sensors are used to sense the water level in the well as well as in the water tank. If the observed value is less than the required value, then alert is given to the user. If the level of the water in the tank is very low, then it directs the command to the nearby well. The pump is activated thereby the valve is opened and the water is pumped out to the well. Sensors in the field sensed the value of temperature and moisture in the field and based on the input value, the system decides either to open the valve or not.

REMOTE MONITORING USING SENSORS

In the irrigation monitoring system, the information regarding soil moisture, temperature etc. are sensed by the Bluetooth wireless transmitters and the time-specific decision for irrigation is made according to the information sensed. The

Figure 9. Pictorial representation of a process of working during monitoring in drip irrigation

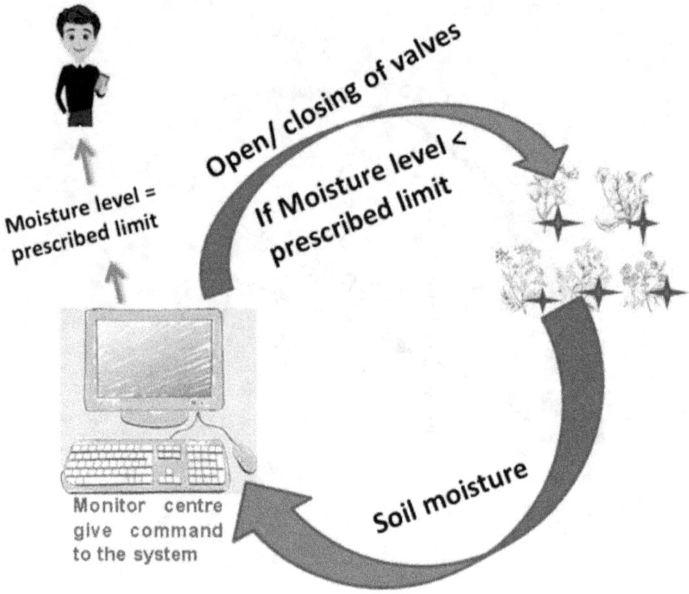

irrigation control unit gets the information and sends the position of the irrigation system using GPS receiver to the base station through real-time monitoring. A base station, in turn, sends control signals to the irrigation control station to operate the device for water usage and distributed irrigation system, automated field-specific irrigation system and sensor-based irrigation system are also provided good irrigation control. Yet, all these do not consider the pollution of water (Majsztrik, et al., 2013).

SOLAR POWERED SMART IRRIGATION SYSTEM

Increasing energy demand and decreasing energy resources have forced scientists to introduce new techniques which will use water resources in an efficient way. It is estimated that 15% of the total energy expended for all crop production is used to pump irrigation water (Martin et al., 2011). Total annual energy consumption is 168913 Giga Watt Hours (GWh) in the agriculture sector in 2014-15 which is 17.81% of total electricity consumption (Central Statistics Office Ministry of Statistics and Programme Implementation Government of India New Delhi, 2016). Solar energy is the most easily available source of energy in the world. The solar-powered irrigation system can be a suitable alternative for farmers in the present state of the energy crisis in India and other countries of the world. Solar powered smart irrigation system (SIS) in agricultural management uses soil moisture sensor.

Components of Solar Powered Smart Irrigation System (Gora and Dulawat, 2017)

- **Solar Photovoltaic Panel:** Solar panel is an assembly of photovoltaic (PV) cells electrically connected and mounted on a supporting structure such as roof or wall. Photovoltaic (PV) cells are made of semiconductors for example silicon. When light incident on the cell, a certain part of it is absorbed on the semiconductor material. This absorbed solar light is coverted and stored in the form of electric current in batteries. The electricity that generated by the semiconductor is in the form of direct current (DC) and can be used immediately or stored in a battery.
- **Battery:** Batteries are used for storage of electricity produced by solar panels in this proposed system. Electrical batteries are mixture of one or more electro- chemical cells, used to transfer chemical energy into electrical energy. The stored energy used for operation of the irrigation system.
- **GSMC (Global System for Mobile Communication):** The GSMC is used for sending a SMS about pump status to farmer (user). Global system for mobile communication (GSMC) is a globally accepted standard for digital cellular communication. A GSM modem can be an external modem device in which a GSMC SIM card inserted and connect the modem to an available serial port.
- **Soil Moisture Sensor:** Soil moisture sensor is used to measure the soil moisture in the system. A cost effective soil moisture sensor is mounted under the ground at particular depth. The moisture sensor just senses the moisture of the soil. The change in moisture will be observed by amount of current water flowing through soil. These sensors required very low power and having high resolution. This gives the capacity to make numerous estimations i.e. hourly over a drawn out stretch of time with insignificant battery utilization.
- **Relay:** Relay is electromagnetic switch use to control the electrical devices. In the system, two relays are used for switching. One relay is used for switching between microcontroller and GSM. Second relay is used for switching between microcontroller and RF. Relay is acts as transmitter. Most of the relay uses an electromagnet to activate a switching mechanism mechanically.
- **Solenoid Valve:** A solenoid valve is an electromechanical device which is mostly used with fluid or gas. The valve is operated by an electric current through a solenoid coil. In this system solenoid valves are used to control the flow of water. Solenoid valves may have at least two ports: on account of a two-port valve the flow is turned on or off; on account of a three-port valve, the outflow is exchanged between the two outlet ports. Solenoid valves are

the most as often as possible utilized control components in fluidics. Their jobs are to release, distribute shut off, or mix fluids. Solenoids is the best option for quick and safe switching, long life, high unwavering quality, great medium similarity of the materials utilized, low control power and compact design.

- **Microcontroller:** Microcontroller is the heart of this system. Every sensor has to send signal to the microcontroller. Then the microcontroller has to take decision according to the situation. ARM-LPC2148 is commonly used IC from ARM -7 family. It is pre loaded with many programmes and manufactured by Phillips. It is useful option for beginners and as well as high end application developer (Gora and Dulawat, 2017).

The system also focused on the use of solar energy utilization by the sensors during communication. A solar panel pumps electricity into a battery that stores it. Soil moisture sensor sensed the moisture content and sends data to a user. The user activates the water supply system and when the moisture level of the field reached the required limit, the water supply stops automatically.

Advantage and Disadvantages of Solar Powered Smart Irrigation System

Solar powered system uses the solar energy as an energy source that conserves the other energy resources. Other advantages of this system are explained as follow:

Figure 10: Flowchart of a process of working of a solar-powered smart irrigation system

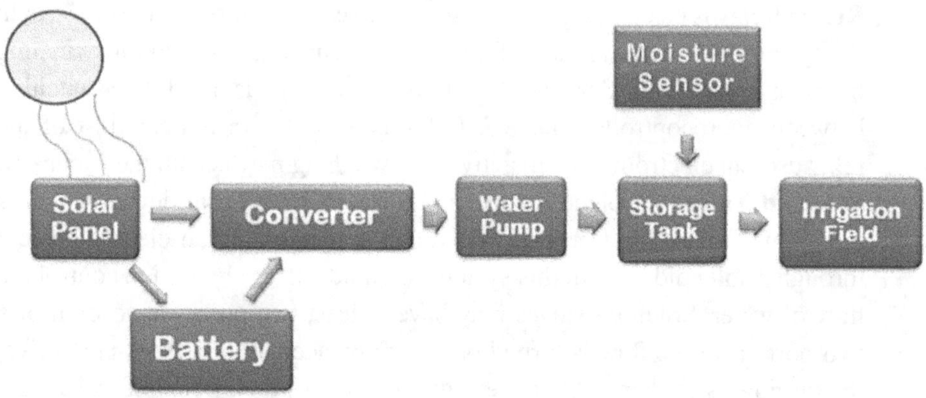

- It provides precise water delivery to the fields as larger amount to the seedling with time intervals and lower quantity to the mature plants as per the requirement.
- This system is programmable and promoting scientific usage of soil and plant additives so that watering pattern is maintained week over weeks.
- This system is flexible in working according to type of crop grow in the fields.
- Cleaner source of energy is used for working so that it is pollution free and efficient system to use.

Excluding above advantages, this system has also some limitations such as

- Depends on the climatic conditions as in rainy day or cloudy or foggy day solar system does not work.
- This system is expensive in installation for small scale projects.

CASE STUDIES ABOUT SMART IRRIGATION SYSTEMS

Utilization of Automatic Irrigation System to the First Time

The first types of sensors adapted to automatic irrigation control were tensiometers and GMS. The first used as soil matric potential sensor to control subsurface drip-irrigated processing in tomatoes (Phene and Howell, 1984). The results showed that yields of the automated system were similar to those from tomatoes irrigated based and had potential to use less water and conserve irrigation water. Switching tensiometers devices are that operate in bypass mode with a timer such that irrigation will be allowed within a timed irrigation window if the soil matric potential exceeds a threshold setting.

Automatic Subsurface Drip Irrigation in Sweet Corn/Peanut Crops

An automatic subsurface drip irrigation control system used in a sweet corn/peanut crop rotation (Nogueira et al., 2003). This system used TDR sensors to control a subsurface drip irrigation system on-demand. During successive analysis of this system, 11% irrigation savings with the on-demand subsurface drip irrigation system (23 cm deep) compared to sprinkler irrigation was reported with similar yields between the systems (Dukes and Scholberg, 2005). A system used to irrigate onion with frequent bypass control using GMS. After study it found that overall water used was slightly lower than calculated crop evapotranspiration with acceptable yields (Shock et al. 2002).

Smart Water Distribution Management Pilot (CBEC, Italy)

The Consorzio di Bonifica Emilia Centrale (CBEC) is one of the reclamation associations of Northern Italy which is responsible for the irrigation and water drainage of an area. Water used for the irrigation is taken from the Po River (on other intake structures operate along the Secchia and Enza rivers, two major Po river tributaries). The water is distributed over the cultivated area by means of a complex irrigation infrastructure. Water losses are due to infiltration through canal banks and bottom, as well as to the management of the irrigation network that requires the filing of long canals stretches and several minor streams to accommodate farmer needs. Efficient water network management based on an appropriate planning of the irrigation scheduling for the farmers served by a given irrigation district. The SWAMP project are aimed to enhance the overall system efficiency by acting at two different levels, one on farmer level which ensures the better estimation of water needs. This will achieve by integration of ground based information with weather forecast. The evaluation in term of water consumptions and crop productivity will be monitored by comparing between pilot areas with same crops. Another level is consortium level; the technological platform will collect request of water needs and management by adjusting the operational management of canal platform. In particular, the SWAMP project will enable the monitoring, automation and remote control of the principal hydraulic infrastructures, through which the CBEC manages the water distribution within the irrigation district.

- In M. Guerbaoui, elafou,a.ed-dahhak " GSM based automated drip irrigation system " proposed a system support to the development of greenhouse production in Morocco. The proposed project involves the development of an integrated automatic drip fertilizing irrigation system in green house. The project adopted involves a data acquisition card PCL-812PG controlled by PC. The irrigation is provided by an electric pump. Water needs are evaluated by measuring soil water status by soil humidity sensor (Sindhuja, 2016).

Automatic Irrigation System Using Solar Photovoltaic System in CSIR, Chennai

The procuring and assembling of Solar Photovoltaic System (SPV) for the purpose of automatic watering of testing field selected in CSIR Chennai, Taramani, and Campus was involved. Solar Powered Automatic Drip Irrigation System (SPADIS) has components such as batteries, solar charge controller, pump, solar panel, inverter, soil moisture sensor, Arduino mega 2560 micro controller. Based on the setup, many experiments were conducted and observation also noted to understand the functional behaviour of soil moisture sensors, batteries charging voltage, batteries discharging voltage while running the pump by drawing the power from battery alone (keeping solar panel disconnected from the circuit) and batteries discharging voltage while running the pump drawing power from batteries and simultaneously it is also kept in charged condition from SPV. In manual irrigation the pump switch ON was done at soil moisture at 45, 35, 40 and 38. The pump was switched OFF at soil moisture at 98, 92 and 90. It has been decided on the basis of experiments to switch ON the submersible pump whenever the soil moisture goes down to 65% and switch OFF the pump when the moisture level reaches to 75%. On the basis of studies it was found that SPADIS can save the irrigation water about 30-70%.

THE ROLE OF NANOTECHNOLOGY IN AGRICULTURE

Nanotechnology has many applications in all stages of agriculture such as in production, storing, processing, packaging and transport of agricultural products (FAO, 2013b). Nanotechnology will modernized and revolutionized agriculture and food industry by addition of new techniques such as: precision farming techniques and technically controlled environment agriculture techniques, increasing the ability of plants to absorb nutrients, more efficient and targeted use of inputs, disease detection and control diseases, withstand environmental pressures and effective systems for processing, storage and packaging. Moisture, temperature, pH and nutrients are soil parameters that will also be detected by using nanomaterial based sensors and then possible corrections may be taken place.

CONCLUSION AND FUTURE PROSPECTS

The smart irrigation system controller automates the irrigation process by sensing primary agricultural parameters such as soil moisture, soil PH, temperature, and humidity. Also, the system allows farm owners to remotely monitor and control the irrigation system. The management of drip irrigation presents additional challenges as a result of high irrigation frequency, limited soil wetting, and technical skills required for proper system operation and care. Management at high efficiencies common under drip irrigation invariably means operating with limited soil water and nutrient storage. Conservation of water sources and limiting wastage of it, done by this effective framework will help for the better profitability of crops. It saves time and attends to the plant's water need at the right time and reduces human labour hours and errors and increases profit for the farmer. Solar powered based smart irrigation system is best option to save energy and water. It is environment-friendly and pollution free technique. By combining the rain harvesting system with smart irrigation system, a huge amount of water can save in future.

REFERENCES

Abbas, A. H., & Mohammed, M. M. (2014). Smart Watering System for Gardens using Wireless Sensor Networks. *Proceedings of International Conference on Engineering and Technology*, 1-5 10.1109/ICEngTechnol.2014.7016780

Central Statistics Office, Ministry of Statistics and Programme Implementation Government of India New Delhi. (2016). *Energy Statistics., 23*, 39–52.

Chavan, C. H., & Karande, M. P. (2014). Wireless monitoring of soil moisture, temperature & humidity using zigbee in agriculture. *Int. J. Eng. Trends Technol, 11*(10), 493–497. doi:10.14445/22315381/IJETT-V11P296

Clean Water Action Council (CWAC). (2014). *Population facts the future*. Retrieved from http://www.cleanwateractioncouncil.org/issues/resource-issues/population

Damas, M., Prados, A. M., Gomez, F., & Oliyares, G. (2001). Hydro Bus system: Fieldbus for integrated management of extensive areas of irrigated land. *Microprocessors and Microsystems, 25*(3), 177–184. doi:10.1016/S0141-9331(01)00110-7

Dillehay, T. D., Eling, H. H., & Rossen, J. (2005). Preceramic irrigation canals in the Peruvian Andes. *Proceedings of the National Academy of Sciences of the United States of America, 102*(47), 17241–17244. doi:10.1073/pnas.0508583102 PMID:16284247

Dukes, M. D., & Scholberg, J. M. (2005). Soil moisture controlled subsurface drip irrigation on sandy soils. *Applied Engineering in Agriculture, 21*(1), 89–101. doi:10.13031/2013.17916

FAO. (2011a). *Climate change, water, and food security*. Rome: FAO.

FAO. (2013b). *Food Balance Sheet*. FAO.

Gora, A., & Dulawat, M. S. (2017). Solar powered smart irrigation. *Research Journal of Agriculture and Forest, 5*(6), 15–19.

Harishankar, S., Kumar, R. S., Sudharsan, K. P., Vignesh, U., & Viveknath, T. (2014). Solar powered smart irrigation system. *Advance in Electronic and Electric Engineering*, *4*(4), 341–346.

Hill, R. W., Patterson, R., & Barnhill, J. V. (2008). Small Acreage Irrigation System Operation and Maintenance. Utah State University -Extension Service. AG/Small Acreage/ 2008-01pr.

Hillel, D., Braimoh, A. K., & Vlek, P. L. G. (2008). Soil Degradation under Irrigation. In A. K. Braimoh & P. L. G. Vlek (Eds.), *Land Use and Soil Resources*. Dordrecht: Springer. doi:10.1007/978-1-4020-6778-5_6

Hillel, D. (1994). *Rivers of Eden. The struggle for water and the quest for peace in the Middle East*. Academic Press.

Jin, Song, Han, Wang, & Yan. (2007). School of Electric and Electronic Engineering, A Remote Measurement and Control System for Greenhouse Based on GSM-SMS. *IEEE 8th International Conference on Electronic Measurement and Instrument*, 45-82.

Majsztrik, J. C., Price, E. W., & King, D. M. (2013). Environmental benefits of wireless sensor-based irrigation networks: Case-study projections and potential adoption rates. *HortTechnology*, *23*(6), 783–793.

Martin, D. L., Dorn, T. W., Melvin, S. R., Corr, A. J., & Kranz, W. L. (2011, February). Evaluating energy use for pumping irrigation water. *Proceedings of the 23rd Annual Central Plains Irrigation Conference*, 22-23.

Nogueira, L. C., Dukes, M. D., Haman, D. Z., Scholberg, J. M., & Cornejo, C. (2003, January). Data acquisition system and irrigation controller based on CR10X datalogger and TDR sensor. *Proceedings - Soil and Crop Science Society of Florida*, *62*, 38–46.

Phene, C. J., & Howell, T. A. (1984). Soil sensor control of high-frequency irrigation systems. *Transactions of the ASAE. American Society of Agricultural Engineers*, *27*(2), 392–0396. doi:10.13031/2013.32798

Rodda, J. C., & Ubertini, L. (2004). *The Basis of Civilization -Water Science*. International Association of Hydrological Sciences.

Schroder, H. J. (2006). *Soil moisture-based irrigation for efficient use of water and nutrients and sustainability of vegetables cropped on coarse soils* (Thesis). University of Florida.

Shock, C. C., Feibert, E. B., Saunders, L. D., & Eldredge, E. P. (2002). Automation of subsurface drip irrigation for crop research. In *World Congress of Computers in Agriculture and Natural Resources, Proceedings of the 2002 Conference* (p. 809). American Society of Agricultural and Biological Engineers.

Sieberta, S., Hoogeveenb, J., Dölla, P., Faurèsb, J. M., Feicka, S., & Frenkenb, K. (2006, October). The Digital Global Map of Irrigation Areas–Development and Validation of Map Version 4. *Conference on International Agricultural Research for Development (Tropentag).*

Sindhuja, M. P., Prathusha, M., Kalaiselvi, M. C., Sudhaa, S. J., & Velthai, R. (2016). *Solar driven Aurdino based automatic irrigation using GSM*. Academic Press.

Singh, S. (1997). *Taming the waters: The political ecology of large dams*. Academic Press.

Snyder, R. L., & de Melo-Abreu, J. P. (2005). *Frost protection: fundamentals, practice and economics*. Academic Press.

Southern Nevada Water Authority (SNWA). (2017). *Subsurface Irrigation*. Retrieved from https://www.snwa.com/land/irrig_subsurface.html

Uddin, J., Reza, S. T., Newaz, Q., Uddin, J., Islam, T., & Kim, J. M. (2012, December). Automated irrigation system using solar power. In *Electrical & Computer Engineering (ICECE), 2012 7th International Conference on* (pp. 228-231). IEEE. 10.1109/ICECE.2012.6471527

Valente, A., Morais, R., Tuli, A., Hopmans, J. W., & Kluitenberg, G. J. (2006). Multi-functional probe for small-scale simultaneous measurements of soil thermal properties, water content, and electrical conductivity. *Sensors and Actuators. A, Physical, 132*(1), 70–77. doi:10.1016/j.sna.2006.05.010

KEY TERMS AND DEFINITIONS

Humidity Sensor: A humidity sensor (or hygrometer) senses, measures and reports the relative humidity in the air. It therefore measures both moisture and air temperature.

Moisture Sensor: Soil moisture sensors measure the volumetric water content indirectly by using some other property of the soil. Soil moisture sensors typically refer to sensors that estimate volumetric water content.

Smart Irrigation: Smart irrigation systems save water by adjusting the watering schedule and amount of water used for irrigation based on a variety of factors and inputs, including weather, plant species, and soil type.

Soil Moisture: Soil moisture is the water that is held in the spaces between soil particles. Surface soil moisture is the water that is in the upper 10 cm of soil, whereas root zone soil moisture is the water that is available to plants, which is generally considered to be in the upper 200 cm of soil.

Sprinkler System: Sprinkler Irrigation is a method of applying irrigation water which is similar to rainfall. Water is distributed through a system of pipes usually by pumping. It is then sprayed into the air and irrigated entire soil surface through spray heads so that it breaks up into small water drops which fall to the ground.

Water Resources: Water resources are sources of water that are useful or potentially useful to humans. Only 2.5% of water on the Earth is fresh water, and over two thirds of this is frozen in glaciers and polar ice caps.

Chapter 10
Environmentally Friendly Slow Release Nano-Chemicals in Agriculture:
A Synoptic Review

Richa Kothari
ITM University, Gwalior, India

Khursheed Ahmad Wani
ITM University, Gwalior, India

ABSTRACT

Agriculture is important for people all over the world in order to obtain food to sustain the ever-growing population. However, the current practices for obtaining more and more food has several environmental challenges. Hence, new environmentally friendly fertilizers, herbicides, and pesticides have been developed that enhance crop yield by facilitating maximum nutrient uptake by the application of nanotechnology that will help in promoting sustainable agriculture by the slow or controlled release fertilizers. This slow discharge encourages improved delivery of nutrients to the plants that further speeds up early germination, fast growth, and high nutritional level. The current study is aimed to review nano-chemicals used in agriculture that have been developed by the researchers all over the world.

DOI: 10.4018/978-1-5225-5909-2.ch010

INTRODUCTION

Agricultural is a backbone to economy of most countries and currently feeding to the entire human race and has no alternative till date. However, it has various challenges both at national and international level. In one hand, agricultural products has to meet various challenges to ever growing population with same limited resources of land and water. Due to this, most of the chemical fertilizers are being used to enhance the productivity of crops (Liu and Lal, 2015). The efficiency of most fertilizers, especially the nitrogen forms, is reduced greatly due to the volatilization and leaching of nutrients. This has not only caused decline in land quality but 40 to 75% leaching losses were reported, they contribute less towards plant growth and more towards environmental issues (Trenkel, 2010) that may various health concerns directly and indirectly not only to humans but a great damage to other organisms (Rodriguez, 2004). The economic loss due to the use of fertilizers can't be ignored at the same time. This is can't be seen as a healthy sign for the developing countries that are continuously struggling for economic development. Hence, there is an urgent need to develop new environmental friendly fertilizers that can increase the yield of crops and manage the environment.

It is reported that application of nanotechnology may assist in promoting sustainable agriculture by slow or controlled release of fertilizers, herbicides and pesticides (Rai, Acharya, & Dey, 2012; Mura, Seddaiu, Bacchini, Roggero, & Greppi, 2013; Baruah & Dutta, 2009). Research in the direction has been already started by various organizations all over the world by making nanoparticles or nano-composites that will help in plant growth either by direct uptake or by slow release of nutrients (Naderi & Shahraki, 2013). Out of 16 nutrients 13 nutrients are present in soil and are being taken by plants from soil ecosystem by slight modifications and it is argued that nano-fertilizers can help in provision of these essential nutrients to the soil by slow release to enhance delivery of nutrients to the plants that will further enhance early germination, fast growth and high nutritional level (Hamid, Mohamad, Hing, Dimin, Azam, Hassan, Mustaq & Ahmad, 2013).

The barrier to inhibit the "burst effect" of fertilizers is generally used to prepare slow-release fertilizers by physically encapsulating the soluble fertilizers with hydrophobic inorganic and/or organic materials. Among those coating materials, The insoluble inorganic materials such as sulfur used to be attractive candidates are the coating materials that aids in retarding the release of nutrients from fertilizers and promote additional functions like secondary plant nutrients and agents that neutralize soil alkalinity as well (Tsai, 1986). The different types of polymers were widely tested as coating materials, however, some developed cracks and pores like sulfur film formed on the surface of fertilizer granules (Bao et al., 2015), and some were also susceptible to microorganisms (Azeem, Kushaari, Man, Basit, & Thanh,

2014). Petroleum-based polymers and renewable biopolymers such as lignin were also used but due to their high cost, process complications and environmental pollution (Mulder, Gosselink, Vingerhoeds, Harmsen, & Eastham, 2011), cellulose and starch are being used as substitutes as coating materials to improve the performance of slow-release fertilizers.

The current study is aimed to review nano-chemicals used in agriculture that have been developed by the researchers all over the world with the pros and cons of synthetic fertilizers and future implications of nano fertilizers.

AGROCHEMICALS AND ENVIRONMENTAL IMBALANCE

Agrochemicals are one of the major contributors for enhanced crop production. However, serious issues has been raised by scientific community for synthetic agrochemicals as these are responsible for various types of environmental in the form of heavy metal pollution of soil, atmospheric pollution due to emission of nitrous oxide and ammonia, eutrophication of surface water, nitrate pollution of groundwater, acid rain, soil erosion and leaching etc.

India which is second largest populous country of the world after china has to produce 380 million tonnes of food grain per annum in order to feed a population of 4 billion by 2025. Here it is to mention that in Essay on the Principle of Population, Malthus proposes the principle that human populations grow up exponentially while food production grows at an arithmetic rate. Thus, while food output was likely to enhance in a series of twenty-five year intervals in the arithmetic progression 1, 2, 3, 4, 5, 6, 7, 8, 9, and so on, population was capable of increasing in the geometric progression 1, 2, 4, 8, 16, 32, 64, 128, 256, and so forth. This is an indication that arithmetic food growth with simultaneous geometric human population growth predicted a future when humans would have no resources to survive on. Hence, there is no scope for horizontal expansion of agricultural land, additional amount of food grain has to be produced vertically in which fertilizers can be used again and again and as such environment will become the soft victim as synthetic fertilizers have implication on the environment. The environmental consequences due to various types of fertilizers are below:

NPK Fertilizers

NPK fertilizer is a fertilizer composed primarily of the three primary nutrients (NPK) essential for healthy plant growth. The agriculture productivity relies heavily on the use of NPK fertilizer to satisfy the global food supply and ensure healthy crops. NPK fertilizers after their application in soil undergo various transformation processes like mineralization, immobilization, urea hydrolysis, nitrification, volatilization and, denitrification. Phosphorus may be converted into various insoluble forms as Fe and Al phosphate in acid soil and Ca-phosphate in alkaline soils. The leaching of phosphrous is a primary cause for eutrophication of water bodies as significant amount of P is lost from the soil through surface runoff.

Nitrate Pollution

The buildup of oxidized nitrogen in the form of nitrates may result in high concentrations (may be upto 50 mg/l) of nitrates in the groundwater, which in some cases may exceed the limit for water consumption. Due to the multiple character of nitrates in groundwater, concentrations higher than 25 mg/l should be treated with concern, as it may lead to Blue baby syndrome i.e., cyanosis in infants: cyanotic heart disease and methemoglobinemia. The most common cyanotic heart defects include transposition of the great arteries, tetralogy of Fallot, persistent truncus arteriosus, tricuspid atresia and total anomalous pulmonary venous return and cancer due to nitrate ingestion in food and water. In surface waters, the presence of nitrogen as well as phosphorus may lead to eutrophication, with adverse impact, due the multipulcation of algae and their byproducts, on fish, oxygen levels and amenity value.

World Health Organization (WHO) prescribes that drinking water should not have more than 10 mg nitrates. The nitrate is converted into nitrite in the intestine and then absorbed in the blood stream can be poisonous. Young infants are incapable of detoxifying the nitrite, thus reducing the capacity of blood to carry oxygen hence; clinical symptoms such as gray or blue skin develop known as "methaemoglobinaemia" or "blue baby syndrome".

Eutrophication

The excessive plant and algal growth due to the increased availability of one or more limiting growth factors required for photosynthesis (Schindler 2006), like sunlight, carbon dioxide, and nutrient fertilizers is known as Eutrophication. Eutrophication does not take place quickly as it takes centuries as lakes age and are filled in with sediments (Carpenter 1981) and is dependent on several factors as mentioned above. However, human activities have accelerated the rate and extent of eutrophication through both The point-source discharges and non-point loadings by human activities has accelerated the process of eutrophication due to continuous use of synthetic fertilizers mostly nitrogen and phosphorus fertilizers that find their way into aquatic ecosystems. The process of eutrophication has several dramatic consequences for drinking water sources, fisheries, and recreational water bodies (Carpenter et al.

1998). The known consequences of eutrophication include blooms of blue-green algae, tainted drinking water supplies, degradation of recreational opportunities, and hypoxia. It is estimated that the cost of damage mediated by eutrophication in the U.S. alone is approximately \$2.2 billion annually (Dodds et al. 2009).

Eutrophication is linked with dramatic changes in aquatic community structure. Few observational studies have attributed anti-herbivore traits of cyanobacteria like toxicity, morphology, and poor food quality due to eutrophication as small-bodied zooplankton tend to dominate plankton communities (Porter 1977). on the other hand, the biomass of planktivorous fish is often positively related to nutrient levels and ecosystem productivity. Furthermore, Piscivorous fishes (e.g., bass, pike) have a tendency to control the fish community of nutrient-poor, oligotrophic lakes, while as, planktivorous fishes (e.g., shad, bream) become increasingly dominant with nutrient enrichment (Jeppesen et al. 1997).

Ammonia Volatilization

It is reported that 94% of the NH_3, released from synthetic fertilizers used in agricultural sources is re-deposited into surrounding ecosystems. The loss of NH_3 due to volatilization is also a cause of concern and may cause environmental pollution. The re-deposition of NH_3 on to the terrestrial ecosystem after it is oxidized to N_2O from the atmosphere acts as a green house gas and is responsible for depletion of ozone in the stratosphere. It also forms salts with acidic gases and these salt particles can be transported to long distances.

Acid Rain

Acid rain is any form of precipitation with high levels of nitric and sulfuric acids that occur in the form of snow, fog, and tiny bits of dry material that settle to earth. Nitrogenous fertilizers used in agriculture contribute significantly towards emissions of ammonia, which is considered as one of the agents causing acid rain. The high concentration of ammonia may cause acidification of land and water surfaces, damage and reduce plant bio-diversity in natural systems. Acid rain has ecological effects on lakes, streams, wetlands, and other aquatic environments and makes waters acidic. This combination makes waters poisonous to crayfish, clams, fish, and other aquatic animals.

Greenhouse Gases

Atmosphere researchers contended that the reason for the enlarged nitrous oxide was nitrogen-based compost, which energize organisms in the soil to change over nitrogen to nitrous oxide at a quicker rate than typical (Sanders, 2012).The Organization for Economic Cooperation and Development (OECD) uncovered that CO_2, CH_4 and N_2O emissions in agriculture at global level are evaluated to represent 14 percent of the aggregate discharge of GHGs. Limiting nitrous oxide emissions could be part of a first step toward decreasing all greenhouse gases and attenuation of global warming. Diminishing nitrous oxide discharges can counterbalance more than what's coming to its of ozone depleting substance outflows by and large, since N_2O traps warm at an unexpected wavelength in comparison to CO_2 and stops up a "window" that enables Earth to chill free of CO_2 levels. A lot of nitrogen gas are discharged to the climate by means of denitrification, including that of nitrogen composts. Release of N2 diminishes nitrogen (N) accessible to crops, yet isn't generally negative to the earth. Both nitric oxide (NO) and nitrogen dioxide (NO_2) respond in daylight with unpredictable natural mixes to frame tropospheric ozone (O_3).Ozone is toxic to crops, even at low concentrations, and detrimental to the health of sensitive individuals.

Trace Element and Heavy Metals Contamination

Huge amount of fertilizers are added to soils to provide adequate N, P, and K for crop growth and yield. The compounds used to supply these elements have small amount of heavy metals (e.g., Cd and Pb) as impurities may significantly increase their content in the soil when applied continuously (Jones and Jarvis, 1981). Metals, such as Cd and Pb, have no known physiological activity. Application of certain phosphatic fertilizers inadvertently adds Cd and other potentially toxic elements to the soil, including F, Hg, and Pb (Raven et al, 1998). A serious concern has been raised about the occurrence of heavy metals and trace elements in the environment in concentrations that may cause potential dangers to the humans. It has been reported that many fertilizers especially phosphatic fertilizers contain different amounts of trace elements such as F, As, Cd, Co, Cr, Hg. Mo, Ni and Pb. The pig and poultry manures added to soil may contaminate agricultural fields by Cu and Zn as these elements are added to their diet for growth promoters. (Sumner, 2000; Chaney and Oliver, 1996). These heavy metals may enter into human body by a process of biomagnification through different food chains. A small amount of Cd has been reported to occur in animal manures along with other heavy metals, although, these concentrations are very low.. Different amounts of trace and heavy elements are transferred to P fertilizers in production processes from phosphate rocks and later find their way into soil on its application, although the concentration of heavy metals depend on its origin. Most of the trace elements and heavy metals present in fertilizers are arsenic (As), chromium (Cr), lead (Pb), mercury (Hg), nickel (Ni), and vanadium (V). However, these metals are of less concern than Cd, either because they are not as readily absorbed by plants from P-fertilized soils or their apparent relative effects on human health are less than that of Cd. Most of the countries have established tolerance limits on heavy-metal additions to soil for the tillage layer (surface 20-30 cm) of soil where most root activity occurs as their long-term effects are unknown. Regulations are also proposed for phased-in limits on maximum heavy metal concentrations allowed in P fertilizers, or they are already in effect. Most of the fertilizer regulations relate Cd limits to P concentrations, so P application rates dictate Cd inputs to soil (Mortvedt, 1995).

Biofertilizers and Implications

The global market for bio fertilizers was expected to be worth about five billion USD in 2011 and is predicated to double by 2017 (Marketsandmarkets 2013), mostly in Latin America, India and China (Fuentes-Ramirez and Caballero-Mellado 2005 ; Bashan and de-Bashan 2010).

Bio fertilizers have the tendency to enhance the growth and productivity of plant life, crop yield and help in minimizing the use synthetic fertilizers. The application of micro organisims and organic compounds has been very useful in this direction. The micro-organisms and organic compounds mobilize the nutrients and help in assimilation of nutrients by plants. Bio fertilizers have the ability to reduce the buildup, leaching or run off nutrients from the agricultural fields, participate in nutrient recycling and improve crop productivity (Singh et al., 2011). Further, bio fertilizers can support economically and are beneficial for the environment as compared to chemical fertilizers (Adesemoye, Torbert, Kloepper, 2009 ; Adesemoye and Kloepper 2009; Malusà and Sas Paszt 2009). The application of different strains of microorganisms cooperating with autochthonous microorganisms like endophytic bacteria (Reinhold-Hurek and Hurek 2011 ; Ryan et al. 2008), or utilizing the synergies with microbial communities (Bernard et al. 2012), and the introduction of protozoa in the formulation of biofertilizers (Bonkowski 2004 ; Ronn, McGraig, Griffi, & Prosser, 2002) will develop new kinds of biofertilizers that may be very important for future use. The formation of the soil web by different microorganisms when dimethyl sulfide is released by legumes to attract nematodes that transport rhizobia toward the roots is an indication that wider network of microorganisms can be involved during the process to act as bio fertilizers. (Horiuchi, Prithiviraj, Bais, Kimball & Vivanco, 2005). Different bacterial species were also successfully tested as bio fertilizers to promote the plant growth by producing phytohormones and are regarded as bio enhancers. These bacterial species can fix atmospheric N, both, in association with plant roots and without it, solubilise insoluble soil phosphates and produces plant growth substances in the soil. Biofertilizers could also be produced invitro grown plant material leading to increased growth of seedlings, being more resistant to biotic and abiotic stresses (Sekar and Kandavel 2010), and for their quantitative or qualitative improvement of plant secondary metabolites content in medicinal plants (Zubek, Mielcarek, & Turnau, 2012). However, several biotic, abiotic and anthropogenic factors cause challenges in successful application of commercial biofertilizer and are responsible for the effectiveness of the biofertilizers in agricultural fields.

The liquid bio fertilizers are also in use and are more advantageous as to compared to solid bio fertilizers as they contain desired microorganisms along with nutrients. Furthermore, they have special cell protectants or chemicals that encourage formation of resting spores or cysts for longer shelf life and tolerance to adverse conditions.

The use of bio fertilizers have become very important as the imbalanced use synthetic fertilizers has lead to considerable damage to soil ecosystem. However, the environmental limitations for application of bio fertilizers can't be ignored. The farmer has to bear considerable initial cost in terms of skill acquisition, trial and failure and risk thus they are reluctant to use these types of fertilizers in their agricultural fields. The producer firms have serious uncertainty about the demand or sale of the product, which discourage investment, particularly if it is irretrievable (Ghosh, 2004). The quantification results are not guaranteed with bio fertilizers as in case of synthetic fertilizers. The effectiveness of bio fertilizers is governed by different factors like pH, temperature, soil moisture and other environmental factors. The money, time and yield of the farmers gets affected heavily when the microorganisms don't multiply sometimes due to various reasons. The different soil conditions, climatic factors and inadequate experienced staff are the other constraints to use bio fertilizers for crop productivity.

NANO FERTILIZERS AND ADVANTAGES

The history of nano technology dates back to 4th century and has been continuously improved (Table 1).

The nutrient deficiency in soils due to anthropogenic factors has caused economic loss for farmers and decreases in nutritional quality that resulted in overall decrease in the quantity of grains for human population and livestock. The imbalance in the soil mineral ratio due of large-scale application of chemical fertilizers to increase the crop productivity is a matter of concern. This has also caused irreparable damage to the soil structure, soil microbial flora, plants, mineral cycles, and to food chains and food webs across different ecosystems. The applications of nano technology is diverse and are considered very important in present century (Figure 1)

In recent years, progress in the field of nanotechnology has produced nanoparticles that are used to improve fertilizer formulations for increased uptake in plant cells and by minimizing nutrient loss (Table 2).

Table 1. Milestones in nanotechnology

Period	Major Breakthrough
4th Century	Dichroic glass: Colloidal gold and silver in the glass enable it to look dark green when lit from outside however translucent red when light radiates through
6th-15th Centuries	Vibrant Stained glass windows: gold chloride and other metal oxides and chlorides impart rich colors to nanoparticles
9th-17th Centuries:	Gleaming, sparkling "gloss" clay coats utilized as a part of the Islamic world, and later in Europe, contained silver or copper or other metallic nanoparticles.
13th-18th Centuries	"Damascus" saber cutting edges contained carbon nanotubes and cementite nanowires.
1857 solutions.	Michael Faraday found colloidal "ruby" gold, exhibiting that nanostructured gold under certain lighting conditions produces distinctive shaded
1936	Erwin Müller, working at Siemens Research Laboratory, invented the field emission microscope, allowing near-atomic-resolution images of materials.
1947	John Bardeen, William Shockley, and Walter Brattain at Bell Labs discovered the semiconductor transistor
1950	Victor La Mer and Robert Dinegar developed the theory and a process for growing monodisperse colloidal materials.
1951	Erwin Müller pioneered the field ion microscope
1956	Arthur von Hippel at MIT introduced molecular engineering
1958	Jack Kilby designed, and built the first integrated circuit, for which he received the Nobel Prize in 2000
1965	Intel co-founder Gordon Moore described "Moore's Law"
1974	Tokyo Science University Professor Norio Taniguchi coined the term nanotechnology to describe precision machining of materials to within atomic-scale dimensional tolerances.
1981	Gerd Binnig and Heinrich Rohrer at IBM's Zurich lab invented the scanning tunneling microscope
1981	Russia's Alexei Ekimov discovered nanocrystalline, semiconducting quantum dots in a glass matrix
1985	Rice University researchers Harold Kroto, Sean O'Brien, Robert Curl, and Richard Smalley discovered the Buckminsterfullerene (C60), The team was awarded the 1996 Nobel Prize
1985	Bell Labs's Louis Brus discovered colloidal semiconductor nanocrystals (quantum dots), for which he shared the 2008 Kavli Prize in Nanotechnology.
1986	Gerd Binnig, Calvin Quate, and Christoph Gerber invented the atomic force microscope
1989	Don Eigler and Erhard Schweizer at IBM's Almaden Research Center manipulated 35 individual xenon atoms to spell out the IBM logo.
1990s	Early nanotechnology companies began to operate, e.g., Nanophase Technologies in 1989, Helix Energy Solutions Group in 1990, Zyvex in 1997, Nano-Tex in 1998
1991	Sumio Iijima of NEC is credited with discovering the carbon nanotube (CNT). Iijima shared the Kavli Prize in Nanoscience in 2008 for this advance and other advances in the field.
1993	Moungi Bawendi of MIT invented a method for controlled synthesis of nanocrystals (quantum dots).
1998	The Interagency Working Group on Nanotechnology (IWGN) was formed under the National Science and Technology Council to investigate the state of the art in nanoscale science and technology and to forecast possible future developments.

continued on following page

Table 1. Continued

Period	Major Breakthrough
1999	Cornell University researchers Wilson Ho and Hyojune Lee probed secrets of chemical bonding by assembling a molecule [iron carbonyl Fe(CO)2] from constituent components [iron (Fe) and carbon monoxide (CO)] with a scanning tunneling microscope.
1999	Chad Mirkin at Northwestern University invented dip-pen nanolithography® (DPN®), leading to manufacturable, reproducible "writing" of electronic circuits as well as patterning of biomaterials for cell biology research, nanoencryption, and other applications.
1999–early 2000's	Consumer products making use of nanotechnology began appearing in the marketplace.
2000	President Clinton launched the National Nanotechnology Initiative (NNI) to coordinate Federal R&D efforts and promote U.S. competitiveness in nanotechnology. .
2003	Congress enacted the 21st Century Nanotechnology Research and Development Act (P.L. 108-153).
2003	Naomi Halas, Jennifer West, Rebekah Drezek, and Renata Pasqualin at Rice University developed gold nanoshells
2004	The European Commission adopted the Communication "Towards a European Strategy for Nanotechnology.
	2004: Britain's Royal Society and the Royal Academy of Engineering published Nanoscience and Nanotechnologies: Opportunities and Uncertainties advocating the need to address potential health, environmental, social, ethical, and regulatory issues associated with nanotechnology.
2004	SUNY Albany launched the first college-level education program in nanotechnology in the United States, the College of Nanoscale Science and Engineering.
2005	Erik Winfree and Paul Rothemund from the California Institute of Technology developed theories for DNA-based computation and "algorithmic self-assembly" in which computations are embedded in the process of nanocrystal growth.
2006	James Tour and colleagues at Rice University built a nanoscale car made of oligo(phenylene ethynylene) with alkynyl axles and four spherical C60 fullerene (buckyball) wheels.
2007	Angela Belcher and colleagues at MIT built a lithium-ion battery with a common type of virus that is nonharmful to humans, using a low-cost and environmentally benign process.
2008	The first official NNI Strategy for Nanotechnology-Related Environmental, Health, and Safety (EHS) Research was published
2009–2010	Nadrian Seeman and colleagues at New York University created several DNA-like robotic nanoscale assembly devices.
2010	IBM used a silicon tip measuring only a few nanometers at its apex .
2011	The NSET Subcommittee updated both the NNI Strategic Plan and the NNI Environmental, Health, and Safety Research Strategy, drawing on extensive input from public workshops and online dialog with stakeholders from Government, academia, NGOs, and the public, and others.
2012	The NNI launched two more Nanotechnology Signature Initiatives (NSIs)--Nanosensors and the Nanotechnology Knowledge Infrastructure (NKI)--bringing the total to five NSIs.
2013	Stanford researchers develop the first carbon nanotube computer.
2014	The NNI releases the updated 2014 Strategic Plan.

Modified from: https://www.nano.gov/timeline

Figure 1.

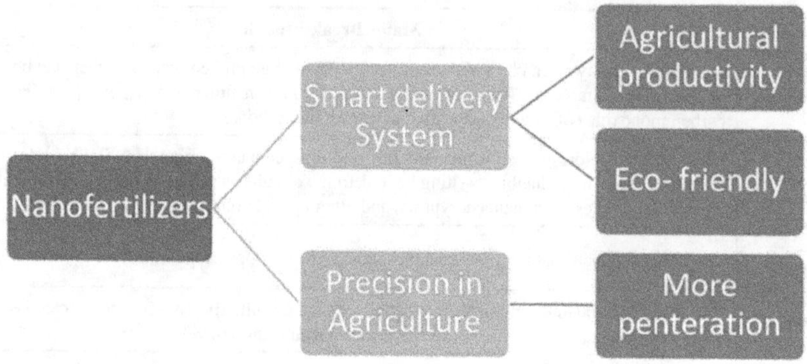

Table 2. Recent breakthroughs in nanotechnology in agriculture

Product	Application	Institution
Nanocides	Pesticides encapsulated in nanoparticles for controlled release	BASF, Ludwigshafen, Germany
	Nanoemulsions for greater efficiency	Syngenta, Greensboro, NC, USA
Buckyball fertilizer	Ammonia from buckyballs	Kyoto University, Kyoto, Japan
Nanoparticles	Adhesion-specific nanoparticles for removal of *Campylobacter jejuni* from poultry	Clemson University, Clemson, SC, USA
Food packaging	Airtight plastic packaging with silicate nanoparticles	Bayer AG, Leverkusen, Germany
Use of agricultural waste	Nanofibers from cotton waste for improved strength of clothing	Cornell University, Ithaca, NY, USA
Nanosensors	Contamination of packaged food	Nestle, Kraft, Chicago, USA
	Pathogen detection	Cornell University, Vevey, Switzerland
Precision farming	Nanosensors linked to a global positioning system tracking unit for real-time monitoring of soil conditions and crop growth	US Department of Agriculture, Washington, DC, USA
Livestock and fisheries	Nanoveterinary medicine (nanoparticles, buckyballs, dendrimers, nanocapsules for drug delivery, nanovaccines; smart herds, cleaning fish ponds (Nanocheck [Nano-Ditech Corp., Cranbury, NJ, USA]), and feed (iron nanoparticles)).	Cornell University NanoVic, Dingley, Australia

Source: Siddhartha S Mukhopadhyay (2014).

The high surface area, sorption capacity, and site specific controlled-release kinetics make nano fertilizers unique as compared to synthetic and bio fertilizers. These nano fertilizers can enhance the nutrient use efficiency through mechanisms like targeted delivery, slow or controlled release and could precisely release their active ingredients in responding to environmental triggers and biological demands. Nano fertilizers can enhance seed germination rates, growth of seedling, photosynthetic activity, nitrogen metabolism, and carbohydrate and protein synthesis.

Nano fertilizers are characterized by large surface area having particle size less than the pore size of root and leaves of the plant that increases the chances of penetration into the plant from applied surface and help in uptake of nutrients. The decrease in particle size results in increased surface area and number of particles per unit area of a fertilizer that increases more chances of contact of nano-fertilizers for uptake of the nutrient (Maurice and Hochella, 2008). The availability and uptake of nutrient to the crop plants is achieved by fertilizer encapsulation (Cui et al, 2010). The penetration capacity, size and very higher surface area which is usually different from the synthetic fertilizers are the prime drivers to achieve goals of target delivery system (Figure 2).

Nanoparticles delivered at safe dose can help in increasing plant growth and overall yield (Gao et al. 2006; Moreno et al. 2010), although, most of the recent studies have reported adverse effects of nanoparticles on plants (Lee, An, Yoon & Kweon et al, 2008; Barrena et al. 2009). Nanosilver particles were found effective at a lower concentration than silver nitrate to prevent leaf abscission. Effect of biosynthesized silver nanoparticles on emergence of seedling and various plant growth parameters of many economically important plant species were studied by Namasivayam and Chitrakala (2011). Multi-walled carbon nanotubes (MWCNTs) have the ability to enhance seed germination and growth of tomato (Khodakovskaya, Silva, Biris, Dervishi, & Villagarica, 2009; 2012). The enhancement of seed germination and plant growth using MWCNTs in mustard plant was revealed by Mondal, Basu, Das, S., & Nandy (2011). TiO_2 nanoparticles have increased 73% dry weight, threefold higher photosynthetic rate, and 45% increment in chlorophyll a after seed was treated in spinach (Zheng, Hong, Lu, & Liu, 2005). Nano silver particles are better than silver nitrate in improving the seed yield and preventing leaf abscission in borage plant (Sahandi, Sorooshzadeh, Rezazadeh, & Naghdibadi, 2011). While as, lower concentration (1.5 ppm) of ZnO nanoparticles has positive effect on *Vigna radiata* and *Cicer arietinum* seedling growth (Mahajan, Dhoke, & Khanna, 2011; Burman, Saini, & Kumar, 2013. On the other hand, Mahmoodzadeh, Nabavi & Kashefi (2013) found that at high concentration (up to 2,000 ppm) TiO_2 nanoparticles enhance seed germination and seedling vigor in *Brassica napus*. The mixture of SiO_2 and TiO_2 nanoparticles in G. max showed productive effect with increase in water and fertilizer uptake capacity and stimulation of nitrate reductase

Figure 2.
Source: https://phys.org/news/2015-11-tomatoes-boost-growth-antioxidants-nano-sized.html

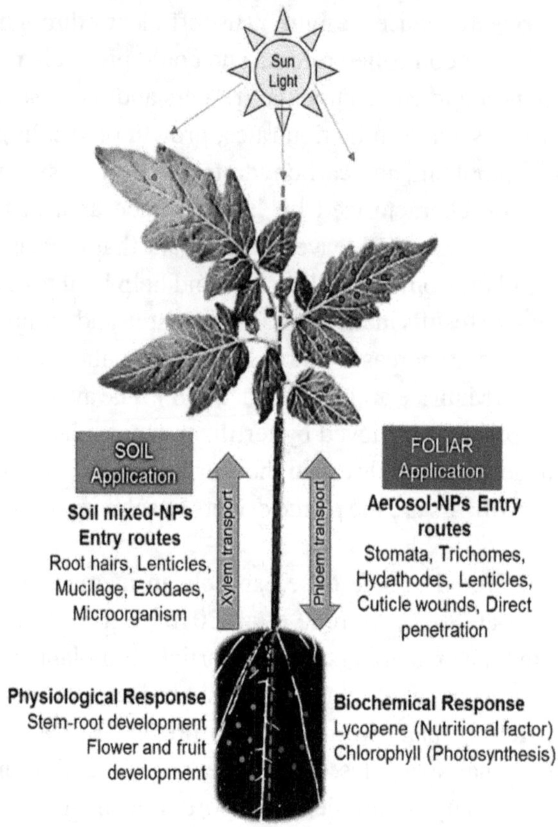

and antioxidant activity (Lu, Zhang, Wu, & Tao. 2002). Moreover, Quoc Buu et al. (2014) reported an enhanced seed germination rate in G. max when treated with nanocrystalline powder of iron, cobalt, and copper at very low concentration. With marked increase in the chlorophyll index, number of nodules and crop yield. Varied effect on lettuce seed germination was observed by Shah and Belozerova (2009) with different metal nanoparticles, Pd and Au at lower concentration, Si and Cu at higher concentration, and Au and Cu in combined mixture showed positive influence. The levels of organic compounds such as proteins, chlorophyll, and phenols in maize plants was significantly improved with SiO_2 nanoparticles (Suriyaprabha et al. 2012).

The nano-fertilizers have higher surface area it is mainly due to very less size of particles which provide more site to facilitate different metabolic process in the plant system result production of more photosynthesis. Due to higher surface area and very less size they have high reactivity with other compound. They have high solubility in different solvent such as water. Particles size of nano-fertilizers is less than 100 nm which facilitates more penetration of nano particles in to the plant from applied surface such as soil or leaves.

CONCLUSION

Nano-fertilizers application to agricultural field have important role in enhancing crop production that will reduce the cost of fertilizer for crop production and also minimize the impact to different ecosystems. Fertilizer nutrient use efficiency in crop production can be increased with effective use of nano-fertilizers. However, inhibitory effect on plants at higher concentration than the optimum may decrease growth and productivity. The efficacy for a nano fertilizer in a particular soil with a specific variety of crop is a complex task that needs to be investigated under different environmental conditions. Manufacturers, agricultural advisors and farmers when designing and applying a specific nano fertilizer for a particular crop need to look into safety concerns and environmental constraints of associated with nano fertilizers. This may be a challenge to transform agriculture by the use of nano technology. However, Agricultural technology should take benefit of the powerful tools of nanotechnology. The tools of nanotechnology can be employed to address the vital issues of environmental protection and pollution.

ACKNOWLEDGMENT

We are highly thankful to management ITM University, Gwalior for providing all the facilities for the present work.

REFERENCES

Acharya, S., & Dey, N. (2012). Implications of nanobiosensors in agriculture. *Journal of Biomaterials and Nanobiotechnology, 3*, 315.

Adesemoye, A. O., & Kloepper, J. W. (2009). Plant-microbes interactions in enhanced fertilizer-use efficiency. *Applied Microbiology and Biotechnology, 85*(1), 1–12. doi:10.100700253-009-2196-0 PMID:19707753

Adesemoye, A. O., Torbert, H. A., & Kloepper, J. W. (2009). Plant growth-promoting rhizobacteria allow reduced application rates of chemical fertilizers. *Microbial Ecology, 58*(4), 921–929. doi:10.100700248-009-9531-y PMID:19466478

Azeem, B. S., Kushaari, K., Man, Z. B., Basit, A., & Thanh, T. H. (2014). Review on materials & methods to produce controlled release coated urea fertilizer. *Journal of Controlled Release*, *181*, 11–21. doi:10.1016/j.jconrel.2014.02.020 PMID:24593892

Bao, X.-Y., Ali, A., Qiao, D.-L., Liu, H.-S., Chen, L., & Yu, L. (2015). Application Of Polymer Materials In Developing Slow/Control Release Fertilizer. *Gaofenzi Xuebao*, *9*, 1010–1019.

Barrena, R., Casals, E., Colon, J., Font, X., Sanchez, A., & Puntes, V. (2009). Evaluation of the ecotoxicity of model nanoparticles. *Chemosphere*, *75*(7), 850–857. doi:10.1016/j.chemosphere.2009.01.078 PMID:19264345

Baruah, S., & Dutta, J. (2009). Nanotechnology applications in pollution sensing and degradation in agriculture: A review. *Environmental Chemistry Letters*, *7*(3), 191–204. doi:10.100710311-009-0228-8

Bashan, Y., & de-Bashan, L. E. (2010). Microbial populations of arid lands and their potential for restoration of deserts. In P. Dion (Ed.), *Soil biology and agriculture in the tropics, Soil biology series* (pp. 109–137). Berlin: Springer. doi:10.1007/978-3-642-05076-3_6

Bernard, E., Larkin, R. P., Tavantzis, S., Erich, M. S., Alyokhin, A., Sewell, G., ... Gross, S. D. (2012). Compost, rapeseed rotation, and biocontrol agents significantly impact soil microbial communities in organic and conventional potato production systems. *Applied Soil Ecology*, *52*, 29–41. doi:10.1016/j.apsoil.2011.10.002

Bonkowski, M. (2004). Protozoa and plant growth: The microbial loop in soil revisited. *The New Phytologist*, *162*(3), 617–631. doi:10.1111/j.1469-8137.2004.01066.x

Boyd, C. E., & Tucker, C. S. (1998). *Pond aquaculture water quality management*. Norwell, MA: Kluwer. doi:10.1007/978-1-4615-5407-3

Burman, U., Saini, M., & Kumar, P. (2013). Effect of zinc oxide nanoparticles on growth and antioxidant system of chickpea seedlings. *Toxicological and Environmental Chemistry*, *95*(4), 605–616. doi:10.1080/02772248.2013.803796

Carpenter, S. R. (1981). Submersed vegetation: An internal factor in lake ecosystem succession. *American Naturalist*, *118*(3), 372–383. doi:10.1086/283829

Carpenter, S. R., Caraco, N. F., Correll, D. L., Howarth, R. W., Sharpley, A. N., & Smith, V. H. (1998). Nonpoint pollution of surface waters with phosphorus and nitrogen. *Ecological Applications*, *8*(3), 559–568. doi:10.1890/1051-0761(1998)008[0559:NPOSWW]2.0.CO;2

Chaney, R. L., & Oliver, D. P. (1996). *Sources, potential adverse effects and remediation of agricultural soil contaminants.* In R. Naidu (Ed.), *Contaminants and the Soil Environments in the Australia-Pacific Region* (pp. 323–359). Dordrecht, The Netherlands: Kluwer Academic Publishers.

Cui, H. X., Sun, C. J., Liu, Q., Jiang, J., & Gu, W. (2010). *Applications of nanotechnology in agrochemical formulation, perspectives, challenges and strategies.* International conference on Nanoagri, Sao Pedro, Brazil.

Dodds, W. K., Bouska, W. W., Eitzmann, J. L., Pilger, T. J., Pitts, K. L., Riley, A. J., ... Thornbrugh, D. J. (2009). Eutrophication of U.S. freshwaters: Analysis of potential economic damages. *Environmental Science & Technology, 43*(1), 12–19. doi:10.1021/es801217q PMID:19209578

Fuentes-Ramirez, L. E., & Caballero-Mellado, J. (2005). Bacterial biofertilizers. In Z. A. Siddiqui (Ed.), *PGPR: biocontrol and biofertilization* (pp. 143–172). Dordrecht: Springer.

Gao, F., Hong, F., Liu, C., Zhangm, L., Su, M., Wu, X., ... Yang, P. (2006). Mechanism of nano- anatase TiO_2 on promoting photosynthetic carbon reaction of spinach. *Biological Trace Element Research, 111*, 239–253. doi:10.1385/BTER:111:1:239 PMID:16943609

Ghosh, J. (2004). *Final Report of the Commission on Farmers' Welfare.* Retrieved from http://www.macroscan.org/pol/apr05/pdf/Full_Report_Commission_Farmer_AP.pdf

Hamid, N. N. A., Mohamad, N., Hing, L. Y., Dimin, M. F., Azam, M. A., Hassan, M. H. C., ... Ahmad, A. S. (2013). The effect of chitosan content to physical and degradation properties of biodegradable urea fertilizer. *International Journal of Innovation Science and Research, 2*, 893–902.

Horiuchi, J., Prithiviraj, B., Bais, H. P., Kimball, B. A., & Vivanco, J. M. (2005). Soil nematodes mediate positive interactions between legume plants and rhizobium bacteria. *Planta, 222*(5), 848–857. doi:10.100700425-005-0025-y PMID:16025342

Jeppesen, E., Jensen, J. P., Søndergaard, M., Lauridsen, T., Pedersen, L. J., & Jensen, L. (1997). Top-down control in freshwater lakes: The role of nutrient state, submerged macrophytes and water depth. *Hydrobiologia, 342/343*, 151–164. doi:10.1023/A:1017046130329

Jones, L. H. P., & Jarvis, S. C. (1981). *The fate of heavy metals in the Chemistry of Soil Processes.* New York: John Wiley & Sons.

Khodakovskaya, M., Dervishi, E., Mahmood, M., Xu, Y., Li, Z., Watanbe, F., & Biris, A. S. (2009). Carbon nanotubes are able to penetrate plant seed coat and dramatically affect seed germination and plant growth. American Chemical Society, 3, 3221–3227.

Khodakovskaya, M., Silva, K. D., Biris, A. S., Dervishi, E., & Villagarica, H. (2012). Carbon nanotubes induce growth enhancement in tobacco cells. *ACS Nano*, *6*(3), 2128–2135. doi:10.1021/nn204643g PMID:22360840

Lee, W., An, Y. J., Yoon, H., & Kweon, H. S. (2008). Toxicity and bioavailability of copper nanoparticles to the terrestrial plants mung bean (Phaseolus radiatus) and wheat (Triticum aestivum): Plant agar test for water-insoluble nanoparticles. *Environmental Toxicology and Chemistry*, *27*(9), 1915–1921. doi:10.1897/07-481.1 PMID:19086317

Liu, R., & Lal, R. (2015). Potentials of engineered nanoparticles as fertilizers for increasing agronomic productions. *The Science of the Total Environment*, *514*, 131–139. doi:10.1016/j.scitotenv.2015.01.104 PMID:25659311

Lu, C. M., Zhang, C. Y., Wu, J. Q., & Tao, M. X. (2002). Research of the effect of nanometer on germination and growth enhancement of Glycine max and its mechanism. *Dadou Kexue*, *21*, 168–172.

Mahajan, P., Dhoke, S. K., & Khanna, A. S. (2011). Effect of nano-ZnO particle suspension on growth of mung (Vigna radiata) and gram (Cicer arietinum) seedlings using plant agar method. *Journal of Nanotechnology*, *2011*, 1–7. doi:10.1155/2011/696535

Mahmoodzadeh, H., Nabavi, M., & Kashefi, H. (2013). Effect of nanoscale titanium dioxide particles on the germination and growth of canola (Brassica napus). *International Journal of Horticultural Science and Ornamental Plants*, *3*, 25–32.

Malusà, E., & Sas Paszt, L. (2009). The development of innovative technologies and products for organic fruit production. An integrated project. *Proceedings of international plant nutrition Coll. XVI*. Retrieved from http://repositories.cdlib.org/ipnc/xvi/1359,1–4

Markets and Markets. (2013). *Global Biofertilizers Market By Types, Applications And Geography – Trends An Forecasts To 2017*. Markets and Markets.

Maurice, P. A., & Hochella, M. F. (2008). Nanoscale particles and processes: A new dimension in soil science. *Advances in Agronomy*, *100*, 123–153. doi:10.1016/S0065-2113(08)00605-6

Mondal, A., Basu, R., Das, S., & Nandy, P. (2011). Beneficial role of carbon nanotubes on mustard plant growth: An agricul tural prospect. *Journal of Nanoparticle Research, 13*(10), 4519–4528. doi:10.100711051-011-0406-z

Moreno, M. L., Rosa, G., Herna'ndez-Viezcas, J. A., Castillo-Michel, H., Botez, C. E., Peralta-Videa, J. R., & Gardea-Torresdey, J. L. (2010). Evidence of the differential biotransformation and genotoxicity of ZnO and CeO2 nanoparticles on soybean (Glycine max) plants. *Environmental Science & Technology, 44*(19), 7315–7320. doi:10.1021/es903891g PMID:20384348

Mortvedt, J.J. (1995). Heavy metal contaminants in inorganic and organic fertilizers. *Fertilizer Research, 43*(1), 55–61.

Mulder, W.J., Gosselink, R.J.A., Vingerhoeds, M.H., Harmsen, P.F.H., & Eastham, D. (2011). Lignin based Controlled Release Coatings. *Industrial Crops and Products, 34*(1), 915-920.

Mura, S., Seddaiu, G., Bacchini, F., Roggero, P. P., & Greppi, G. F. (2013). Advances of nanotechnology in agro-environmental studies. *Italian Journal of Agronomy, 8*(18).

Naderi, M., & Shahraki, A. (2013). Nano technologies and their role in sustainable agriculture. *International Journal of Agriculture and Crop Sciences, 5*, 2229–2232.

Namasivayam, S. K. R., & Chitrakala, K. (2011). Ecotoxicological effect of Lecanicillium lecanii (Ascomycota: Hypocreales) based silver nanoparticles on growth parameters of economically important plants. *Journal of Biopesticides, 4*, 97–101.

Porter, K. G. (1977). The plant-animal interface in freshwater ecosystems. *American Scientist, 65*, 159–170.

Quoc Buu, N., Hien, D. T., Chau, H. N., Tin, X. T., Van, N. T., Duong, T. K., & Ha, T. H. (2014). Effects of nanocrystalline powders (Fe, Co and Cu) on the germination, growth, crop yield and product quality of soybean (Vietnamese species DT-51). *Advances in Natural Sciences: Nanoscience and Nanotechnology, 5*, 015016.

Raven, P. H., Berg, L. R., & Johnson, G. B. (1998). *Environment* (2nd ed.). New York: Saunders College Publishing.

Reinhold-Hurek, B., & Hurek, T. (2011). Living inside plants: Bacterial endophytes. *Current Opinion in Plant Biology, 14*(4), 435–443. doi:10.1016/j.pbi.2011.04.004 PMID:21536480

Rodriguez, E., Sultan, R., & Hilliker, A. (2004). Negative Effects of Agriculture on Our Environment. *The Traprock, 3*, 28–32.

Ronn, R., & McGraig, A. E. (2002). Impact of protozoan grazing on the bacterial community structure in soil microcosms. *Applied and Environmental Microbiology*, *68*(12), 6094–6105. doi:10.1128/AEM.68.12.6094-6105.2002 PMID:12450833

Ryan, R. P., Germaine, K., Franks, A., Ryan, D. J., & Dowling, D. N. (2008). Bacterial endophytes: Recent developments and applications. *FEMS Microbiology Letters*, *278*(1), 1–9. doi:10.1111/j.1574-6968.2007.00918.x PMID:18034833

Sahandi, S. M., Sorooshzadeh, A., Rezazadeh, H. S., & Naghdibadi, H. A. (2011). Effect of nanosilver and silver nitrate on seed yield of borage. *Journal of Medicinal Plants Research*, *5*, 706–710.

Sanders, R. (2012). *Fertilizer use responsible for increase in nitrous oxide in atmosphere*. Research, Science & Environment. Retrieved from http://news.berkeley.edu/2012/04/02/fertilizer-use-responsible-for-increase-in-nitrous-oxide-in-atmosphere

Schindler, D. W. (1974). Eutrophication and recovery in experimental lakes: Implications for lake management. *Science*, *174*(4139), 897–899. doi:10.1126cience.184.4139.897 PMID:17782381

Schindler, D. W. (2006). Recent advances in the understanding and management of eutrophication. *Limnology and Oceanography*, *51*(1part2), 356–363. doi:10.4319/lo.2006.51.1_part_2.0356

Sekar, S., & Kandavel, D. (2010). Interaction of plant growth promoting rhizobacteria (PGPR) and endophytes with medicinal plants– new avenues for phytochemicals. *Journal of Phytology*, *2*, 91–100.

Shah, V., & Belozerova, I. (2009). Influence of metal nanoparticles on the soil microbial community and germination of lettuce seeds. *Water, Air, and Soil Pollution*, *197*(1-4), 143–148. doi:10.100711270-008-9797-6

Siddhartha, S. (2014). Mukhopadhyay (2014). Nanotechnology in agriculture: Prospects and constraints. *Nanotechnology, Science and Applications*, *7*, 63–71. doi:10.2147/NSA.S39409 PMID:25187699

Singh, M D, Chirag, G., Om Prakash, P., Mohan, M.h., & Prakash, G., & Vishwajit. (2017). Nano fertilizers is a new way to increase nutrients use efficiency in crop production. *International Journal of Agricultural Sciences*, 3831–3833.

Sumner, M. E. (2000). Beneficial use of effluents, wastes, and biosolids. *Communications in Soil Science and Plant Analysis*, *31*(11), 1701–1715. doi:10.1080/00103620009370532

Suriyaprabha, R., Karunakaran, G., Yuvakkumar, R., Prabu, P., Rajendran, V., & Kannan, N. (2012). Growth and physiological responses of maize (Zea mays L.) to porous silica nanoparticles in soil. *Journal of Nanoparticle Research, 14*(12), 1294–1308. doi:10.100711051-012-1294-6

Trenkel, M.E. (2010). *Slow and Controlled-Release and Stabilized Fertilizers: An Option for Enhancing Nutrient Use Efficiency in Agriculture*. International Fertilizer Industry Association.

Tsai, B. S. (1986). *Continuous spouted bed process for sulphur-coating urea. Chemical And Biological Engineering (Vol. Master Of Applied Science)*. Vancouver: University Of British Columbia.

Zheng, L., Hong, F., Lu, S., & Liu, C. (2005). Effect of nano-TiO_2 on strength of naturally aged seeds and growth of spinach. *Biological Trace Element Research, 104*(1), 83–92. doi:10.1385/BTER:104:1:083 PMID:15851835

Zubek, S., Mielcarek, S., & Turnau, K. (2012). Hypericin and pseudohypericin concentrations of a valuable medicinal plant Hypericum perforatum L are enhanced by arbuscular mycorrhizal fungi. *Mycorrhiza, 22*(2), 149–156. doi:10.100700572-011-0391-1 PMID:21626142

ADDITIONAL READING

Arora, S., Sharma, P., Kumar, S., Nayan, R., Khanna, P. K., & Zaidi, M. G. H. (2012). Gold-nanopar ticle induced enhancement in growth and seed yield of Brassica juncea. *Plant Growth Regulation, 66*(3), 303–310. doi:10.100710725-011-9649-z

Barber, S. A. (1995). *Soil nutrient bioavailability: a mechanistic approach* (2nd ed.). New Jersey: Wiley Blackwell.

Borm, P. J. A., Robbins, D., Haubold, S., Kuhlbusch, T., Fissan, H., Donaldson, K., ... Oberdorster, E. (2006). The potential risks of nanomaterials: A review carried out for ECETOC. *Particle and Fibre Toxicology, 3*(1), 11. doi:10.1186/1743-8977-3-11 PMID:16907977

Brady, N. R., & Weil, R. R. (1999). In N. R. Brady & R. R. Weil (Eds.), *The nature and properties of soils* (pp. 415–473). New Jersey: Prentice Hall.

Comerford, N. B. (2005). Soil factors affecting nutrient bioavailability ecological studies. In H. Bassiri Rad (Ed.), *Nutrient acquisition by plants an ecological perspective* (pp. 1–14). Berlin: Springer. doi:10.1007/3-540-27675-0_1

Gao, F., Hong, F., Liu, C., Zhang, L., Su, M., Wu, X., ... Yang, P. (2006). Mechanism of nano- anatase TiO 2 on promoting photosynthetic carbon reaction of spinach. *Biological Trace Element Research*, *111*, 239–253. doi:10.1385/BTER:111:1:239 PMID:16943609

Gericke, W. F. (1937). Hydroponics – crop production in liquid culture media. *Science*, *85*(2198), 177–178. doi:10.1126cience.85.2198.177 PMID:17732930

Ghormade, V., Deshpande, M. V., & Paknikar, K. M. (2011). Perspectives for nano-biotechnology enabled protection and nutrition of plants. *Biotechnology Advances*, *29*(6), 792–803. doi:10.1016/j.biotechadv.2011.06.007 PMID:21729746

Larue, C., Khodja, H., Herlin-Boime, N., Brisset, F., Flank, A. M., Fayard, B., ... Carrière, M. (2011). Investigation of titanium dioxide nanoparticles toxicity and uptake by plants. *Journal of Physics: Conference Series*, *304*, 012057. doi:10.1088/1742-6596/304/1/012057

Loomis, R. S., & Connor, D. J. (1992). *Crop ecology: productivity and management in agricultural systems* (p. 538). Cambridge: Cambridge University Press. doi:10.1017/CBO9781139170161

Schulte, P. A., & Salamanca-Buentello, F. (2007). Ethical and scientific issues of nanotechnology in the workplace. *Environmental Health Perspectives*, *115*(1), 5–12. doi:10.1289/ehp.9456 PMID:17366812

KEY TERMS AND DEFINITIONS

Fertilizer Encapsulation: Encapsulation of fertilizers in polymeric coatings is a method used to reduce fertilizer losses and to minimize environmental pollution.

Nanotechnology: Originally used to define any work done on the molecular scale, or one billionth of a meter. This term is now used broadly (and loosely) for anything that is really small (usually smaller than a micrometer).

Smart Materials and Products: Here, materials and products capable of relatively complex behavior due to the incorporation of nanocomputers and nanomachines. Also used for products having some ability to respond to the environment.

Compilation of References

Abbas, A. H., & Mohammed, M. M. (2014). Smart Watering System for Gardens using Wireless Sensor Networks. *Proceedings of International Conference on Engineering and Technology*, 1-5 10.1109/ICEngTechnol.2014.7016780

Abdel Salam, H. S., & Olariu, S. (2011). Toward efficient task management in wireless sensor networks. *IEEE Transactions on Computers*, *60*(11), 1638–1651. doi:10.1109/TC.2010.264

Abdullahi, H. S., Mahieddine, F., & Sheriff, R. E. (2015, July). Technology impact on agricultural productivity: a review of precision agriculture using unmanned aerial vehicles. In *International Conference on Wireless and Satellite Systems* (pp. 388-400). Springer. 10.1007/978-3-319-25479-1_29

Abidine, A. Z., Heidman, B. C., Upadhyaya, S. K., & Hills, D. J. (2002). Application of RTK GPS based auto-guidance system in agricultural production. The American Society of Agricultural and Biological Engineers Paper, (21152).

Acharya, S., & Dey, N. (2012). Implications of nanobiosensors in agriculture. *Journal of Biomaterials and Nanobiotechnology, 3*, 315.

Adamchuk, V. I., & Rossel, R. A. V. (2011). Precision agriculture: proximal soil sensing. In *Encyclopedia of Agrophysics* (pp. 650–656). Springer. doi:10.1007/978-90-481-3585-1_126

Adesemoye, A. O., & Kloepper, J. W. (2009). Plant-microbes interactions in enhanced fertilizer-use efficiency. *Applied Microbiology and Biotechnology*, *85*(1), 1–12. doi:10.100700253-009-2196-0 PMID:19707753

Adesemoye, A. O., Torbert, H. A., & Kloepper, J. W. (2009). Plant growth-promoting rhizobacteria allow reduced application rates of chemical fertilizers. *Microbial Ecology*, *58*(4), 921–929. doi:10.100700248-009-9531-y PMID:19466478

Agency, I. A. (2008). Field Estimation of Soil. Vienna: Academic Press.

Aguilar, J., Rogers, D., & Kisekka, I. (2015). Irrigation Scheduling Based on Soil Moisture Sensors and Evapotranspiration. *Kansas Agricultural Experiment Station Research Reports*, *1*(5), 20. doi:10.4148/2378-5977.1087

Akshay, S., Aishwarya, B. P., & Dr, T. H. (2016). Srinivas (2016), Smart Farm - An Android Based Monitoring System for Agricultural Land. *International Journal of Modern Trends in Engineering and Research*, *03*(04).

Akyildiz, I. F., Su, W., Sankarasubramaniam, Y., & Cayirci, E. (2002). Wireless sensor networks: A survey. *Computer Networks*, *38*(4), 393–422. doi:10.1016/S1389-1286(01)00302-4

Alam, K. M., Kamruzzaman, J., Karmakar, G., & Murshed, M. (2011, August). Dynamic event coverage in hybrid wireless sensor networks. In *Network Computing and Applications (NCA), 2011 10th IEEE International Symposium on* (pp. 348-353). IEEE. 10.1109/NCA.2011.62

Al-Fuqaha, A., Guizani, M., Mohammadi, M., Aledhari, M., & Ayyash, M. (2015). Internet of things: A survey on enabling technologies, protocols, and applications. *IEEE Communications Surveys and Tutorials*, *17*(4), 2347–2376. doi:10.1109/COMST.2015.2444095

Ammad Uddin, M., Mansour, A., & Le Jeune, D. (2017). Agriculture internet of things: AG-IoT. In *2017- 27th International Telecommunication Networks and Applications Conference (ITNAC)*. IEEE. DOI: 10.1109/ATNAC.2017.8215399

Andrews, L. J. B., & Raja, L. (2017, December). Remote based patient monitoring system using wearable sensors through online and offline mode for android based mobile platforms. In *Infocom Technologies and Unmanned Systems (Trends and Future Directions) (ICTUS), 2017 International Conference on* (pp. 602-606). IEEE. 10.1109/ICTUS.2017.8286080

Andrews, L. J. B., & Raja, L. (2017). A study on m-health inline with the sensors applying for a real time environment. *Journal of Statistics and Management Systems*, *20*(4), 659–667. doi:10.1080/09720510.2017.1395186

ASBAE Standards, . (2009). *Agricultural Machinery Management Data*. St. Joseph, MI: Author.

Ashton. (2009). Internet of Things. *RFiD Journal*.

Atzori, L., Iera, A., & Morabito, G. (2010). The internet of things: A survey. *Computer Networks*, *54*(15), 2787–2805. doi:10.1016/j.comnet.2010.05.010

Austin, R. F., DiSera, D. P., & Brooks, T. J. (2015). *GIS for Critical Infrastructure Protection*. CRC Press. doi:10.1201/b18839

Awasthi, A., & Reddy, S. R. N. (2013). *Monitoring for precision agriculture using wireless sensor network-a review*. Global Journal of Computer Science and Technology.

Azeem, B. S., Kushaari, K., Man, Z. B., Basit, A., & Thanh, T. H. (2014). Review on materials & methods to produce controlled release coated urea fertilizer. *Journal of Controlled Release*, *181*, 11–21. doi:10.1016/j.jconrel.2014.02.020 PMID:24593892

Azzola, F. (2017). *IoT System To Monitor Soil Moisture With Arduino*. Retrieved from https://create.arduino.cc/projecthub/jfrankie/iot-system-to-monitor-soil-moisture-with-arduino-5e370c

Backman, J., Oksanen, T., & Visala, A. (2013). Applicability of the ISO 11783 network in a distributed combined guidance system for agricultural machines. *Biosystems Engineering*, *114*(3), 306–317. doi:10.1016/j.biosystemseng.2012.12.017

Balafoutis, A. T., Beck, B., Fountas, S., Tsiropoulos, Z., Vangeyte, J., van der Wal, T., … Pedersen, S. M. (2017). Smart Farming Technologies–Description, Taxonomy and Economic Impact. In Precision Agriculture: Technology and Economic Perspectives (pp. 21–77). Springer.

Balasaheb. (2011). *A study of export marketing management of grapes with special reference to Nasik district* (Doctoral Thesis). Retrieved from http://shodhganga.inflibnet.ac.in/handle/10603/92452

Bandyopadhyay, D., & Sen, J. (2011). Internet of things: Applications and challenges in technology and standardization. *Wireless Personal Communications*, *58*(1), 49–69. doi:10.100711277-011-0288-5

Bao, X.-Y., Ali, A., Qiao, D.-L., Liu, H.-S., Chen, L., & Yu, L. (2015). Application Of Polymer Materials In Developing Slow/Control Release Fertilizer. *Gaofenzi Xuebao*, *9*, 1010–1019.

Barrena, R., Casals, E., Colon, J., Font, X., Sanchez, A., & Puntes, V. (2009). Evaluation of the ecotoxicity of model nanoparticles. *Chemosphere*, *75*(7), 850–857. doi:10.1016/j.chemosphere.2009.01.078 PMID:19264345

Baruah, S., & Dutta, J. (2009). Nanotechnology applications in pollution sensing and degradation in agriculture: A review. *Environmental Chemistry Letters*, *7*(3), 191–204. doi:10.100710311-009-0228-8

Bashan, Y., & de-Bashan, L. E. (2010). Microbial populations of arid lands and their potential for restoration of deserts. In P. Dion (Ed.), *Soil biology and agriculture in the tropics, Soil biology series* (pp. 109–137). Berlin: Springer. doi:10.1007/978-3-642-05076-3_6

Beckerman, W. (1994). 'Sustainable development': Is it a useful concept? *Environmental Values*, *3*(3), 191–209. doi:10.3197/096327194776679700

Bernard, E., Larkin, R. P., Tavantzis, S., Erich, M. S., Alyokhin, A., Sewell, G., ... Gross, S. D. (2012). Compost, rapeseed rotation, and biocontrol agents significantly impact soil microbial communities in organic and conventional potato production systems. *Applied Soil Ecology*, *52*, 29–41. doi:10.1016/j.apsoil.2011.10.002

Bishop, A. A., Jensen, M. E., & Hall, W. A. (1967). *Surface irrigation systems*. Academic Press.

Bochtis, D. D., & Sørensen, C. G. (2009). The vehicle routing problem in field logistics part I. *Biosystems Engineering*, *104*(4), 447–457. doi:10.1016/j.biosystemseng.2009.09.003

Boisot, M. H. (1998). *Knowledge assets: Securing competitive advantage in the information economy*. OUP Oxford.

Bongiovanni, R., & Lowenberg-DeBoer, J. (2004). Precision agriculture and sustainability. *Precision Agriculture*, *5*(4), 359–387. doi:10.1023/B:PRAG.0000040806.39604.aa

Bonkowski, M. (2004). Protozoa and plant growth: The microbial loop in soil revisited. *The New Phytologist*, *162*(3), 617–631. doi:10.1111/j.1469-8137.2004.01066.x

Bornstein, D. (2007). *How to change the world: Social entrepreneurs and the power of new ideas*. Oxford University Press.

Bouabdallah, F., Bouabdallah, N., & Boutaba, R. (2009). On balancing energy consumption in wireless sensor networks. *IEEE Transactions on Vehicular Technology*, *58*(6), 2909–2924. doi:10.1109/TVT.2008.2008715

Boyd, C. E., & Tucker, C. S. (1998). *Pond aquaculture water quality management*. Norwell, MA: Kluwer. doi:10.1007/978-1-4615-5407-3

Brock. (2002). *Integrating the Electronic Product Code (EPC) and the Global Trade Item Number (GTIN), Updated edition*. MIT Auto-Id center.

Bui, T., & Lee, J. (1999). An agent-based framework for building decision support systems. *Decision Support Systems*, *25*(3), 225–237. doi:10.1016/S0167-9236(99)00008-1

Burman, U., Saini, M., & Kumar, P. (2013). Effect of zinc oxide nanoparticles on growth and antioxidant system of chickpea seedlings. *Toxicological and Environmental Chemistry*, *95*(4), 605–616. doi:10.1080/02772248.2013.803796

Byerlee, D., & Alex, G. E. (1998). *Strengthening national agricultural research systems: policy issues and good practice* (Vol. 24). World Bank Publications. doi:10.1596/0-8213-4173-1

Caffrey, K., & Veal, M. (2013). Conducting an Agricultural Life Cycle Assessment: Challenges and Perspectives. *The Scientific World Journal*, *2013*, 1–13. doi:10.1155/2013/472431 PMID:24391463

Calderón, R., Navas-Cortés, J. A., Lucena, C., & Zarco-Tejada, P. J. (2013). High-resolution airborne hyperspectral and thermal imagery for early detection of Verticillium wilt of olive using fluorescence, temperature and narrow-band spectral indices. *Remote Sensing of Environment*, *139*, 231–245. doi:10.1016/j.rse.2013.07.031

Carpenter, S. R. (1981). Submersed vegetation: An internal factor in lake ecosystem succession. *American Naturalist, 118*(3), 372–383. doi:10.1086/283829

Carpenter, S. R., Caraco, N. F., Correll, D. L., Howarth, R. W., Sharpley, A. N., & Smith, V. H. (1998). Nonpoint pollution of surface waters with phosphorus and nitrogen. *Ecological Applications, 8*(3), 559–568. doi:10.1890/1051-0761(1998)008[0559:NPOSWW]2.0.CO;2

Central Statistics Office, Ministry of Statistics and Programme Implementation Government of India New Delhi. (2016). *Energy Statistics., 23*, 39–52.

Chakravarti, A. K. (1973). Green revolution in India. *Annals of the Association of American Geographers, 63*(3), 319–330. doi:10.1111/j.1467-8306.1973.tb00929.x

Chandra, A., McNamara, K. E., & Dargusch, P. (2018). Climate-smart agriculture: Perspectives and framings. *Climate Policy, 18*(4), 526–541. doi:10.1080/14693062.2017.1316968

Chaney, R. L., & Oliver, D. P. (1996). *Sources, potential adverse effects and remediation of agricultural soil contaminants.* In R. Naidu (Ed.), *Contaminants and the Soil Environments in the Australia-Pacific Region* (pp. 323–359). Dordrecht, The Netherlands: Kluwer Academic Publishers.

Chang, H., Zhou, N., Zhao, X., Cao, Q., Tan, M., & Zhang, Y. (2014). *A new agriculture monitoring system based on WSNs*. Paper presented at 2014 12th International Conference on Signal Processing (ICSP), Hangzhou, China. 10.1109/ICOSP.2014.7015295

Charan, S. (2017). *Enabling Innovation in Agriculture and Improving Access to Key Service by IOT (Internet of Things).* Gateway.

Chatterjee, A., & Mukherjee, D. (2014, January). Reliable multipath wireless sensor network routing protocol scheme for network lifetime maximization. In *Business and Information Management (ICBIM), 2014 2nd International Conference on* (pp. 64-68). IEEE. 10.1109/ICBIM.2014.6970935

Chavan, C. H., & Karande, M. P. (2014). Wireless monitoring of soil moisture, temperature & humidity using zigbee in agriculture. *Int. J. Eng. Trends Technol, 11*(10), 493–497. doi:10.14445/22315381/IJETT-V11P296

Chen, H., Jia, X., & Li, H. (2011, October). A brief introduction to IoT gateway. In *Communication Technology and Application (ICCTA 2011), IET International Conference on* (pp. 610-613). IET.

Chen, J., Du, R., Wang, Q., & Yao, S. (2013, July). Secure routing based on network coding in wireless sensor networks. In *Trust, Security and Privacy in Computing and Communications (TrustCom), 2013 12th IEEE International Conference on* (pp. 58-64). IEEE. 10.1109/TrustCom.2013.11

Chen, J., Zhou, M., Li, D., & Sun, T. (2008, November). A priority based dynamic adaptive routing protocol for wireless sensor networks. In *Intelligent Networks and Intelligent Systems, 2008. ICINIS'08. First International Conference on* (pp. 160-164). IEEE. 10.1109/ICINIS.2008.171

Chen, Y. (2015). *Routing algorithm dedicated to environmental data collection: precision agriculture* (Doctoral dissertation). University Blaise Pascal-Clermont-Ferrand II.

Cheng, J. (2009). *Biomass to Renewable Energy Processes*. CRC Press.

Christian, D. G. (2000). Biomass for Renewable Energy, Fuels, and Chemicals. *Journal of Environmental Quality*, *29*(2), 662. doi:10.2134/jeq2000.00472425002900020040x

Clean Water Action Council (CWAC). (2014). *Population facts the future*. Retrieved from http://www.cleanwateractioncouncil.org/issues/resource-issues/population

Cooklev, T. (2004). *Wireless Communication Standards: A Study of IEEE 802.11, 802.15, 802.16*. IEEE Standards Association.

Countries with their GDP Components. (2017). Retrieved from https://www.cia.gov/library/publications/the-world-factbook/fields/2012.html

Csikszentmihalyi, M. (2004). *Good business: Leadership, flow, and the making of meaning*. Penguin.

Cui, H. X., Sun, C. J., Liu, Q., Jiang, J., & Gu, W. (2010). *Applications of nanotechnology in agrochemical formulation, perspectives, challenges and strategies*. International conference on Nanoagri, Sao Pedro, Brazil.

Dadhich, M., Rathore, V. S., & Vyas, S. (2017, December). Comparison and selection of SaaS service providers using ASMAN framework. In Infocom Technologies and Unmanned Systems (Trends and Future Directions) (ICTUS), 2017 International Conference on (pp. 568-572). IEEE.

Damas, M., Prados, A. M., Gomez, F., & Oliyares, G. (2001). Hydro Bus system: Fieldbus for integrated management of extensive areas of irrigated land. *Microprocessors and Microsystems*, *25*(3), 177–184. doi:10.1016/S0141-9331(01)00110-7

Daniele Zaccaria, S. O. (2014). *Advantages and Constraints of Soil Moisture Monitoring*. Academic Press.

Dargie, W., & Poellabauer, C. (2010). *Fundamentals of Wireless Sensor Networks*. Chichester, UK: John Wiley & Sons Ltd. doi:10.1002/9780470666388

Dar, K., Bakhouya, M., Gaber, J., Wack, M., & Lorenz, P. (2010). Wireless communication technologies for ITS applications. *IEEE Communications Magazine*, *48*(5), 156–162. doi:10.1109/MCOM.2010.5458377

Davis, S. L., & Dukes, M. D. (2014). Methodologies for successful implementation of smart irrigation controllers. *Journal of Irrigation and Drainage Engineering*, *141*(3), 04014055. doi:10.1061/(ASCE)IR.1943-4774.0000804

de Lima, G. H., Silva, L. C., & Neto, P. F. (2010, March). WSN as a Tool for Supporting Agriculture in the Precision Irrigation. In *Networking and Services (ICNS), 2010 Sixth International Conference on* (pp. 137-142). IEEE 10.1109/ICNS.2010.26

Demir, A. (2017). *Remote Monitoring of the Soil Quality with the Internet of Things*. Retrieved from https://sites.tufts.edu/eeseniordesignhandbook/files/2017/05/Azure_Demir_F1.pdf

Demirbas, A. (2008). Biofuels sources, biofuel policy, biofuel economy and global biofuel projections. *Energy Conversion and Management*, *49*(8), 2106–2116. doi:10.1016/j.enconman.2008.02.020

Dhaka, V. S., Poonia, R. C., & Raja, L. (2013). A literature review on dedicated short range communication for intelligent transport. *International Journal of Advanced Research in Computer Science and Software Engineering*, *3*(9), 1066–1071.

Dhaka, V. S., Poonia, R. C., & Raja, L. (2014). The Realistic Mobility Evaluation of Vehicular Ad-Hoc Network for Indian Automotive Networks. International Journal of Ad hoc. *Sensor & Ubiquitous Computing*, *5*(2), 1. doi:10.5121/ijasuc.2014.5201

Dhaka, V. S., & Vyas, S. (2014). Analysis of Server Performance with Different Techniques of Virtual Databases. *Journal of Emerging Trends in Computing and Information Sciences*, *5*(10).

Dillehay, T. D., Eling, H. H., & Rossen, J. (2005). Preceramic irrigation canals in the Peruvian Andes. *Proceedings of the National Academy of Sciences of the United States of America*, *102*(47), 17241–17244. doi:10.1073/pnas.0508583102 PMID:16284247

Dodds, W. K., Bouska, W. W., Eitzmann, J. L., Pilger, T. J., Pitts, K. L., Riley, A. J., ... Thornbrugh, D. J. (2009). Eutrophication of U.S. freshwaters: Analysis of potential economic damages. *Environmental Science & Technology*, *43*(1), 12–19. doi:10.1021/es801217q PMID:19209578

dsPIC® Digital Signal Controllers. (n.d.). Access on 13 April 2018, from http://www.mouser.com/ds/2/268/00001032Q-1131382.pdf

Dukes, M. D., Zotarelli, L., Liu, G. D., & Simonne, E. H. (2012). *Principles and practices of irrigation management for vegetables*. Academic Press.

Dukes, M. D., & Scholberg, J. M. (2005). Soil moisture controlled subsurface drip irrigation on sandy soils. *Applied Engineering in Agriculture*, *21*(1), 89–101. doi:10.13031/2013.17916

Dukes, M. D., Shedd, M., & Cardenas-Lailhacar, B. (2009). *Smart Irrigation Controllers: How Do Soil Moisture Sensor (SMS) Irrigation Controllers Work?* University of Florida, Institute of Food and Agricultural Sciences.

Dunkels, Gronvall, & Voigt. (2004). Contiki - A lightweight and flexible operating system for tiny networked sensors. *Local Computer Networks, 2004. 29th Annual IEEE International Conference on*, 455 – 462.

Ekpeni, L. (2014). Energy Diversity through Renewable Energy Source (RES) – A Case Study of Biomass. *Energy Procedia*, 61, 1740–1747. 10.1016/j.egypro.2014.12.202

European Comission. (2007). Renewable Energy Road Map. Renewable energies in the 21st century: building a more sustainable future. Brussels, Belgium: Author.

Export and Import of Indian Agriculture Commodities. (2018). Retrieved from https://www.export.gov/article?id=India-Agricultural-Commodities

Factor Affect Agriculture. (2018). Retrieved from http://eagri.tnau.ac.in/eagri50/AGRO101/lec09.pdf.9

FAO. (2011a). *Climate change, water, and food security*. Rome: FAO.

FAO. (2013b). *Food Balance Sheet*. FAO.

Ferreira, D., Corista, P., Gião, J., Ghimire, S., Sarraipa, J., & Jardim-Gonçalves, R. (2017, June). Towards smart agriculture using FIWARE enablers. In *Engineering, Technology and Innovation (ICE/ITMC), 2017 International Conference on* (pp. 1544-1551). IEEE. 10.1109/ICE.2017.8280066

Finck, A. (1982). *Fertilizers and fertilization: introduction and practical guide to crop fertilization*. Verlag Chemie.

Fountas, S., Blackmore, S., Ess, D., Hawkins, S., Blumhoff, G., Lowenberg-Deboer, J., & Sorensen, C. G. (2005). Farmer experience with precision agriculture in Denmark and the US Eastern Corn Belt. *Precision Agriculture*, *6*(2), 121–141. doi:10.100711119-004-1030-z

Fountas, S., Wulfsohn, D., Blackmore, B. S., Jacobsen, H. L., & Pedersen, S. M. (2006). A model of decision-making and information flows for information-intensive agriculture. *Agricultural Systems*, *87*(2), 192–210. doi:10.1016/j.agsy.2004.12.003

Fritz Westover, K. B. (2014). *Using Soil Moisture Sensors for Vineyard Irrigation Management*. Atascadero, CA: Vineyard Team.

Fuentes-Ramirez, L. E., & Caballero-Mellado, J. (2005). Bacterial biofertilizers. In Z. A. Siddiqui (Ed.), *PGPR: biocontrol and biofertilization* (pp. 143–172). Dordrecht: Springer.

Gadgil, S. (2013). *The Indian monsoon and its variability Clouds, climate and Tropical Meteorology. ICTS*. CAOS.

Gao, F., Hong, F., Liu, C., Zhangm, L., Su, M., Wu, X., ... Yang, P. (2006). Mechanism of nano- anatase TiO_2 on promoting photosynthetic carbon reaction of spinach. *Biological Trace Element Research, 111*, 239–253. doi:10.1385/BTER:111:1:239 PMID:16943609

Geography of Rajasthan. (2017). Retrieved from http://niti.gov.in/writereaddata/files/Rajasthan_Presentation_0.pdf

Gerber, A. (2017). *Choosing the best hardware for your next IoT project.* Retrieved from https://www.ibm.com/developerworks/library/iot-lp101-best-hardware-devices-iot-project/index.html

Gerber, A. (2017). *Making sense of IoT data.* Retrieved from https://www.ibm.com/developerworks/library/iot-lp301-iot-manage-data/index.html

Ghosh, J. (2004). *Final Report of the Commission on Farmers' Welfare.* Retrieved from http://www.macroscan.org/pol/apr05/pdf/Full_Report_Commission_Farmer_AP.pdf

Goh, H. G., Kwong, K. H., Michie, C., & Andonovic, I. (2008, August). Performance evaluation of priority packet for wireless sensor network. In *Sensor Technologies and Applications, 2008. SENSORCOMM'08. Second International Conference on* (pp. 494-499). IEEE. 10.1109/SENSORCOMM.2008.90

Goldsmith, A. (2005). *Wireless communications.* Cambridge University Press. doi:10.1017/CBO9780511841224

Gondchawar & Kawitkar. (2016). IoT based Smart Agriculture. *International Journal of Advanced Research in Computer and Communication Engineering*, *5*(6).

Gora, A., & Dulawat, M. S. (2017). Solar powered smart irrigation. *Research Journal of Agriculture and Forest*, *5*(6), 15–19.

Government of India. (2017). *Department of Fertilization.* Monthly Bulletin September.

Government of Rajasthan, Department of Agriculture. (2016-17). *Administrative Progress Report 2016-17.* Author.

Gubbi, J., Buyya, R., Marusic, S., & Palaniswami, M. (2013). Internet of Things (IoT): A vision, architectural elements, and future directions. *Future Generation Computer Systems*, *29*(7), 1645–1660. doi:10.1016/j.future.2013.01.010

Guerrini, F. (2015). *The Future of Agriculture? Smart Farming*. Retrieved from https://www.forbes.com/

Guharkata, P. M. (2006). Trends in the Rainfall Pattern. Pune: National Climate Centre.

Gupta, H. P., Rao, S. V., & Venkatesh, T. (2013, January). Sleep scheduling for partial coverage in heterogeneous wireless sensor networks. In *Communication Systems and Networks (COMSNETS), 2013 Fifth International Conference on* (pp. 1-10). IEEE. 10.1109/COMSNETS.2013.6465580

Gutiérrez, J. F., Villa-Medina, A., & Porta-Gándara, M. A. (2014). Automated irrigation system using a wireless sensor network and GPRS Module. *IEEE Transactions on Instrumentation and Measurement*, *63*(1), 166–176. doi:10.1109/TIM.2013.2276487

Gu, X., Chai, Y., Liu, Y., Shen, J., Huang, Y., & Nan, Y. (2017). A MCIN-based architecture of smart agriculture. *International Journal of Crowd Science*, *1*(3), 237–248. doi:10.1108/IJCS-08-2017-0017

Hac, A. (2003). *Wireless sensor network designs*. John Wiley & Sons Ltd. doi:10.1002/0470867388

Hamelinck, C., Suurs, R., & Faaij, A. (2005). International bioenergy transport costs and energy balance. *Biomass and Bioenergy*, *29*(2), 114–134. doi:10.1016/j.biombioe.2005.04.002

Hamid, N. N. A., Mohamad, N., Hing, L. Y., Dimin, M. F., Azam, M. A., Hassan, M. H. C., ... Ahmad, A. S. (2013). The effect of chitosan content to physical and degradation properties of biodegradable urea fertilizer. *International Journal of Innovation Science and Research*, *2*, 893–902.

Harishankar, S., Kumar, R. S., Sudharsan, K. P., Vignesh, U., & Viveknath, T. (2014). Solar powered smart irrigation system. *Advance in Electronic and Electric Engineering*, *4*(4), 341–346.

Havlak, R. D. (2012). *Lawn Irrigation: How, When and Where to Water!* Texas AgriLife Extension.

Helgadottir, A., & Hopkins, A. (Eds.). (2013). The role of grasslands in a green future: threats and perspectives in less favoured areas. In *Proceedings of the 17th Symposium of the European Grassland Federation*. Borgarnes: Agricultural University of Iceland (Grassland Science in Europe).

Herrera, J. L. (1999). *Handbook of Agriculture* (1st ed.). Taylor And Francis.

Hess, K. (2014). *Five security considerations for IoT implementations*. Retrieved from http://www.zdnet.com/article/five-security-considerations-for-iot-implementations/

Hill, R. W., Patterson, R., & Barnhill, J. V. (2008). Small Acreage Irrigation System Operation and Maintenance. Utah State University -Extension Service. AG/Small Acreage/ 2008-01pr.

Hillel, D. (1994). *Rivers of Eden. The struggle for water and the quest for peace in the Middle East*. Academic Press.

Hillel, D., Braimoh, A. K., & Vlek, P. L. G. (2008). Soil Degradation under Irrigation. In A. K. Braimoh & P. L. G. Vlek (Eds.), *Land Use and Soil Resources*. Dordrecht: Springer. doi:10.1007/978-1-4020-6778-5_6

Honess, D. (2015). Raspberry Pi Weather Station for schools [Blog post]. Retrieved from https://www.raspberrypi.org/blog/school-weather-station-project/

Horiuchi, J., Prithiviraj, B., Bais, H. P., Kimball, B. A., & Vivanco, J. M. (2005). Soil nematodes mediate positive interactions between legume plants and rhizobium bacteria. *Planta*, *222*(5), 848–857. doi:10.100700425-005-0025-y PMID:16025342

How robotics is transforming 21st century farming. (n.d.). Retrieved from http://www.wired.co.uk/article/farmville-gets-an-upgrade

Hsieh, C. M., Samie, F., Srouji, M. S., Wang, M., Wang, Z., & Henkel, J. (2014, October). Hardware/software co-design for a wireless sensor network platform. In *Hardware/Software Codesign and System Synthesis (CODES+ ISSS), 2014 International Conference on* (pp. 1-10). IEEE.

Huisman, W., Venturi, P., & Molenaar, J. (1997). Costs of supply chains of *Miscanthus giganteus*. *Industrial Crops and Products*, *6*(3), 353–366. doi:10.1016/S0926-6690(97)00026-5

IEEE 802.15.4. (2011). IEEE Standard for Local and metropolitan area networks--Part 15.4: Low-Rate Wireless Personal Area Networks (LR-WPANs), (pp.314).

India, Government. (2014). *Agricultural Statistics*. New Delhi: Oxford University Press.

India, Government. (2015). *Statistics Related to Climate Change India - 2015*. New Delhi: Ministry of Statistics and Programme Implementation.

Indian Fertilization Pattern. (2017). Retrieved from http://fert.nic.in/sites/default/files/Indian%20Fertilizer%20SCENARIO-2014.pdf

Indian Union Budget. (2018). Retrieved from http://www.indiabudget.gov.in/

Internet of Things Irrigation System. (2018). Retrieved from http://www.microchip.com/design-centers/internet-of-things/applications-pages/irrigation-system

Islam, M. A., & Huosong, X. (2016). Can Corporate Social Responsibility (CSR) Ensure the Sustainable Growth of Ready Made Garments (RMG) Sector in Bangladesh. *World Journal of Social Science Research*, *3*(4), 615. doi:10.22158/wjssr.v3n4p615

Izadi, D., Abawajy, J., & Ghanavati, S. (2015). An alternative node deployment scheme for WSNs. *IEEE Sensors Journal*, *15*(2), 667–675. doi:10.1109/JSEN.2014.2351405

Jahnavi, V. S., & Ahamed, S. F. (2015). Smart wireless sensor network for automated greenhouse. *Journal of the Institution of Electronics and Telecommunication Engineers*, *61*(2), 180–185. doi:10.1080/03772063.2014.999834

Jarchow, M. E., Kubiszewski, I., Larsen, G. D., Zdorkowski, G., Costanza, R., Gailans, S. R., & Petrehn, M. R. (2012). The future of agriculture and society in Iowa: Four scenarios. *International Journal of Agricultural Sustainability*, *10*(1), 76–92. doi:10.1080/14735903.2012.646730

Jeppesen, E., Jensen, J. P., Søndergaard, M., Lauridsen, T., Pedersen, L. J., & Jensen, L. (1997). Top-down control in freshwater lakes: The role of nutrient state, submerged macrophytes and water depth. *Hydrobiologia*, *342/343*, 151–164. doi:10.1023/A:1017046130329

Jiber, Y., Harroud, H., & Karmouch, A. (2011, July). Precision agriculture monitoring framework based on WSN. In *Wireless Communications and Mobile Computing Conference (IWCMC), 2011 7th International* (pp. 2015-2020). IEEE. 10.1109/IWCMC.2011.5982844

Jin, Song, Han, Wang, & Yan. (2007). School of Electric and Electronic Engineering, A Remote Measurement and Control System for Greenhouse Based on GSM-SMS. *IEEE 8th International Conference on Electronic Measurement and Instrument*, 45-82.

Jones, L. H. P., & Jarvis, S. C. (1981). *The fate of heavy metals in the Chemistry of Soil Processes*. New York: John Wiley & Sons.

Jose, L., & Chavez, S. R. (2012). using soil water sensors to improve to improve irrigation management. *Proceedings of the 24th Annual Central Plains Irrigation Conference*.

Jungbluth, N., Büsser, S., & Frischknecht, R. (2008). *Life Cycle Assessment of Biomass to Liquid Fuels*. ESU-services Ltd.

Kainrath, D. (2009). *Ecopreneurship in theory and practice: A proposed emerging framework for ecopreneurship*. Academic Press.

Kanjilal, Singh, & Reddy. (2014). Smart Farm: Extending Automation To The Farm Level. *International Journal of Scientific & Technology Research*, *3*(7).

Kaplan, S., & Kaplan, R. (2009). Creating a larger role for environmental psychology: The Reasonable Person Model as an integrative framework. *Journal of Environmental Psychology*, *29*(3), 329–339. doi:10.1016/j.jenvp.2008.10.005

Karagiannis, V., & Chatzimisios, P. (2015). A survey on application layer protocols for the internet of things. *Transaction on IoT and Cloud Computing*, *3*(1), 11–17.

Karan, K. (2015). Sensor-based automated irrigation system with IoT: A technical review. *International Journal of Computer Science and Information Technologies*, *6*, 520–524.

Karl, H., & Willig, A. (2007). *Protocols and architectures for wireless sensor networks*. John Wiley & Sons.

Kavita, B., & Chaudhari, D. (2016). Solar powered sensor base irrigation system. *Internationnal Research Journal of Engineering and Technology*, *3*, 678–682.

Keshtgari, M., & Deljoo, A. (2011). *A wireless sensor network solution for precision agriculture based on zigbee technology*. Academic Press.

Khan, R., Khan, S. U., Zaheer, R., & Khan, S. (2012, December). Future internet: the internet of things architecture, possible applications and key challenges. In *Frontiers of Information Technology (FIT), 2012 10th International Conference on* (pp. 257-260). IEEE. Retrieved from https://techbeacon.com/4-stages-iot-architecture

Khodakovskaya, M., Dervishi, E., Mahmood, M., Xu, Y., Li, Z., Watanbe, F., & Biris, A. S. (2009). Carbon nanotubes are able to penetrate plant seed coat and dramatically affect seed germination and plant growth. American Chemical Society, 3, 3221–3227.

Khodakovskaya, M., Silva, K. D., Biris, A. S., Dervishi, E., & Villagarica, H. (2012). Carbon nanotubes induce growth enhancement in tobacco cells. *ACS Nano*, *6*(3), 2128–2135. doi:10.1021/nn204643g PMID:22360840

Kim, H. (2015). Wireless Communications Systems Design and Considerations. *Wireless Communications Systems Design*, 349-378.

Kim, Y., Evans, R. G., & Iversen, W. M. (2008). Remote sensing and control of an irrigation system using a distributed wireless sensor network. *IEEE Transactions on Instrumentation and Measurement*, *57*(7), 1379–1387. doi:10.1109/TIM.2008.917198

Krishna, K. L., Silver, O., Malende, W. F., & Anuradha, K. (2017). *Internet of Things application for implementation of smart agriculture system*. Paper presented at 2017 International Conference on I-SMAC (IoT in Social, Mobile, Analytics and Cloud) (I-SMAC), Palladam, India. 10.1109/I-SMAC.2017.8058236

Lacombe, G., & McCartney, M. (2014). Uncovering consistencies in Indian rainfall trends observed over the last half century. *Climatic Change*, *123*(2), 287–299. doi:10.100710584-013-1036-5

Latchem, C., & Jung, I. (2009). *Distance and blended learning in Asia*. Routledge.

Lee, J. S., Su, Y. W., & Shen, C. C. (2007, November). A comparative study of wireless protocols: Bluetooth, UWB, ZigBee, and Wi-Fi. In *Industrial Electronics Society, 2007. IECON 2007. 33rd Annual Conference of the IEEE* (pp. 46-51). IEEE.

Lee, W., An, Y. J., Yoon, H., & Kweon, H. S. (2008). Toxicity and bioavailability of copper nanoparticles to the terrestrial plants mung bean (Phaseolus radiatus) and wheat (Triticum aestivum): Plant agar test for water-insoluble nanoparticles. *Environmental Toxicology and Chemistry*, *27*(9), 1915–1921. doi:10.1897/07-481.1 PMID:19086317

Liu, Q., & Zhao, D. (2009, December). Priority polling strategy for wireless sensor networks. In *Information Science and Engineering (ICISE), 2009 1st International Conference on* (pp. 4062-4065). IEEE. 10.1109/ICISE.2009.852

Liu, D. (2016). Wireless Sensor Network in Precision agriculture application. *International conference on computer, Information and telecommunication systems (CITS)*, *8*, 345-350.

Liu, R., & Lal, R. (2015). Potentials of engineered nanoparticles as fertilizers for increasing agronomic productions. *The Science of the Total Environment*, *514*, 131–139. doi:10.1016/j.scitotenv.2015.01.104 PMID:25659311

Lokesh, K. K., Silver, O., Malende, W. F., & Anuradha, K. (2017). Internet of things application for implementation of smart agriculture system. *International conference on I-SMAC (IoT in Social, Mobile, Analytics and Cloud)*, 54-59.

Lu, C. M., Zhang, C. Y., Wu, J. Q., & Tao, M. X. (2002). Research of the effect of nanometer on germination and growth enhancement of Glycine max and its mechanism. *Dadou Kexue*, *21*, 168–172.

Luo, H., Yang, P. L., Li, Y. K., & Xu, F. P. (2016). An intelligent controlling system for greenhouse environment based on the architecture of the internet of things. *Sensor Letters*, *10*(1), 514–522. doi:10.1166l.2012.1850

Macleod, N. A. (1951). *U.S. Patent No. 2,536,196*. Washington, DC: U.S. Patent and Trademark Office.

Mafakheri, F., & Nasiri, F. (2014). Modeling of biomass-to-energy supply chain operations: Applications, challenges and research directions. *Energy Policy*, *67*, 116–126. doi:10.1016/j.enpol.2013.11.071

Mahajan, P., Dhoke, S. K., & Khanna, A. S. (2011). Effect of nano-ZnO particle suspension on growth of mung (Vigna radiata) and gram (Cicer arietinum) seedlings using plant agar method. *Journal of Nanotechnology*, *2011*, 1–7. doi:10.1155/2011/696535

Maharjan. (2010). *RFID and IOT: An Overview*. White Paper. Simula Research Laboratory, University of Oslo.

Maheshwari, B. (2016). Understanding the performance of irrigation systems around homes. *Journal of Environmental Engineering and Landscape Management*, *24*(4), 278–292. doi:10.3846/16486897.2016.1176575

Mahmoodzadeh, H., Nabavi, M., & Kashefi, H. (2013). Effect of nanoscale titanium dioxide particles on the germination and growth of canola (Brassica napus). *International Journal of Horticultural Science and Ornamental Plants*, *3*, 25–32.

Maina, M. M., Amin, M. S. M., & Yazid, M. A. (2014). Web geographic information system decision support system for irrigation water management: A review. Acta Agriculturae Scandinavica, Section B—Soil &. *Plant Science*, *64*(4), 283–293.

Mainetti, L., Patrono, L., & Vilei, A. (2011, September). Evolution of wireless sensor networks towards the internet of things: A survey. In *Software, Telecommunications and Computer Networks (SoftCOM), 2011 19th International Conference on* (pp. 1-6). IEEE.

Majsztrik, J. C., Price, E. W., & King, D. M. (2013). Environmental benefits of wireless sensor-based irrigation networks: Case-study projections and potential adoption rates. *HortTechnology*, *23*(6), 783–793.

Malusà, E., & Sas Paszt, L. (2009). The development of innovative technologies and products for organic fruit production. An integrated project. *Proceedings of international plant nutrition Coll. XVI*. Retrieved from http://repositories.cdlib.org/ipnc/xvi/1359,1–4

Mao, X., Liu, Y., Tang, S., Liu, H., Han, J., & Li, X. Y. (2013). Finding best and worst k-coverage paths in multihop wireless sensor networks. *IEEE Transactions on Parallel and Distributed Systems*, *24*(12), 2396–2406. doi:10.1109/TPDS.2012.329

Markets and Markets. (2013). *Global Biofertilizers Market By Types, Applications And Geography – Trends An Forecasts To 2017*. Markets and Markets.

Martin, D. L., Dorn, T. W., Melvin, S. R., Corr, A. J., & Kranz, W. L. (2011, February). Evaluating energy use for pumping irrigation water. *Proceedings of the 23rd Annual Central Plains Irrigation Conference*, 22-23.

Maughan, T., Allen, L. N., & Drost, D. (2015). *Soil Moisture Measurement and Sensors for Irrigation Management*. Academic Press.

Maurice, P. A., & Hochella, M. F. (2008). Nanoscale particles and processes: A new dimension in soil science. *Advances in Agronomy*, *100*, 123–153. doi:10.1016/S0065-2113(08)00605-6

Medaglia. (2011). Services, Use Cases and Future Challenges for Near Field Communication. In The Stolpan Project, Deploying Rfid - Challenges, Solutions, and Open Issues. Academic Press.

Mehoudar, R. (1980). *U.S. Patent No. 4,210,287*. Washington, DC: U.S. Patent and Trademark Office.

Miller, H. G., & Mork, P. (2013). From data to decisions: A value chain for Big Data. *IT Professional*, *15*, 57–59.

Minbo, L., Zhu, Z., & Guangyu, C. (2013). Information service system of agriculture IoT. *Automatika, 54*(4), 415-426.

Miorandi, D., Sicari, S., De Pellegrini, F., & Chlamtac, I. (2012). Internet of things: Vision, applications and research challenges. *Ad Hoc Networks*, *10*(7), 1497–1516. doi:10.1016/j.adhoc.2012.02.016

Mishra, D. P., & Kumar, R. (2015). Qualitative Analysis of Wireless Sensor Network Simulator. *International Journal of Computer Applications*.

Mohamed, R., K., & Ibrahim, M. (2016). *Wireless Sensor Network in Precision Agriculture Application*. Academic Press.

Mondal, A., Basu, R., Das, S., & Nandy, P. (2011). Beneficial role of carbon nanotubes on mustard plant growth: An agricul tural prospect. *Journal of Nanoparticle Research*, *13*(10), 4519–4528. doi:10.100711051-011-0406-z

Moreno, M. L., Rosa, G., Herna'ndez-Viezcas, J. A., Castillo-Michel, H., Botez, C. E., Peralta-Videa, J. R., & Gardea-Torresdey, J. L. (2010). Evidence of the differential biotransformation and genotoxicity of ZnO and CeO2 nanoparticles on soybean (Glycine max) plants. *Environmental Science & Technology*, *44*(19), 7315–7320. doi:10.1021/es903891g PMID:20384348

Mortensen, H. (2014). *Sustainable recovery of biomass on low-lying areas Agromek 2014*. Academic Press.

Mortvedt, J.J. (1995). Heavy metal contaminants in inorganic and organic fertilizers. *Fertilizer Research, 43*(1), 55–61.

Mosaic Crop Nutrition. (n.d.). *Soil Acidity*. Retrieved from http://www.cropnutrition.com/efu-soil-ph

Mulder, W.J., Gosselink, R.J.A., Vingerhoeds, M.H., Harmsen, P.F.H., & Eastham, D. (2011). Lignin based Controlled Release Coatings. *Industrial Crops and Products*, *34*(1), 915-920.

Mura, S., Seddaiu, G., Bacchini, F., Roggero, P. P., & Greppi, G. F. (2013). Advances of nanotechnology in agro-environmental studies. *Italian Journal of Agronomy*, *8*(18).

Naderi, M., & Shahraki, A. (2013). Nano technologies and their role in sustainable agriculture. *International Journal of Agriculture and Crop Sciences*, *5*, 2229–2232.

Namasivayam, S. K. R., & Chitrakala, K. (2011). Ecotoxicological effect of Lecanicillium lecanii (Ascomycota: Hypocreales) based silver nanoparticles on growth parameters of economically important plants. *Journal of Biopesticides*, *4*, 97–101.

Nandhagopal, S. M., & Sivanandam, S. N. (2013). Reliable Data Delivery in Mobile Adhoc Networks Using Light Weight Verification Algorithm with High Node Mobility. *IJITEE*, 2278-3075.

Narayut, P., Sasimanee, P., & Anupong, C. (2016). A control system in an intelligent farming using Arduino technology. *International Student Project Conference ICT-ISPC*, *7*, 345-349.

Natale, F., Gibin, M., Alessandrini, A., Vespe, M., & Paulrud, A. (2015). Mapping Agriculture effort through AIS data. *PLoS One*, 10.

NEC. (2014). *NEC and Dacom collaborate on precision farming solution to maximize yields and reduce costs*. Retrieved from https://www.nec-enterprise.com/news/Latest-press-releases/NECand-Dacom-collaborate-on-precision-farming-solution-to-maximize-yields-andreduce-costs-791

Nelson, S., & Artur, A. (2015). Wireless sensor and actuator system for smart irrigation on the cloud. *2nd World forum on Internet of Things (WF-IoT)*, 150-155.

Nemecek, T. (2007). Life cycle inventories of agricultural production systems. *Data v2. 0, Ecoinvent report*, (15).

Nikesh, G., & Kawitkar, R. S. (2016). IoT based smart agriculture. *International Journal of Advanced Research Computer and Communication Engineering*, *5*, 450–459.

Nogueira, L. C., Dukes, M. D., Haman, D. Z., Scholberg, J. M., & Cornejo, C. (2003, January). Data acquisition system and irrigation controller based on CR10X datalogger and TDR sensor. *Proceedings - Soil and Crop Science Society of Florida*, *62*, 38–46.

Noyes, K. (2014). Big Data Poised to Change the Face of Agriculture. *Fortune Data*. Retrieved from http:// fortune.com/2014/05/30/cropping-up-on-every-farm-big-data-technology/

Nuthall, P. L. (2017). *Farm Business Management: Analysis of Farming Systems*. CABI.

Omonode, R., & Vyn, T. (2006). Vertical distribution of soil organic carbon and nitrogen under warm-season native grasses relative to croplands in west-central Indiana, USA. *Agriculture, Ecosystems & Environment*, *117*(2–3), 159–170. doi:10.1016/j.agee.2006.03.031

Omta, S., Trienekens, J., & Beers, G. (2001). Chain and network science: A research framework. *Journal on Chain and Network Science*, *1*, 1–6.

Orfanou, A., Busato, P., Bochtis, D. D., Edwards, G., Pavlou, D., Sørensen, C. G., & Berruto, R. (2013). Scheduling for machinery fleets in biomass multiple-field operations. *Computers and Electronics in Agriculture*, *94*, 12–19. doi:10.1016/j.compag.2013.03.002

Orts, E., & Spigonardo, J. (2014). *Sustainability in the Age of Big Data. IGEL/Wharton*. Pennsylvania: University of Pennsylvania.

Osterwalder, A. (2014). *The Business Model Ontology: A Proposition in a Design Science Approach, l'Ecole des Hautes Etudes Commerciales*. Université de Lausanne.

Paige, G. B., & Keefer, T. O. (2008). Comparison of Field Performance of Multiple Soil Moisture Sensors in a Semi-Arid Rangeland. *JAWRA Journal of the American Water Resources Association*, *44*(1), 121–135. doi:10.1111/j.1752-1688.2007.00142.x

Park, M. (2015). IEEE 802.11ah: Sub-1-GHz license-exempt operation for the internet of things. *IEEE Communications Magazine*, *53*(9), 145–151. doi:10.1109/MCOM.2015.7263359

Parmenter, J., Jensen, A. N., & Chiu, S. (2014, June). Smart irrigation controller. In *Electro/Information Technology (EIT), 2014 IEEE International Conference on* (pp. 394-398). IEEE 10.1109/EIT.2014.6871796

Patil, K. A., & Kale, N. R. (2016, December). A model for smart agriculture using IoT. In *Global Trends in Signal Processing, Information Computing and Communication (ICGTSPICC), 2016 International Conference on* (pp. 543-545). IEEE 10.1109/ICGTSPICC.2016.7955360

Pavlou, D., Orfanou, A., Busato, P., Berruto, R., Sørensen, C., & Bochtis, D. (2016). Functional modeling for green biomass supply chains. *Computers and Electronics in Agriculture*, *122*, 29–40. doi:10.1016/j.compag.2016.01.014

Peeters, A. (2009). Importance, evolution, environmental impact and future challenges of grasslands and grassland-based systems in Europe. *Grassland Science*, *55*(3), 113–125. doi:10.1111/j.1744-697X.2009.00154.x

Pereira, A., Shock, C. A., Wang, F. X., Feibert, E. B. G., Shock, C. C., & Flock, R. (2013). *Irrigation monitoring using soil water tension*. Academic Press.

Peters, R. T., Desta, K. G., & Nelson, L. (2013). *Practical use of soil moisture sensors and their data for irrigation scheduling*. Academic Press.

Phene, C. J., & Howell, T. A. (1984). Soil sensor control of high-frequency irrigation systems. *Transactions of the ASAE. American Society of Agricultural Engineers*, *27*(2), 392–0396. doi:10.13031/2013.32798

Plume, K. (2014). The Big Data Bounty: U.S. Startups Challenge Agribusiness Giants. *CloudTweaks*. Retrieved from http://www.reuters.com/article/us-usa-farming-startupsidUSKCN0HX0C620141008

Polygon. (n.d.). How Humidity Affects the Growth of Plants [Blog post]. Retrieved from https://www.polygongroup.com/en-US/blog/how-humidity-affects-the-growth-of-plants/

Poppe, K. J., Wolfert, J., Verdouw, C. N., & Renwick, A. (2015). A European perspective on the economics of Big Data. *Farm Policy Journal*, *12*, 11–19.

Porter, M.E., & Heppelmann, J.E. (2014, November). How smart, connected products are transforming competition. *Harv. Bus. Rev.*, 65–88.

Porter, K. G. (1977). The plant-animal interface in freshwater ecosystems. *American Scientist*, *65*, 159–170.

Porter, M. E. (1985). *Competitive Advantage: Creating and Sustaining Superior Performance. 1985*. New York: Free Press.

Prathibha, Hongal, & Jyothi. (2017). IOT Based Monitoring System in Smart Agriculture. *2017 International Conference on Recent Advances in Electronics and Communication Technology*. Retrieved from https://ieeexplore.ieee.org/document/8081906/authors

Prathibha, S. R., Hongal, A., & Jyothi, M. P. (2017, March). IOT Based Monitoring System in Smart Agriculture. In *Recent Advances in Electronics and Communication Technology (ICRAECT), 2017 International Conference on* (pp. 81-84). IEEE 10.1109/ICRAECT.2017.52

Provan, K. G., & Kenis, P. (2008). Modes of network governance: Structure, management, and effectiveness. *Journal of Public Administration: Research and Theory*, *18*(2), 229–252. doi:10.1093/jopart/mum015

Puri, V., Nayyar, A., & Raja, L. (2017). Agriculture drones: A modern breakthrough in precision agriculture. *Journal of Statistics and Management Systems*, *20*(4), 507–518. doi:10.1080/09720510.2017.1395171

Quoc Buu, N., Hien, D. T., Chau, H. N., Tin, X. T., Van, N. T., Duong, T. K., & Ha, T. H. (2014). Effects of nanocrystalline powders (Fe, Co and Cu) on the germination, growth, crop yield and product quality of soybean (Vietnamese species DT-51). *Advances in Natural Sciences: Nanoscience and Nanotechnology*, *5*, 015016.

Raghavendra, C. S., Sivalingam, K. M., & Znati, T. (Eds.). (2006). *Wireless sensor networks*. Springer.

Rahuri, M. P. (2017). *Soil testing Analysis*. Rahuri: MPKU Rahuri.

Rajalakshmi, P., & Devi, M. (2016). IOT based crop-field monitoring and irrigation automation. *10th International conference on intelligent systems and control (ISCO)*, *7*, 89 – 95. 10.1109/ISCO.2016.7726900

Rajasthan Budget. (2018). Retrieved from http://plan.rajasthan.gov.in/

Rappaport, T. S. (1996). *Wireless communications: principles and practice* (Vol. 2). Prentice Hall.

Raven, P. H., Berg, L. R., & Johnson, G. B. (1998). *Environment* (2nd ed.). New York: Saunders College Publishing.

Ray, B. (2015). 5 Types of Wireless Technology For The IoT [Blog post]. Retrieved from https://www.link-labs.com/blog/types-of-wireless-technology

Reinhold-Hurek, B., & Hurek, T. (2011). Living inside plants: Bacterial endophytes. *Current Opinion in Plant Biology*, *14*(4), 435–443. doi:10.1016/j.pbi.2011.04.004 PMID:21536480

Remya, K. (2016). Greenhouse monitoring and control based on IoT using WSN. *ITSI-TEEE*, *4*, 2016–2020.

Report: Global Agriculture Towards 2050. (n.d.). Retrieved from http://www.fao.org/

Report: How to Feed the World in 2050. (n.d.). Retrieved from http://www.fao.org/

Ritaban, D., Ahsan, M., Jagannath, A., D'Este, C., & Aruneema, D. (2014). Development of an intelligent environmental knowledge system for sustainable agricultural decision support. *Environmental Modelling & Software*, *52*, 264–272. doi:10.1016/j.envsoft.2013.10.004

Rodda, J. C., & Ubertini, L. (2004). *The Basis of Civilization -Water Science.* International Association of Hydrological Sciences.

Rodriguez, E., Sultan, R., & Hilliker, A. (2004). Negative Effects of Agriculture on Our Environment. *The Traprock*, *3*, 28–32.

Ronn, R., & McGraig, A. E. (2002). Impact of protozoan grazing on the bacterial community structure in soil microcosms. *Applied and Environmental Microbiology*, *68*(12), 6094–6105. doi:10.1128/AEM.68.12.6094-6105.2002 PMID:12450833

Rotter, S., & Rohrhofer, C. (2013). *Biomass based energy intermediates boosting biofuel production*. Academic Press.

Royse, R. (2014). The Growing Field of Agriculture Technology. *Vator News*. Retrieved from http://vator.tv/ news/2014-05-14-the-growing-field-of-agriculture-technology

Ruby, R. A., & Jawahar, A. (2017). Smart agro system using wireless sensor networks. *International Conference on Intelligent Computing and Control Systems*, 400-403.

Ruiz-Garcia, L., Lunadei, L., Barreiro, P., & Robla, I. (2009). A review of wireless sensor technologies and applications in agriculture and food industry: State of the art and current trends. *Sensors (Basel)*, *9*(6), 4728–4750. doi:10.339090604728 PMID:22408551

Ryan, R. P., Germaine, K., Franks, A., Ryan, D. J., & Dowling, D. N. (2008). Bacterial endophytes: Recent developments and applications. *FEMS Microbiology Letters*, *278*(1), 1–9. doi:10.1111/j.1574-6968.2007.00918.x PMID:18034833

Sahandi, S. M., Sorooshzadeh, A., Rezazadeh, H. S., & Naghdibadi, H. A. (2011). Effect of nanosilver and silver nitrate on seed yield of borage. *Journal of Medicinal Plants Research*, *5*, 706–710.

Salarian, H., Chin, K. W., & Naghdy, F. (2014). An energy-efficient mobile-sink path selection strategy for wireless sensor networks. *IEEE Transactions on Vehicular Technology*, *63*(5), 2407–2419. doi:10.1109/TVT.2013.2291811

Sanders, R. (2012). *Fertilizer use responsible for increase in nitrous oxide in atmosphere*. Research, Science & Environment. Retrieved from http://news.berkeley.edu/2012/04/02/fertilizer-use-responsible-for-increase-in-nitrous-oxide-in-atmosphere

Sasikala, C. (2013). Cloud Computing as Value Addition and Cost Effective Solution to Library and Information Centres. *Electronic Resources Management in Libraries*, *1*, 177.

Schindler, D. W. (1974). Eutrophication and recovery in experimental lakes: Implications for lake management. *Science*, *174*(4139), 897–899. doi:10.1126cience.184.4139.897 PMID:17782381

Schindler, D. W. (2006). Recent advances in the understanding and management of eutrophication. *Limnology and Oceanography*, *51*(1part2), 356–363. doi:10.4319/lo.2006.51.1_part_2.0356

Schribe, S. (n.d.). *Smart Agriculture Sensors: Helping Small Farmers and Positively Impacting Global Issues, Too.* Retrieved from https://www.mouser.in/applications/smart-agriculture-sensors/

Schroder, H. J. (2006). *Soil moisture-based irrigation for efficient use of water and nutrients and sustainability of vegetables cropped on coarse soils* (Thesis). University of Florida.

Schroder, J. H. (2006). *Soil moisture-based drip irrigation for efficient use of water, nutrients, and sustainability of vegetables cropped on coarse soils* (Doctoral dissertation). University of Florida.

Schweier, J., & Becker, G. (2013). Economics of poplar short rotation coppice plantations on marginal land in Germany. *Biomass and Bioenergy*, *59*, 494–502. doi:10.1016/j.biombioe.2013.10.020

SEGES. (2015). *The cost for using belts and tyres*. Author.

Sekar, S., & Kandavel, D. (2010). Interaction of plant growth promoting rhizobacteria (PGPR) and endophytes with medicinal plants– new avenues for phytochemicals. *Journal of Phytology*, *2*, 91–100.

Semantic Community. (2015). *Big Data Science for Precision Farming Business*. Retrieved from http://semanticommunity.info/Data_Science/Big_Data_Science_for_Precision_Farming_Business

Series Y: Global Information Infrastructure, Internet Protocol Aspects and Next-Generation Networks,Overview of the Internet of things, ITU-T Y.2060.

Service, N. R. (2014). *Irrigation Water Management*. National Resource Conservation Service, 01-04.

Service, N. R. (2012). *Irrigation Water Management*. NRCS Montana.

Sethi, P., & Sarangi, S. R. (2017). Internet of things: Architectures, protocols, and applications. *Journal of Electrical and Computer Engineering*.

Shah, V., & Belozerova, I. (2009). Influence of metal nanoparticles on the soil microbial community and germination of lettuce seeds. *Water, Air, and Soil Pollution*, *197*(1-4), 143–148. doi:10.100711270-008-9797-6

Sharma, B., Ingalls, R. G., Jones, C. L., & Khanchi, A. (2013). Biomass supply chain design and analysis: Basis, overview, modeling, challenges, and future. *Renewable & Sustainable Energy Reviews*, *24*, 608–627. doi:10.1016/j.rser.2013.03.049

Shelby, Z., & Bormann, C. (2010). *6LoWPAN: The Wireless Embedded Internet*. Wiley.

Shock, C. C., Feibert, E. B., Saunders, L. D., & Eldredge, E. P. (2002). Automation of subsurface drip irrigation for crop research. In *World Congress of Computers in Agriculture and Natural Resources, Proceedings of the 2002 Conference* (p. 809). American Society of Agricultural and Biological Engineers.

Short, M. (2010, January). The case for non-preemptive, deadline-driven scheduling in real-time embedded systems. *Lecture notes in engineering and computer science: Proceedings of the World Congress on Engineering.*

Siddhartha, S. (2014). Mukhopadhyay (2014). Nanotechnology in agriculture: Prospects and constraints. *Nanotechnology, Science and Applications, 7*, 63–71. doi:10.2147/NSA.S39409 PMID:25187699

Sieberta, S., Hoogeveenb, J., Dölla, P., Faurèsb, J. M., Feicka, S., & Frenkenb, K. (2006, October). The Digital Global Map of Irrigation Areas–Development and Validation of Map Version 4. *Conference on International Agricultural Research for Development (Tropentag).*

Sindhuja, M. P., Prathusha, M., Kalaiselvi, M. C., Sudhaa, S. J., & Velthai, R. (2016). *Solar driven Aurdino based automatic irrigation using GSM*. Academic Press.

Singh, R. (2017). *Indian Economy*. McGraw-Hill.

Singh, S. (1997). *Taming the waters: The political ecology of large dams*. Academic Press.

Singh, M D, Chirag, G., Om Prakash, P., Mohan, M.h., & Prakash, G., & Vishwajit. (2017). Nano fertilizers is a new way to increase nutrients use efficiency in crop production. *International Journal of Agricultural Sciences*, 3831–3833.

Smart Irrigation. (n.d.). Retrieved from https://www.hydropoint.com/what-is-smart-irrigation/

Smeets, E. M. W., Lewandowski, I. M., & Faaij, A. P. C. (2009). The economical and environmental performance of miscanthus and switchgrass production and supply chains in a European setting. *Renewable & Sustainable Energy Reviews, 13*(6–7), 1230–1245. doi:10.1016/j.rser.2008.09.006

Snyder, R. L., & de Melo-Abreu, J. P. (2005). *Frost protection: fundamentals, practice and economics*. Academic Press.

Sohraby, K., Minoli, D., & Znati, T. (2007). *Wireless sensor networks: technology, protocols, and applications*. John Wiley & Sons. doi:10.1002/047011276X

Sokhansanj, S., Kumar, A., & Turhollow, A. (2006). Development and implementation of integrated biomass supply analysis and logistics model (IBSAL). *Biomass and Bioenergy*, *30*(10), 838–847. doi:10.1016/j.biombioe.2006.04.004

Sonka, S. (2014). Big Data and the Ag sector: More than lots of numbers. *The International Food and Agribusiness Management Review*, *17*, 1.

Sonka, S. (2015). Big Data: From hype to agricultural tool. *Farm Policy Journal*, *12*, 1–9.

Sørensen, C. G., & Bochtis, D. D. (2010). Conceptual model of fleet management in agriculture. *Biosystems Engineering*, *105*(1), 41–50. doi:10.1016/j.biosystemseng.2009.09.009

Sørensen, C., & Nielsen, V. (2005). Operational Analyses and Model Comparison of Machinery Systems for Reduced Tillage. *Biosystems Engineering*, *92*(2), 143–155. doi:10.1016/j.biosystemseng.2005.06.014

Sørensen, G., Halberg, N., Oudshoorn, F. W., Petersen, B. M., & Dalgaard, R. (2014). Energy inputs and GHG emissions of tillage systems. *Biosystems Engineering*, *120*, 2–14. doi:10.1016/j.biosystemseng.2014.01.004

Southern Nevada Water Authority (SNWA). (2017). *Subsurface Irrigation*. Retrieved from https://www.snwa.com/land/irrig_subsurface.html

SparkFun. (n.d.). *What is an Arduino?* Retrieved on 13 April 2018 from https://learn.sparkfun.com/tutorials/what-is-an-arduino

Spijker, A. (2014). *The New Oil - Using Innovative Business Models to Turn Data into Profit*. Basking Ridge: Technics Publications.

Stojmenovic, I. (Ed.). (2003). *Handbook of wireless networks and mobile computing* (Vol. 27). John Wiley & Sons.

Suma, Samson, Saranya, Shanmugapriya, & Subhashri. (n.d.). IOT Based Smart Agriculture Monitoring System. *International Journal on Recent and Innovation Trends in Computing and Communication, 5*(2), 177 – 181.

Sumner, M. E. (2000). Beneficial use of effluents, wastes, and biosolids. *Communications in Soil Science and Plant Analysis, 31*(11), 1701–1715. doi:10.1080/00103620009370532

Sundmaeker, H., Verdouw, C., Wolfert, S., & Pérez Freire, L. (2016). Internet of food and farm 2020. In O. Vermesan & P. Friess (Eds.), *Digitising the Industry. Internet of Things Connecting Phy*.

Sun, J., Zhou, Z., Bu, Y., Zhuo, J., Chen, Y., & Li, D. (2013a). Research and development for potted flowers automated grading system based on internet of things. *Shenyang Nongye Daxue Xuebao, 44*, 687–691.

Sun, Z., Du, K., Zheng, F., & Yin, S. (2013b). Perspectives of research and application of Big Data on smart agriculture. *J. Agric. Sci. Technol., 15*, 63–71.

Suriyaprabha, R., Karunakaran, G., Yuvakkumar, R., Prabu, P., Rajendran, V., & Kannan, N. (2012). Growth and physiological responses of maize (Zea mays L.) to porous silica nanoparticles in soil. *Journal of Nanoparticle Research, 14*(12), 1294–1308. doi:10.100711051-012-1294-6

Swami, A., Zhao, Q., Hong, Y. W., & Tong, L. (Eds.). (2007). *Wireless sensor networks: signal processing and communications perspectives*. John Wiley & Sons. doi:10.1002/9780470061794

Tanmay, B., & Pushpendra, K. P. (2016). Development of IoT based Smart Security and Monitoring Devices for Agriculture. *6th International Conference - Cloud System and Big Data Engineering*, 978 – 982.

The essential building blocks of an IoT architecture. (2017). Retrieved from https://mobilebusinessinsights.com/2017/09/the-essential-building-blocks-of-an-iot-architecture/

The European Commission. (2000). Towards a European strategy for the security of energy supply. *EUR-Lex*, pp. 1–111. doi: 92-894-0319-5

Tien, J. M. (2013). Big Data: Unleashing information. *Journal of Systems Science and Systems Engineering*, *22*, 127–151.

Tolentino. (2013). *How QR Codes Can Deliver the Internet of Everything*. News.

Tong, L., & Hong, T., & JingHua, Z. (2015). Research on the Big Data-based government decision and public information service model of food safety and nutrition industry. *Shipin Anquan Zhiliang Jiance Xuebao*, *6*, 366–371.

Trenkel, M.E. (2010). *Slow and Controlled-Release and Stabilized Fertilizers: An Option for Enhancing Nutrient Use Efficiency in Agriculture*. International Fertilizer Industry Association.

Tsai, B. S. (1986). *Continuous spouted bed process for sulphur-coating urea. Chemical And Biological Engineering (Vol. Master Of Applied Science)*. Vancouver: University Of British Columbia.

Tse, D., & Viswanath, P. (2005). *Fundamentals of wireless communication*. Cambridge University Press.

Types of Irrigation. (2017). Retrieved from http://nptel.ac.in/courses/105105110/pdf/m3l01.pdf

Uddin, J., Reza, S. T., Newaz, Q., Uddin, J., Islam, T., & Kim, J. M. (2012, December). Automated irrigation system using solar power. In *Electrical & Computer Engineering (ICECE), 2012 7th International Conference on* (pp. 228-231). IEEE. 10.1109/ICECE.2012.6471527

Umamaheswari, S., & Vanitha, K. (2016). *Interaction Model of Service Discovery Using Visa Processing Algorithm and Constrained Application Protocol (CoAP)*. Springer.

UNECE. (2013). *Classification of types of Big Data*. Retrieved from http://www1.unece.org/stat/platform/ display/bigdata/Classification+of+Types+of+Big+Data

Unger, R. M. (1998). *Democracy realized: The progressive alternative*. Verso.

Vaibhavraj, Roham, Pawar, Patil, & Rupnar. (2015). Smart Farm using Wireless Sensor Network. *International Journal of Computer Applications, National Conference on Advances in Computing.*

Valente, A., Morais, R., Tuli, A., Hopmans, J. W., & Kluitenberg, G. J. (2006). Multi-functional probe for small-scale simultaneous measurements of soil thermal properties, water content, and electrical conductivity. *Sensors and Actuators. A, Physical*, *132*(1), 70–77. doi:10.1016/j.sna.2006.05.010

Van der Vorst, J., & Beulens, B. (2005). Innovations in logistics and ICT in food supply chain networks. In W. M. F. Jongen & M. T. G. Meulenberg (Eds.), *Innovations in Agri-food Systems: Product Quality and Consumer Acceptance. Wageningen Academic Publishers.*

Van Rijmenam, M. (2015). *John Deere is revolutionizing farming with Big Data.* Retrieved from https:// datafloq.com/read/john-deere-revolutionizing-farming-big-data/511

Verdouw, C. N., Beulens, A. J. M., Reijers, H. A., & van der Vorst, J. G. A. J. (2015). A control model for object virtualization in supply chain management. *Computers in Industry*, *68*, 116–131. doi:10.1016/j.compind.2014.12.011

Verdouw, C. N., Beulens, A. J. M., & van der Vorst, J. G. A. J. (2013). Virtualisation of floricultural supply chains: A review from an internet of things perspective. *Computers and Electronics in Agriculture*, *99*, 160–175. doi:10.1016/j.compag.2013.09.006

Verhoosel, J., van Bekkum, M., Verwaart, T. (2016). HortiCube: a platform for transparent, trusted data sharing in the food supply chain. *Proceedings in Food System Dynamics*, 384–388.

Vilmos, Medaglia, & Moroni. (2011). *NFC Technology and Its Application Scenarios in a Future Of IoT*. Stolpan Project.

Vogt, W. (2013). Looking at Big Data One Plant at a Time. *Farm Industry News.* Retrieved from http:// farmindustrynews.com/blog/looking-big-data-one-plant-time

Walter, A., Finger, R., Huber, R., & Buchmann, N. (2017). Opinion: Smart farming is key to developing sustainable agriculture. *Proceedings of the National Academy of Sciences of the United States of America, 114*(24), 6148–6150. doi:10.1073/pnas.1707462114 PMID:28611194

Wamba, S. F., & Wicks, A. (2010). RFID deployment and use in the dairy value chain: applications, current issues and future research directions. *Technology and Society (ISTAS), 2010 IEEE International Symposium on. IEEE,* 172–179.

Weber, R. H. (2010). Internet of Things–New security and privacy challenges. *Computer Law & Security Review, 26*(1), 23–30. doi:10.1016/j.clsr.2009.11.008

Welte, J. T., Ault, A. C., Bowman, C., Ellis, S., Buckmaster, D. R., Ess, D., & Krogmeier, J. V. (2013). An Approach to Farm Management Information Systems Using Task-Specific, Collaborative Mobile Apps and Cloud Storage Services. *American Society of Agricultural and Biological Engineers,* 1.

Werner, H. (1992). *Measuring soil moisture for irrigation water management.* Cooperative Extension Service, South Dakota State University, US Department of Agriculture.

Wigboldus, S., Klerkx, L., Leeuwis, C., Schut, M., Muilerman, S., & Jochemsen, H. (2016). Systemic perspectives on scaling agricultural innovations. A review. *Agronomy for Sustainable Development, 36*(3), 1–20. doi:10.100713593-016-0380-z

Wightman, P. M., & Labrador, M. A. (2009). *Topology Control in Wireless Sensor Networks.* Academic Press.

Wikipedia. (2016). *Big Data.* Retrieved from https://en.wikipedia.org/wiki/Big_data

Williamson, O. E. (2016). *The Mechanisms of Governance.* Oxford University Press.

Winter. (2012). *IPv6 routing protocol for low power and lossy networks.* Internet Engineering Task Force, RFC 6550.

Winter, T., & Thubert, P.ROLL Team. (2011). *RPL, Internet-Draft.* RPL.

Wolfert, J., Sørensen, C. G., & Goense, D. (2014). A Future Internet Collaboration Platform for Safe and Healthy Food from Farm to Fork. *Global Conference (SRII), 2014 Annual SRII,* 266–273.

Wright, J., Wildung, D., & Nennich, T. (2012). *Irrigation Considerations and Soil Moisture Monitoring Tools*. University of Minnesota.

Wu, M., Lu, T. J., Ling, F. Y., Sun, J., & Du, H. Y. (2010, August). Research on the architecture of Internet of things. In *Advanced Computer Theory and Engineering (ICACTE), 2010 3rd International Conference on* (Vol. 5, pp. V5-484). IEEE.

Wu, Y., Li, X. Y., Li, Y., & Lou, W. (2010). Energy-efficient wake-up scheduling for data collection and aggregation. *IEEE Transactions on Parallel and Distributed Systems*, *21*(2), 275–287. doi:10.1109/TPDS.2009.45

Xia, F., Yang, L. T., Wang, L., & Vinel, A. (2012). Internet of things. *International Journal of Communication Systems*, *25*(9), 1101–1102. doi:10.1002/dac.2417

Xiao, K., Xiao, D., & Luo, X. (2010). Smart water-saving irrigation system in precision agriculture based on wireless sensor network. *Nongye Gongcheng Xuebao (Beijing)*, *26*(11), 170–175.

Xie, T., Huang, Z., Chi, Z., & Zhu, T. (2017, April). Minimizing amortized cost of the on-demand irrigation system in smart farms. In *Proceedings of the 3rd International Workshop on Cyber-Physical Systems for Smart Water Networks* (pp. 43-46). ACM 10.1145/3055366.3055370

Yan, B., Shi, P., & Huang, G. (2013). Development of traceability system of aquatic foods supply chain based on RFID and EPC internet of things. *Nongye Gongcheng Xuebao (Beijing)*, *29*, 172–183.

Yang, C. (2014). Big Data and its potential applications on agricultural production. *Zuowu, Huanjing Yu Shengwu Zixun*, *11*, 51–56.

Yoe, H. (2017, June). Smart farming: ICT based agriculture: Keynote address. In *Software Engineering, Artificial Intelligence, Networking and Parallel/Distributed Computing (SNPD), 2017 18th IEEE/ACIS International Conference on* (pp. 3-3). IEEE.

Yoo, S. E., Kim, J. E., Kim, T., Ahn, S., Sung, J., & Kim, D. (2007, June). A 2 S: automated agriculture system based on WSN. In *Consumer Electronics, 2007. ISCE 2007. IEEE International Symposium on* (pp. 1-5). IEEE

Young, L. (2016). *Farmobile takes ag data to the next level.* Retrieved from http://www.agrinews.com/news/minnesota_news/farmobile-takes-ag-data-to-the-next-level/article_5074de27-ecdf-5dd3-a89e-e3b28815475c.html

Zhang, L., & Hu, G. (2013). Supply chain design and operational planning models for biomass to drop-in fuel production. *Biomass and Bioenergy, 58*, 238–250. doi:10.1016/j.biombioe.2013.08.016

Zheng, L., Hong, F., Lu, S., & Liu, C. (2005). Effect of nano-TiO_2 on strength of naturally aged seeds and growth of spinach. *Biological Trace Element Research, 104*(1), 83–92. doi:10.1385/BTER:104:1:083 PMID:15851835

Zhou, Y., Evans, G. H. Jr, Chowdhury, M., Wang, K. C., & Fries, R. (2011). Wireless communication alternatives for intelligent transportation systems: A case study. *Journal of Intelligent Transport Systems, 15*(3), 147–160. doi:10.1080/15472450.2011.594681

Zhu, W., Dai, C., & Huang, P. (2012). Environmental control system based on IOT for nursery pig house. *Nongye Gongcheng Xuebao (Beijing), 28*, 177–182.

Zong, Z. L., Fares, R., Romoser, B., & Wood, J. (2014). FastStor: Improving the performance of a large scale hybrid storage system via caching and prefetching. *Cluster Computing, 17*, 593–604.

Zubek, S., Mielcarek, S., & Turnau, K. (2012). Hypericin and pseudohypericin concentrations of a valuable medicinal plant Hypericum perforatum L are enhanced by arbuscular mycorrhizal fungi. *Mycorrhiza, 22*(2), 149–156. doi:10.100700572-011-0391-1 PMID:21626142

Related References

To continue our tradition of advancing academic research, we have compiled a list of recommended IGI Global readings. These references will provide additional information and guidance to further enrich your knowledge and assist you with your own research and future publications.

Abresch, J. (2003). Geographic Information Systems Research and Data Centers. In A. Hanson & B. Levin (Eds.), *Building a Virtual Library* (pp. 52–64). Hershey, PA: IGI Global. doi:10.4018/978-1-59140-106-3.ch004

Ajiboye, O. E., & Yusuff, O. S. (2017). Foreign Land Acquisition: Food Security and Food Chains – The Nigerian Experience. In I. Management Association (Ed.), Natural Resources Management: Concepts, Methodologies, Tools, and Applications (pp. 1524-1545). Hershey, PA: IGI Global. doi:10.4018/978-1-5225-0803-8.ch072

Alapiki, H. E., & Amadi, L. A. (2018). Sustainable Food Consumption in the Neoliberal Order: Challenges and Policy Implications. In A. Obayelu (Ed.), *Food Systems Sustainability and Environmental Policies in Modern Economies* (pp. 90–123). Hershey, PA: IGI Global. doi:10.4018/978-1-5225-3631-4.ch005

Altaş, A. (2018). Geographical Information System Applications Utilized in Museums in Turkey Within the Scope of the Cultural Heritage Tourism: A Case Study of Mobile Application of Müze Asist. In S. Chaudhuri & N. Ray (Eds.), *GIS Applications in the Tourism and Hospitality Industry* (pp. 42–60). Hershey, PA: IGI Global. doi:10.4018/978-1-5225-5088-4.ch002

Andreea, I. R. (2018). Beyond Macroeconomics of Food and Nutrition Security. *International Journal of Sustainable Economies Management*, *7*(1), 13–22. doi:10.4018/IJSEM.2018010102

Anwar, J. (2017). Reproductive and Mental Health during Natural Disaster: Implications and Issues for Women in Developing Nations – A Case Example. In I. Management Association (Ed.), Gaming and Technology Addiction: Breakthroughs in Research and Practice (pp. 446-472). Hershey, PA: IGI Global. doi:10.4018/978-1-5225-0778-9.ch021

Awadh, H., Aksissou, M., Benhardouze, W., Darasi, F., & Snaiki, J. (2018). Socioeconomic Status of Artisanal Fishers in the West Part of Moroccan Mediterranean. *International Journal of Social Ecology and Sustainable Development*, *9*(1), 40–52. doi:10.4018/IJSESD.2018010104

Aye, G. C., & Haruna, R. F. (2018). Effect of Climate Change on Crop Productivity and Prices in Benue State, Nigeria: Implications for Food Security. In V. Erokhin (Ed.), *Establishing Food Security and Alternatives to International Trade in Emerging Economies* (pp. 244–268). Hershey, PA: IGI Global. doi:10.4018/978-1-5225-2733-6.ch012

Azizan, S. A., & Suki, N. M. (2017). Consumers' Intentions to Purchase Organic Food Products. In T. Esakki (Ed.), *Green Marketing and Environmental Responsibility in Modern Corporations* (pp. 68–82). Hershey, PA: IGI Global. doi:10.4018/978-1-5225-2331-4.ch005

Barakabitze, A. A., Fue, K. G., Kitindi, E. J., & Sanga, C. A. (2017). Developing a Framework for Next Generation Integrated Agro Food-Advisory Systems in Developing Countries. In I. Management Association (Ed.), Agri-Food Supply Chain Management: Breakthroughs in Research and Practice (pp. 47-67). Hershey, PA: IGI Global. doi:10.4018/978-1-5225-1629-3.ch004

Beachcroft-Shaw, H., & Ellis, D. (2018). Using Successful Cases to Promote Environmental Sustainability: A Social Marketing Approach. In I. Management Association (Ed.), Sustainable Development: Concepts, Methodologies, Tools, and Applications (pp. 936-953). Hershey, PA: IGI Global. doi:10.4018/978-1-5225-3817-2.ch042

Behnassi, M., Kahime, K., Boussaa, S., Boumezzough, A., & Messouli, M. (2017). Infectious Diseases and Climate Vulnerability in Morocco: Governance and Adaptation Options. In I. Management Association (Ed.), Public Health and Welfare: Concepts, Methodologies, Tools, and Applications (pp. 91-109). Hershey, PA: IGI Global. doi:10.4018/978-1-5225-1674-3.ch005

Bekele, F., & Bekele, I. (2017). Social and Environmental Impacts on Agricultural Development. In W. Ganpat, R. Dyer, & W. Isaac (Eds.), *Agricultural Development and Food Security in Developing Nations* (pp. 21–56). Hershey, PA: IGI Global. doi:10.4018/978-1-5225-0942-4.ch002

Benaouda, A., & García-Peñalvo, F. J. (2018). Towards an Intelligent System for the Territorial Planning: Agricultural Case. In F. García-Peñalvo (Ed.), *Global Implications of Emerging Technology Trends* (pp. 158–178). Hershey, PA: IGI Global. doi:10.4018/978-1-5225-4944-4.ch010

Bhaskar, A., Rao, G. B., & Vencatesan, J. (2017). Characterization and Management Concerns of Water Resources around Pallikaranai Marsh, South Chennai. In P. Rao & Y. Patil (Eds.), *Reconsidering the Impact of Climate Change on Global Water Supply, Use, and Management* (pp. 102–121). Hershey, PA: IGI Global. doi:10.4018/978-1-5225-1046-8.ch007

Bianchini, E., & Sicurella, S. (2010). GIS: A New Tool for Criminology and Victimology's Studies. In R. Sette (Ed.), *Cases on Technologies for Teaching Criminology and Victimology: Methodologies and Practices* (pp. 87–110). Hershey, PA: IGI Global. doi:10.4018/978-1-60566-872-7.ch006

Bogataj, D., & Drobne, D. (2017). Control of Perishable Goods in Cold Logistic Chains by Bionanosensors. In S. Joo (Ed.), *Applying Nanotechnology for Environmental Sustainability* (pp. 376–402). Hershey, PA: IGI Global. doi:10.4018/978-1-5225-0585-3.ch016

Bogataj, D., & Drobne, D. (2017). Control of Perishable Goods in Cold Logistic Chains by Bionanosensors. In I. Management Association (Ed.), Materials Science and Engineering: Concepts, Methodologies, Tools, and Applications (pp. 471-497). Hershey, PA: IGI Global. doi:10.4018/978-1-5225-1798-6.ch019

Bogueva, D., & Marinova, D. (2018). What Is More Important: Perception of Masculinity or Personal Health and the Environment? In D. Bogueva, D. Marinova, & T. Raphaely (Eds.), *Handbook of Research on Social Marketing and Its Influence on Animal Origin Food Product Consumption* (pp. 148–162). Hershey, PA: IGI Global. doi:10.4018/978-1-5225-4757-0.ch010

Bouzid, M. (2017). Waterborne Diseases and Climate Change: Impact and Implications. In M. Bouzid (Ed.), *Examining the Role of Environmental Change on Emerging Infectious Diseases and Pandemics* (pp. 89–108). Hershey, PA: IGI Global. doi:10.4018/978-1-5225-0553-2.ch004

Bowles, D. C. (2017). Climate Change-Associated Conflict and Infectious Disease. In M. Bouzid (Ed.), *Examining the Role of Environmental Change on Emerging Infectious Diseases and Pandemics* (pp. 68–88). Hershey, PA: IGI Global. doi:10.4018/978-1-5225-0553-2.ch003

Buck, J. J., & Lowry, R. K. (2017). Oceanographic Data Management: Quills and Free Text to the Digital Age and "Big Data". In P. Diviacco, A. Leadbetter, & H. Glaves (Eds.), *Oceanographic and Marine Cross-Domain Data Management for Sustainable Development* (pp. 1–22). Hershey, PA: IGI Global. doi:10.4018/978-1-5225-0700-0.ch001

Buse, C. G. (2017). Are Climate Change Adaptation Policies a Game Changer?: A Case Study of Perspectives from Public Health Officials in Ontario, Canada. In M. Bouzid (Ed.), *Examining the Role of Environmental Change on Emerging Infectious Diseases and Pandemics* (pp. 230–257). Hershey, PA: IGI Global. doi:10.4018/978-1-5225-0553-2.ch010

Calderon, F. A., Giolo, E. G., Frau, C. D., Rengel, M. G., Rodriguez, H., Tornello, M., ... Gallucci, R. (2018). Seismic Microzonation and Site Effects Detection Through Microtremors Measures: A Review. In N. Ceryan (Ed.), *Handbook of Research on Trends and Digital Advances in Engineering Geology* (pp. 326–349). Hershey, PA: IGI Global. doi:10.4018/978-1-5225-2709-1.ch009

Carfì, D., Donato, A., & Panuccio, D. (2018). A Game Theory Coopetitive Perspective for Sustainability of Global Feeding: Agreements Among Vegan and Non-Vegan Food Firms. In I. Management Association (Ed.), Game Theory: Breakthroughs in Research and Practice (pp. 71-104). Hershey, PA: IGI Global. doi:10.4018/978-1-5225-2594-3.ch004

Carr, D. A., & Carr, T. R. (2007). Geographic Information System Applications in the Public Sector. In G. Garson (Ed.), *Modern Public Information Technology Systems: Issues and Challenges* (pp. 293–311). Hershey, PA: IGI Global. doi:10.4018/978-1-59904-051-6.ch015

Castagnolo, V. (2018). Analyzing, Classifying, Safeguarding: Drawing for the Borgo Murattiano Neighbourhood of Bari. In G. Carlone, N. Martinelli, & F. Rotondo (Eds.), *Designing Grid Cities for Optimized Urban Development and Planning* (pp. 93–108). Hershey, PA: IGI Global. doi:10.4018/978-1-5225-3613-0.ch006

Chekima, B. (2018). The Dilemma of Purchase Intention: A Conceptual Framework for Understanding Actual Consumption of Organic Food. *International Journal of Sustainable Economies Management*, 7(2), 1–13. doi:10.4018/IJSEM.2018040101

Chen, Y. (2017). Sustainable Supply Chains and International Soft Landings: A Case of Wetland Entrepreneurship. In B. Christiansen & F. Kasarcı (Eds.), *Corporate Espionage, Geopolitics, and Diplomacy Issues in International Business* (pp. 232–247). Hershey, PA: IGI Global. doi:10.4018/978-1-5225-1031-4.ch013

Çıtak, L., Akel, V., & Ersoy, E. (2018). Investors' Reactions to the Announcement of New Constituents of BIST Sustainability Index: An Analysis by Event Study and Mean-Median Tests. In M. Risso & S. Testarmata (Eds.), *Value Sharing for Sustainable and Inclusive Development* (pp. 270–289). Hershey, PA: IGI Global. doi:10.4018/978-1-5225-3147-0.ch012

Crossland, M. D. (2009). Geographic Information Systems as Decision Tools. In M. Khosrow-Pour, D.B.A. (Ed.), Encyclopedia of Information Science and Technology, Second Edition (pp. 1630-1633). Hershey, PA: IGI Global. doi:10.4018/978-1-60566-026-4.ch257

Crossland, M. D., Herschel, R. T., Perkins, W. C., & Scudder, J. N. (2002). Geographic Information Systems: How Cognitive Style Impacts Decision-Making Effectiveness. In M. Mahmood (Ed.), *Advanced Topics in End User Computing* (Vol. 1, pp. 285–301). Hershey, PA: IGI Global. doi:10.4018/978-1-930708-42-6.ch017

D'Aleo, V., D'Aleo, F., & Bonanno, R. (2018). New Food Industries Toward a New Level of Sustainable Supply: Success Stories, Business Models, and Strategies. In V. Erokhin (Ed.), *Establishing Food Security and Alternatives to International Trade in Emerging Economies* (pp. 74–97). Hershey, PA: IGI Global. doi:10.4018/978-1-5225-2733-6.ch004

Dagevos, H., & Reinders, M. J. (2018). Flexitarianism and Social Marketing: Reflections on Eating Meat in Moderation. In D. Bogueva, D. Marinova, & T. Raphaely (Eds.), *Handbook of Research on Social Marketing and Its Influence on Animal Origin Food Product Consumption* (pp. 105–120). Hershey, PA: IGI Global. doi:10.4018/978-1-5225-4757-0.ch007

Deenapanray, P. N., & Ramma, I. (2017). Adaptations to Climate Change and Climate Variability in the Agriculture Sector in Mauritius: Lessons from a Technical Needs Assessment. In I. Management Association (Ed.), Natural Resources Management: Concepts, Methodologies, Tools, and Applications (pp. 655-680). Hershey, PA: IGI Global. doi:10.4018/978-1-5225-0803-8.ch030

Deenapanray, P. N., & Ramma, I. (2017). Adaptations to Climate Change and Climate Variability in the Agriculture Sector in Mauritius: Lessons from a Technical Needs Assessment. In I. Management Association (Ed.), Natural Resources Management: Concepts, Methodologies, Tools, and Applications (pp. 655-680). Hershey, PA: IGI Global. doi:10.4018/978-1-5225-0803-8.ch030

Deshpande, S., Basu, S. K., Li, X., & Chen, X. (2017). Smart, Innovative and Intelligent Technologies Used in Drug Designing. In I. Management Association (Ed.), Pharmaceutical Sciences: Breakthroughs in Research and Practice (pp. 1175-1191). Hershey, PA: IGI Global. doi:10.4018/978-1-5225-1762-7.ch045

Dlamini, P. N. (2017). Use of Information and Communication Technologies Tools to Capture, Store, and Disseminate Indigenous Knowledge: A Literature Review. In P. Ngulube (Ed.), *Handbook of Research on Theoretical Perspectives on Indigenous Knowledge Systems in Developing Countries* (pp. 225–247). Hershey, PA: IGI Global. doi:10.4018/978-1-5225-0833-5.ch010

Dolejsova, M., & Kera, D. (2017). The Fermentation GutHub Project and the Internet of Microbes. In S. Konomi & G. Roussos (Eds.), *Enriching Urban Spaces with Ambient Computing, the Internet of Things, and Smart City Design* (pp. 25–46). Hershey, PA: IGI Global. doi:10.4018/978-1-5225-0827-4.ch002

Dolunay, O. (2018). A Paradigm Shift: Empowering Farmers to Eliminate the Waste in the Form of Fresh Water and Energy Through the Implementation of 4R+T. In I. Management Association (Ed.), Sustainable Development: Concepts, Methodologies, Tools, and Applications (pp. 882-892). Hershey, PA: IGI Global. doi:10.4018/978-1-5225-3817-2.ch039

Dube, P., Heijman, W. J., Ihle, R., & Ochieng, J. (2018). The Potential of Traditional Leafy Vegetables for Improving Food Security in Africa. In V. Erokhin (Ed.), *Establishing Food Security and Alternatives to International Trade in Emerging Economies* (pp. 220–243). Hershey, PA: IGI Global. doi:10.4018/978-1-5225-2733-6.ch011

Duruji, M. M., & Urenma, D. F. (2017). The Environmentalism and Politics of Climate Change: A Study of the Process of Global Convergence through UNFCCC Conferences. In I. Management Association (Ed.), *Natural Resources Management: Concepts, Methodologies, Tools, and Applications* (pp. 77-108). Hershey, PA: IGI Global. 10.4018/978-1-5225-0803-8.ch004

Dutta, U. (2017). Agro-Geoinformatics, Potato Cultivation, and Climate Change. In S. Londhe (Ed.), *Sustainable Potato Production and the Impact of Climate Change* (pp. 247–271). Hershey, PA: IGI Global. doi:10.4018/978-1-5225-1715-3.ch012

Edirisinghe, R., Stranieri, A., & Wickramasinghe, N. (2017). A Taxonomy for mHealth. In N. Wickramasinghe (Ed.), *Handbook of Research on Healthcare Administration and Management* (pp. 596–615). Hershey, PA: IGI Global. doi:10.4018/978-1-5225-0920-2.ch036

Ekpeni, N. M., & Ayeni, A. O. (2018). Global Natural Hazard and Disaster Vulnerability Management. In A. Eneanya (Ed.), *Handbook of Research on Environmental Policies for Emergency Management and Public Safety* (pp. 83–104). Hershey, PA: IGI Global. doi:10.4018/978-1-5225-3194-4.ch005

Ene, C., Voica, M. C., & Panait, M. (2017). Green Investments and Food Security: Opportunities and Future Directions in the Context of Sustainable Development. In M. Mieila (Ed.), *Measuring Sustainable Development and Green Investments in Contemporary Economies* (pp. 163–200). Hershey, PA: IGI Global. doi:10.4018/978-1-5225-2081-8.ch007

Escamilla, I., Ruíz, M. T., Ibarra, M. M., Soto, V. L., Quintero, R., & Guzmán, G. (2018). Geocoding Tweets Based on Semantic Web and Ontologies. In M. Lytras, N. Aljohani, E. Damiani, & K. Chui (Eds.), *Innovations, Developments, and Applications of Semantic Web and Information Systems* (pp. 372–392). Hershey, PA: IGI Global. doi:10.4018/978-1-5225-5042-6.ch014

Escribano, A. J. (2018). Marketing Strategies for Trendy Animal Products: Sustainability as a Core. In F. Quoquab, R. Thurasamy, & J. Mohammad (Eds.), *Driving Green Consumerism Through Strategic Sustainability Marketing* (pp. 169–203). Hershey, PA: IGI Global. doi:10.4018/978-1-5225-2912-5.ch010

Eudoxie, G., & Roopnarine, R. (2017). Climate Change Adaptation and Disaster Risk Management in the Caribbean. In W. Ganpat & W. Isaac (Eds.), *Environmental Sustainability and Climate Change Adaptation Strategies* (pp. 97–125). Hershey, PA: IGI Global. doi:10.4018/978-1-5225-1607-1.ch004

Farmer, L. S. (2017). Data Analytics for Strategic Management: Getting the Right Data. In V. Wang (Ed.), *Encyclopedia of Strategic Leadership and Management* (pp. 810–822). Hershey, PA: IGI Global. doi:10.4018/978-1-5225-1049-9.ch056

Fattal, L. R. (2017). Catastrophe: An Uncanny Catalyst for Creativity. In R. Shin (Ed.), *Convergence of Contemporary Art, Visual Culture, and Global Civic Engagement* (pp. 244–262). Hershey, PA: IGI Global. doi:10.4018/978-1-5225-1665-1.ch014

Forti, I. (2017). A Cross Reading of Landscape through Digital Landscape Models: The Case of Southern Garda. In A. Ippolito (Ed.), *Handbook of Research on Emerging Technologies for Architectural and Archaeological Heritage* (pp. 532–561). Hershey, PA: IGI Global. doi:10.4018/978-1-5225-0675-1.ch018

Gharbi, A., De Runz, C., & Akdag, H. (2017). Urban Development Modelling: A Survey. In S. Faiz & K. Mahmoudi (Eds.), *Handbook of Research on Geographic Information Systems Applications and Advancements* (pp. 96–124). Hershey, PA: IGI Global. doi:10.4018/978-1-5225-0937-0.ch004

Ghosh, I., & Ghoshal, I. (2018). Implications of Trade Liberalization for Food Security Under the ASEAN-India Strategic Partnership: A Gravity Model Approach. In V. Erokhin (Ed.), *Establishing Food Security and Alternatives to International Trade in Emerging Economies* (pp. 98–118). Hershey, PA: IGI Global. doi:10.4018/978-1-5225-2733-6.ch005

Glaves, H. M. (2017). Developing a Common Global Framework for Marine Data Management. In P. Diviacco, A. Leadbetter, & H. Glaves (Eds.), *Oceanographic and Marine Cross-Domain Data Management for Sustainable Development* (pp. 47–68). Hershey, PA: IGI Global. doi:10.4018/978-1-5225-0700-0.ch003

Godulla, A., & Wolf, C. (2018). Future of Food: Transmedia Strategies of National Geographic. In R. Gambarato & G. Alzamora (Eds.), *Exploring Transmedia Journalism in the Digital Age* (pp. 162–182). Hershey, PA: IGI Global. doi:10.4018/978-1-5225-3781-6.ch010

Gomes, P. P. (2018). Food and Environment: A Review on the Sustainability of Six Different Dietary Patterns. In A. Obayelu (Ed.), *Food Systems Sustainability and Environmental Policies in Modern Economies* (pp. 15–31). Hershey, PA: IGI Global. doi:10.4018/978-1-5225-3631-4.ch002

Gonzalez-Feliu, J. (2018). Sustainability Evaluation of Green Urban Logistics Systems: Literature Overview and Proposed Framework. In A. Paul, D. Bhattacharyya, & S. Anand (Eds.), *Green Initiatives for Business Sustainability and Value Creation* (pp. 103–134). Hershey, PA: IGI Global. doi:10.4018/978-1-5225-2662-9.ch005

Goodland, R. (2017). A Fresh Look at Livestock Greenhouse Gas Emissions and Mitigation. In I. Management Association (Ed.), Natural Resources Management: Concepts, Methodologies, Tools, and Applications (pp. 124-139). Hershey, PA: IGI Global. doi:10.4018/978-1-5225-0803-8.ch006

Goundar, S., & Appana, S. (2018). Mainstreaming Development Policies for Climate Change in Fiji: A Policy Gap Analysis and the Role of ICTs. In I. Management Association (Ed.), Sustainable Development: Concepts, Methodologies, Tools, and Applications (pp. 402-432). Hershey, PA: IGI Global. doi:10.4018/978-1-5225-3817-2.ch020

Granell-Canut, C., & Aguilar-Moreno, E. (2018). Geospatial Influence in Science Mapping. In M. Khosrow-Pour, D.B.A. (Ed.), Encyclopedia of Information Science and Technology, Fourth Edition (pp. 3473-3483). Hershey, PA: IGI Global. doi:10.4018/978-1-5225-2255-3.ch302

Grigelis, A., Blažauskas, N., Gelumbauskaitė, L. Ž., Gulbinskas, S., Suzdalev, S., & Ferrarin, C. (2017). Marine Environment Data Management Related to the Human Activity in the South-Eastern Baltic Sea (The Lithuanian Sector). In P. Diviacco, A. Leadbetter, & H. Glaves (Eds.), *Oceanographic and Marine Cross-Domain Data Management for Sustainable Development* (pp. 282–302). Hershey, PA: IGI Global. doi:10.4018/978-1-5225-0700-0.ch012

Guma, I. P., Rwashana, A. S., & Oyo, B. (2018). Food Security Policy Analysis Using System Dynamics: The Case of Uganda. *International Journal of Information Technologies and Systems Approach*, *11*(1), 72–90. doi:10.4018/IJITSA.2018010104

Guma, I. P., Rwashana, A. S., & Oyo, B. (2018). Food Security Indicators for Subsistence Farmers Sustainability: A System Dynamics Approach. *International Journal of System Dynamics Applications*, *7*(1), 45–64. doi:10.4018/IJSDA.2018010103

Gupta, P., & Goyal, S. (2017). Wildlife Habitat Evaluation. In A. Santra & S. Mitra (Eds.), *Remote Sensing Techniques and GIS Applications in Earth and Environmental Studies* (pp. 258–264). Hershey, PA: IGI Global. doi:10.4018/978-1-5225-1814-3.ch013

Hanson, T., & Hildebrand, E. (2018). GPS Travel Diaries in Rural Transportation Research: A Focus on Older Drivers. In I. Management Association (Ed.), Intelligent Transportation and Planning: Breakthroughs in Research and Practice (pp. 609-625). Hershey, PA: IGI Global. doi:10.4018/978-1-5225-5210-9.ch027

Hartman, M. B. (2017). Research-Based Climate Change Public Education Programs. In I. Management Association (Ed.), Natural Resources Management: Concepts, Methodologies, Tools, and Applications (pp. 992-1003). Hershey, PA: IGI Global. doi:10.4018/978-1-5225-0803-8.ch046

Hashim, N. (2017). Zanzibari Seaweed: Global Climate Change and the Promise of Adaptation. In I. Management Association (Ed.), Natural Resources Management: Concepts, Methodologies, Tools, and Applications (pp. 365-391). Hershey, PA: IGI Global. doi:10.4018/978-1-5225-0803-8.ch019

Herrera, J. E., Argüello, L. V., Gonzalez-Feliu, J., & Jaimes, W. A. (2017). Decision Support System Design Requirements, Information Management, and Urban Logistics Efficiency: Case Study of Bogotá, Colombia. In G. Jamil, A. Soares, & C. Pessoa (Eds.), *Handbook of Research on Information Management for Effective Logistics and Supply Chains* (pp. 223–238). Hershey, PA: IGI Global. doi:10.4018/978-1-5225-0973-8.ch012

Huizinga, T., Ayanso, A., Smoor, M., & Wronski, T. (2017). Exploring Insurance and Natural Disaster Tweets Using Text Analytics. *International Journal of Business Analytics*, *4*(1), 1–17. doi:10.4018/IJBAN.2017010101

Hung, K., Kalantari, M., & Rajabifard, A. (2017). An Integrated Method for Assessing the Text Content Quality of Volunteered Geographic Information in Disaster Management. *International Journal of Information Systems for Crisis Response and Management*, *9*(2), 1–17. doi:10.4018/IJISCRAM.2017040101

Husnain, A., & Avdic, A. (2018). Identifying the Contemporary Status of E-Service Sustainability Research. In I. Management Association (Ed.), Sustainable Development: Concepts, Methodologies, Tools, and Applications (pp. 467-485). Hershey, PA: IGI Global. doi:10.4018/978-1-5225-3817-2.ch022

Iarossi, M. P., & Ferro, L. (2017). "The Past is Never Dead. It's Not Even Past": Virtual Archaeological Promenade. In A. Ippolito & M. Cigola (Eds.), *Handbook of Research on Emerging Technologies for Digital Preservation and Information Modeling* (pp. 228–255). Hershey, PA: IGI Global. doi:10.4018/978-1-5225-0680-5.ch010

Ignjatijević, S., & Cvijanović, D. (2018). Analysis of Serbian Honey Production and Exports. In *Exploring the Global Competitiveness of Agri-Food Sectors and Serbia's Dominant Presence: Emerging Research and Opportunities* (pp. 109–139). Hershey, PA: IGI Global. doi:10.4018/978-1-5225-2762-6.ch005

Jana, S. K., & Karmakar, A. K. (2017). Globalization, Governance, and Food Security: The Case of BRICS. In I. Management Association (Ed.), Natural Resources Management: Concepts, Methodologies, Tools, and Applications (pp. 692-712). Hershey, PA: IGI Global. doi:10.4018/978-1-5225-0803-8.ch032

Jana, S. K., & Karmakar, A. K. (2017). Food Security in Asia: Is There Convergence? In I. Management Association (Ed.), Natural Resources Management: Concepts, Methodologies, Tools, and Applications (pp. 109-123). Hershey, PA: IGI Global. doi:10.4018/978-1-5225-0803-8.ch005

John, J., & Kumar, S. (2018). A Locational Decision Making Framework for Shipbreaking Under Multiple Criteria. In I. Management Association (Ed.), Operations and Service Management: Concepts, Methodologies, Tools, and Applications (pp. 504-527). Hershey, PA: IGI Global. doi:10.4018/978-1-5225-3909-4.ch024

John, J., & Srivastava, R. K. (2018). Decision Insights for Shipbreaking using Environmental Impact Assessment: Review and Perspectives. *International Journal of Strategic Decision Sciences*, *9*(1), 45–62. doi:10.4018/IJSDS.2018010104

Joshi, Y., & Rahman, Z. (2018). Determinants of Sustainable Consumption Behaviour: Review and Conceptual Framework. In A. Paul, D. Bhattacharyya, & S. Anand (Eds.), *Green Initiatives for Business Sustainability and Value Creation* (pp. 239–262). Hershey, PA: IGI Global. doi:10.4018/978-1-5225-2662-9.ch011

Juma, D. W., Reuben, M., Wang, H., & Li, F. (2018). Adaptive Coevolution: Realigning the Water Governance Regime to the Changing Climate. In I. Management Association (Ed.), Hydrology and Water Resource Management: Breakthroughs in Research and Practice (pp. 346-357). Hershey, PA: IGI Global. doi:10.4018/978-1-5225-3427-3.ch014

K., S., & Tripathy, B. K. (2018). Neighborhood Rough-Sets-Based Spatial Data Analytics. In M. Khosrow-Pour, D.B.A. (Ed.), *Encyclopedia of Information Science and Technology, Fourth Edition* (pp. 1835-1844). Hershey, PA: IGI Global. doi:10.4018/978-1-5225-2255-3.ch160

Kabir, F. (2018). Towards a More Gender-Inclusive Climate Change Policy. In N. Mahtab, T. Haque, I. Khan, M. Islam, & I. Wahid (Eds.), *Handbook of Research on Women's Issues and Rights in the Developing World* (pp. 354–369). Hershey, PA: IGI Global. doi:10.4018/978-1-5225-3018-3.ch022

Kabir, F. (2018). Towards a More Gender-Inclusive Climate Change Policy. In I. Management Association (Ed.), Climate Change and Environmental Concerns: Breakthroughs in Research and Practice (pp. 525-540). Hershey, PA: IGI Global. doi:10.4018/978-1-5225-5487-5.ch027

Kanyamuka, J. S., Jumbe, C. B., & Ricker-Gilbert, J. (2018). Making Agricultural Input Subsidies More Effective and Profitable in Africa: The Role of Complementary Interventions. In A. Obayelu (Ed.), *Food Systems Sustainability and Environmental Policies in Modern Economies* (pp. 172–187). Hershey, PA: IGI Global. doi:10.4018/978-1-5225-3631-4.ch008

Karimi, H., & Gholamrezafahimi, F. (2017). Study of Integrated Coastal Zone Management and Its Environmental Effects: A Case of Iran. In R. Singh, A. Singh, & V. Srivastava (Eds.), *Environmental Issues Surrounding Human Overpopulation* (pp. 64–88). Hershey, PA: IGI Global. doi:10.4018/978-1-5225-1683-5.ch004

Kaya, I. R., Hutabarat, J., & Bambang, A. N. (2018). "Sasi": A New Path to Sustain Seaweed Farming From Up-Stream to Down-Stream in Kotania Bay, Molucass. *International Journal of Social Ecology and Sustainable Development, 9*(2), 28–36. doi:10.4018/IJSESD.2018040103

Khader, V. (2018). Technologies for Food, Health, Livelihood, and Nutrition Security. In I. Management Association (Ed.), Food Science and Nutrition: Breakthroughs in Research and Practice (pp. 94-112). Hershey, PA: IGI Global. doi:10.4018/978-1-5225-5207-9.ch005

Kocadağlı, A. Y. (2017). The Temporal and Spatial Development of Organic Agriculture in Turkey. In W. Ganpat, R. Dyer, & W. Isaac (Eds.), *Agricultural Development and Food Security in Developing Nations* (pp. 130–156). Hershey, PA: IGI Global. doi:10.4018/978-1-5225-0942-4.ch006

Koundouri, P., Giannouli, A., & Souliotis, I. (2017). An Integrated Approach for Sustainable Environmental and Socio-Economic Development Using Offshore Infrastructure. In I. Management Association (Ed.), Renewable and Alternative Energy: Concepts, Methodologies, Tools, and Applications (pp. 1581-1601). Hershey, PA: IGI Global. doi:10.4018/978-1-5225-1671-2.ch056

Kumar, A., & Dash, M. K. (2017). Sustainability and Future Generation Infrastructure on Digital Platform: A Study of Generation Y. In N. Ray (Ed.), *Business Infrastructure for Sustainability in Developing Economies* (pp. 124–142). Hershey, PA: IGI Global. doi:10.4018/978-1-5225-2041-2.ch007

Kumar, A., Mukherjee, A. B., & Krishna, A. P. (2017). Application of Conventional Data Mining Techniques and Web Mining to Aid Disaster Management. In A. Kumar (Ed.), *Web Usage Mining Techniques and Applications Across Industries* (pp. 138–167). Hershey, PA: IGI Global. doi:10.4018/978-1-5225-0613-3.ch006

Kumar, C. P. (2017). Impact of Climate Change on Groundwater Resources. In I. Management Association (Ed.), Natural Resources Management: Concepts, Methodologies, Tools, and Applications (pp. 1094-1120). Hershey, PA: IGI Global. doi:10.4018/978-1-5225-0803-8.ch052

Kumari, S., & Patil, Y. (2017). Achieving Climate Smart Agriculture with a Sustainable Use of Water: A Conceptual Framework for Sustaining the Use of Water for Agriculture in the Era of Climate Change. In P. Rao & Y. Patil (Eds.), *Reconsidering the Impact of Climate Change on Global Water Supply, Use, and Management* (pp. 122–143). Hershey, PA: IGI Global. doi:10.4018/978-1-5225-1046-8.ch008

Kumari, S., & Patil, Y. (2018). Achieving Climate Smart Agriculture With a Sustainable Use of Water: A Conceptual Framework for Sustaining the Use of Water for Agriculture in the Era of Climate Change. In I. Management Association (Ed.), Climate Change and Environmental Concerns: Breakthroughs in Research and Practice (pp. 111-133). Hershey, PA: IGI Global. doi:10.4018/978-1-5225-5487-5.ch006

Kursah, M. B. (2017). Least-Cost Pipeline using Geographic Information System: The Limit to Technicalities. *International Journal of Applied Geospatial Research*, *8*(3), 1–15. doi:10.4018/ijagr.2017070101

Lahiri, S., Ghosh, D., & Bhakta, J. N. (2017). Role of Microbes in Eco-Remediation of Perturbed Aquatic Ecosystem. In J. Bhakta (Ed.), *Handbook of Research on Inventive Bioremediation Techniques* (pp. 70–107). Hershey, PA: IGI Global. doi:10.4018/978-1-5225-2325-3.ch004

Lallo, C. H., Smalling, S., Facey, A., & Hughes, M. (2018). The Impact of Climate Change on Small Ruminant Performance in Caribbean Communities. In I. Management Association (Ed.), Climate Change and Environmental Concerns: Breakthroughs in Research and Practice (pp. 193-218). Hershey, PA: IGI Global. doi:10.4018/978-1-5225-5487-5.ch010

Laurini, R. (2017). Nature of Geographic Knowledge Bases. In S. Faiz & K. Mahmoudi (Eds.), *Handbook of Research on Geographic Information Systems Applications and Advancements* (pp. 29–60). Hershey, PA: IGI Global. doi:10.4018/978-1-5225-0937-0.ch002

Lawrence, J., Simpson, L., & Piggott, A. (2017). Protected Agriculture: A Climate Change Adaptation for Food and Nutrition Security. In I. Management Association (Ed.), Natural Resources Management: Concepts, Methodologies, Tools, and Applications (pp. 140-158). Hershey, PA: IGI Global. doi:10.4018/978-1-5225-0803-8.ch007

Leadbetter, A., Cheatham, M., Shepherd, A., & Thomas, R. (2017). Linked Ocean Data 2.0. In P. Diviacco, A. Leadbetter, & H. Glaves (Eds.), *Oceanographic and Marine Cross-Domain Data Management for Sustainable Development* (pp. 69–99). Hershey, PA: IGI Global. doi:10.4018/978-1-5225-0700-0.ch004

Leipnik, M. R., & Mehta, S. S. (2005). Geographic Information Systems (GIS) in E-Marketing. In I. Clarke III & T. Flaherty (Eds.), *Advances in Electronic Marketing* (pp. 193–210). Hershey, PA: IGI Global. doi:10.4018/978-1-59140-321-0.ch011

Lucas, M. R., Rego, C., Vieira, C., & Vieira, I. (2017). Proximity and Cooperation for Innovative Regional Development: The Case of the Science and Technology Park of Alentejo. In L. Carvalho (Ed.), *Handbook of Research on Entrepreneurial Development and Innovation Within Smart Cities* (pp. 199–228). Hershey, PA: IGI Global. doi:10.4018/978-1-5225-1978-2.ch010

Ma, X., Beaulieu, S. E., Fu, L., Fox, P., Di Stefano, M., & West, P. (2017). Documenting Provenance for Reproducible Marine Ecosystem Assessment in Open Science. In P. Diviacco, A. Leadbetter, & H. Glaves (Eds.), *Oceanographic and Marine Cross-Domain Data Management for Sustainable Development* (pp. 100–126). Hershey, PA: IGI Global. doi:10.4018/978-1-5225-0700-0.ch005

Mabe, L. K., & Oladele, O. I. (2017). Application of Information Communication Technologies for Agricultural Development through Extension Services: A Review. In T. Tossy (Ed.), *Information Technology Integration for Socio-Economic Development* (pp. 52–101). Hershey, PA: IGI Global. doi:10.4018/978-1-5225-0539-6.ch003

Malomo, B. I. (2018). A Review of Psychological Resilience as a Response to Natural Hazards in Nigeria. In A. Eneanya (Ed.), *Handbook of Research on Environmental Policies for Emergency Management and Public Safety* (pp. 147–165). Hershey, PA: IGI Global. doi:10.4018/978-1-5225-3194-4.ch008

Manchiraju, S. (2018). Predicting Behavioral Intentions Toward Sustainable Fashion Consumption: A Comparison of Attitude-Behavior and Value-Behavior Consistency Models. In I. Management Association (Ed.), Fashion and Textiles: Breakthroughs in Research and Practice (pp. 1-21). Hershey, PA: IGI Global. doi:10.4018/978-1-5225-3432-7.ch001

Manzella, G. M., Bartolini, R., Bustaffa, F., D'Angelo, P., De Mattei, M., Frontini, F., ... Spada, A. (2017). Semantic Search Engine for Data Management and Sustainable Development: Marine Planning Service Platform. In P. Diviacco, A. Leadbetter, & H. Glaves (Eds.), *Oceanographic and Marine Cross-Domain Data Management for Sustainable Development* (pp. 127–154). Hershey, PA: IGI Global. doi:10.4018/978-1-5225-0700-0.ch006

Mbonigaba, J. (2018). Comparing the Effects of Unsustainable Production and Consumption of Food on Health and Policy Across Developed and Less Developed Countries. In A. Obayelu (Ed.), *Food Systems Sustainability and Environmental Policies in Modern Economies* (pp. 124–158). Hershey, PA: IGI Global. doi:10.4018/978-1-5225-3631-4.ch006

McKeown, A. E. (2017). Nurses, Healthcare, and Environmental Pollution and Solutions: Breaking the Cycle of Harm. In I. Management Association (Ed.), Natural Resources Management: Concepts, Methodologies, Tools, and Applications (pp. 392-415). Hershey, PA: IGI Global. doi:10.4018/978-1-5225-0803-8.ch020

Mennecke, B. E., & West, L. A. Jr. (2001). Geographic Information Systems in Developing Countries: Issues in Data Collection, Implementation and Management. *Journal of Global Information Management*, *9*(4), 44–54. doi:10.4018/jgim.2001100103

Mili, B., Barua, A., & Katyaini, S. (2017). Climate Change and Adaptation through the Lens of Capability Approach: A Case Study from Darjeeling, Eastern Himalaya. In I. Management Association (Ed.), Natural Resources Management: Concepts, Methodologies, Tools, and Applications (pp. 1351-1365). Hershey, PA: IGI Global. doi:10.4018/978-1-5225-0803-8.ch064

Mir, S. A., Shah, M. A., Mir, M. M., & Iqbal, U. (2017). New Horizons of Nanotechnology in Agriculture and Food Processing Industry. In B. Nayak, A. Nanda, & M. Bhat (Eds.), *Integrating Biologically-Inspired Nanotechnology into Medical Practice* (pp. 230–258). Hershey, PA: IGI Global. doi:10.4018/978-1-5225-0610-2.ch009

Mujere, N., & Moyce, W. (2017). Climate Change Impacts on Surface Water Quality. In W. Ganpat & W. Isaac (Eds.), *Environmental Sustainability and Climate Change Adaptation Strategies* (pp. 322–340). Hershey, PA: IGI Global. doi:10.4018/978-1-5225-1607-1.ch012

Mukherjee, A. B., Krishna, A. P., & Patel, N. (2018). Geospatial Technology for Urban Sciences. In *Geospatial Technologies in Urban System Development: Emerging Research and Opportunities* (pp. 99–120). Hershey, PA: IGI Global. doi:10.4018/978-1-5225-3683-3.ch005

Nagarajan, S. K., & Sangaiah, A. K. (2017). Vegetation Index: Ideas, Methods, Influences, and Trends. In N. Kumar, A. Sangaiah, M. Arun, & S. Anand (Eds.), *Advanced Image Processing Techniques and Applications* (pp. 347–386). Hershey, PA: IGI Global. doi:10.4018/978-1-5225-2053-5.ch016

Naraine, L., & Meehan, K. (2017). Strengthening Food Security with Sustainable Practices by Smallholder Farmers in Lesser Developed Economies. In W. Ganpat, R. Dyer, & W. Isaac (Eds.), *Agricultural Development and Food Security in Developing Nations* (pp. 57–81). Hershey, PA: IGI Global. doi:10.4018/978-1-5225-0942-4.ch003

Naraine, L., & Meehan, K. (2017). Strengthening Food Security with Sustainable Practices by Smallholder Farmers in Lesser Developed Economies. In W. Ganpat, R. Dyer, & W. Isaac (Eds.), *Agricultural Development and Food Security in Developing Nations* (pp. 57–81). Hershey, PA: IGI Global. doi:10.4018/978-1-5225-0942-4.ch003

Naraine, L., & Meehan, K. (2017). Strengthening Food Security with Sustainable Practices by Smallholder Farmers in Lesser Developed Economies. In W. Ganpat, R. Dyer, & W. Isaac (Eds.), *Agricultural Development and Food Security in Developing Nations* (pp. 57–81). Hershey, PA: IGI Global. doi:10.4018/978-1-5225-0942-4.ch003

Neufeld, D. J., & Griffith, S. (2006). Isobord's Geographic Information System (GIS) Solution. In M. Khosrow-Pour, D.B.A. (Ed.), Cases on Database Technologies and Applications (pp. 64-80). Hershey, PA: IGI Global. doi:10.4018/978-1-59904-399-9.ch004

Nikolaou, K., Tsakiridou, E., Anastasiadis, F., & Mattas, K. (2017). Exploring Alternative Distribution Channels of Agricultural Products. *International Journal of Food and Beverage Manufacturing and Business Models*, *2*(2), 36–66. doi:10.4018/IJFBMBM.2017070103

Nishat, K. J., & Rahman, M. S. (2018). Disaster, Vulnerability, and Violence Against Women: Global Findings and a Research Agenda for Bangladesh. In N. Mahtab, T. Haque, I. Khan, M. Islam, & I. Wahid (Eds.), *Handbook of Research on Women's Issues and Rights in the Developing World* (pp. 235–250). Hershey, PA: IGI Global. doi:10.4018/978-1-5225-3018-3.ch014

O'Hara, S., Jones, D., & Trobman, H. B. (2018). Building an Urban Food System Through UDC Food Hubs. In A. Burtin, J. Fleming, & P. Hampton-Garland (Eds.), *Changing Urban Landscapes Through Public Higher Education* (pp. 116–143). Hershey, PA: IGI Global. doi:10.4018/978-1-5225-3454-9.ch006

Obayelu, A. E. (2018). Integrating Environment, Food Systems, and Sustainability in Feeding the Growing Population in Developing Countries. In A. Obayelu (Ed.), *Food Systems Sustainability and Environmental Policies in Modern Economies* (pp. 1–14). Hershey, PA: IGI Global. doi:10.4018/978-1-5225-3631-4.ch001

Othman, R., Nath, N., & Laswad, F. (2018). Environmental Reporting and Accounting: Sustainability Hybridisation. In G. Azevedo, J. da Silva Oliveira, R. Marques, & A. Ferreira (Eds.), *Handbook of Research on Modernization and Accountability in Public Sector Management* (pp. 130–158). Hershey, PA: IGI Global. doi:10.4018/978-1-5225-3731-1.ch007

Padigala, B. S. (2018). Traditional Water Management System for Climate Change Adaptation in Mountain Ecosystems. In I. Management Association (Ed.), Climate Change and Environmental Concerns: Breakthroughs in Research and Practice (pp. 630-655). Hershey, PA: IGI Global. doi:10.4018/978-1-5225-5487-5.ch033

Panda, C. K. (2018). Mobile Phone Usage in Agricultural Extension in India: The Current and Future Perspective. In F. Mtenzi, G. Oreku, D. Lupiana, & J. Yonazi (Eds.), *Mobile Technologies and Socio-Economic Development in Emerging Nations* (pp. 1–21). Hershey, PA: IGI Global. doi:10.4018/978-1-5225-4029-8.ch001

Pandian, S. L., Yarrakula, K., & Chaudhury, P. (2018). GIS-Based Decision Support System for Village Level: A Case Study in Andhra Pradesh. In S. Chaudhuri & N. Ray (Eds.), *GIS Applications in the Tourism and Hospitality Industry* (pp. 275–295). Hershey, PA: IGI Global. doi:10.4018/978-1-5225-5088-4.ch012

Quaranta, G., & Salvia, R. (2017). Social-Based Product Innovation and Governance in The Milk Sector: The Case of Carciocacio and Innonatura. In T. Tarnanidis, M. Vlachopoulou, & J. Papathanasiou (Eds.), *Driving Agribusiness With Technology Innovations* (pp. 293–310). Hershey, PA: IGI Global. doi:10.4018/978-1-5225-2107-5.ch015

Rahman, M. K., Schmidlin, T. W., Munro-Stasiuk, M. J., & Curtis, A. (2017). Geospatial Analysis of Land Loss, Land Cover Change, and Landuse Patterns of Kutubdia Island, Bangladesh. *International Journal of Applied Geospatial Research*, *8*(2), 45–60. doi:10.4018/IJAGR.2017040104

Rajack-Talley, T. A. (2017). Agriculture, Trade Liberalization and Poverty in the ACP Countries. In W. Ganpat, R. Dyer, & W. Isaac (Eds.), *Agricultural Development and Food Security in Developing Nations* (pp. 1–20). Hershey, PA: IGI Global. doi:10.4018/978-1-5225-0942-4.ch001

Rajamanickam, S. (2018). Exploring Landscapes in Regional Convergence: Environment and Sustainable Development in South Asia. In I. Management Association (Ed.), Sustainable Development: Concepts, Methodologies, Tools, and Applications (pp. 1051-1087). Hershey, PA: IGI Global. doi:10.4018/978-1-5225-3817-2.ch047

Rizvi, S. M., & Dearden, A. (2018). KHETI: ICT Solution for Agriculture Extension and Its Replication in Open and Distance Learning. In U. Pandey & V. Indrakanti (Eds.), *Open and Distance Learning Initiatives for Sustainable Development* (pp. 163–174). Hershey, PA: IGI Global. doi:10.4018/978-1-5225-2621-6.ch008

Roşu, L., & Macarov, L. I. (2017). Management of Drought and Floods in the Dobrogea Region. In I. Management Association (Ed.), Agri-Food Supply Chain Management: Breakthroughs in Research and Practice (pp. 372-403). Hershey, PA: IGI Global. doi:10.4018/978-1-5225-1629-3.ch016

Rouzbehani, K., & Rouzbehani, S. (2018). Mapping Women's World: GIS and the Case of Breast Cancer in the US. *International Journal of Public Health Management and Ethics*, *3*(1), 14–25. doi:10.4018/IJPHME.2018010102

Roy, D. (2018). Success Factors of Adoption of Mobile Applications in Rural India: Effect of Service Characteristics on Conceptual Model. In M. Khosrow-Pour, D.B.A. (Ed.), Green Computing Strategies for Competitive Advantage and Business Sustainability (pp. 211-238). Hershey, PA: IGI Global. doi:10.4018/978-1-5225-5017-4.ch010

Sajeva, M., Lemon, M., & Sahota, P. S. (2017). Governance for Food Security: A Framework for Social Learning and Scenario Building. *International Journal of Food and Beverage Manufacturing and Business Models*, *2*(2), 67–84. doi:10.4018/IJFBMBM.2017070104

Sallam, G. A., Youssef, T., Embaby, M. E., & Shaltot, F. (2011). Using Geographic Information System to Infollow the Fertilizers Pollution Migration. In I. Management Association (Ed.), Green Technologies: Concepts, Methodologies, Tools and Applications (pp. 564-586). Hershey, PA: IGI Global. doi:10.4018/978-1-60960-472-1.ch312

Sambhanthan, A., & Potdar, V. (2017). A Study of the Parameters Impacting Sustainability in Information Technology Organizations. *International Journal of Knowledge-Based Organizations*, *7*(3), 27–39. doi:10.4018/IJKBO.2017070103

Sanga, C., Kalungwizi, V. J., & Msuya, C. P. (2017). Bridging Gender Gaps in Provision of Agricultural Extension Service Using ICT: Experiences from Sokoine University of Agriculture (SUA) Farmer Voice Radio (FVR) Project in Tanzania. In I. Management Association (Ed.), Discrimination and Diversity: Concepts, Methodologies, Tools, and Applications (pp. 682-697). Hershey, PA: IGI Global. doi:10.4018/978-1-5225-1933-1.ch031

Santra, A., & Mitra, D. (2017). Role of Remote Sensing in Potential Fishing Zone Forecast. In A. Santra & S. Mitra (Eds.), *Remote Sensing Techniques and GIS Applications in Earth and Environmental Studies* (pp. 243–257). Hershey, PA: IGI Global. doi:10.4018/978-1-5225-1814-3.ch012

Schaap, D. (2017). SeaDataNet: Towards a Pan-European Infrastructure for Marine and Ocean Data Management. In P. Diviacco, A. Leadbetter, & H. Glaves (Eds.), *Oceanographic and Marine Cross-Domain Data Management for Sustainable Development* (pp. 155–177). Hershey, PA: IGI Global. doi:10.4018/978-1-5225-0700-0.ch007

Seckin-Celik, T. (2017). Sustainability Reporting and Sustainability in the Turkish Business Context. In U. Akkucuk (Ed.), *Ethics and Sustainability in Global Supply Chain Management* (pp. 115–132). Hershey, PA: IGI Global. doi:10.4018/978-1-5225-2036-8.ch006

Segbefia, A. Y., Barnes, V. R., Akpalu, L. A., & Mensah, M. (2018). Environmental Location Assessment for Seaweed Cultivation in Ghana: A Spatial Multi-Criteria Approach. *International Journal of Applied Geospatial Research*, *9*(1), 51–64. doi:10.4018/IJAGR.2018010104

Sen, Y. (2018). How to Manage Sustainability: A Framework for Corporate Sustainability Tools. In I. Management Association (Ed.), Sustainable Development: Concepts, Methodologies, Tools, and Applications (pp. 568-589). Hershey, PA: IGI Global. doi:10.4018/978-1-5225-3817-2.ch026

Shamshiry, E., Abdulai, A. M., Mokhtar, M. B., & Komoo, I. (2015). Regional Landfill Site Selection with GIS and Analytical Hierarchy Process Techniques: A Case Study of Langkawi Island, Malaysia. In P. Thomas, M. Srihari, & S. Kaur (Eds.), *Handbook of Research on Cultural and Economic Impacts of the Information Society* (pp. 248–282). Hershey, PA: IGI Global. doi:10.4018/978-1-4666-8598-7.ch011

Sharma, Y. K., Mangla, S. K., Patil, P. P., & Uniyal, S. (2018). Analyzing Sustainable Food Supply Chain Management Challenges in India. In M. Ram & J. Davim (Eds.), *Soft Computing Techniques and Applications in Mechanical Engineering* (pp. 162–180). Hershey, PA: IGI Global. doi:10.4018/978-1-5225-3035-0.ch008

Silvestrelli, P. (2018). The Impact of Events: To Which Extent Are They Sustainable for Tourist Destinations? Some Evidences From Expo Milano 2015. In M. Risso & S. Testarmata (Eds.), *Value Sharing for Sustainable and Inclusive Development* (pp. 185–204). Hershey, PA: IGI Global. doi:10.4018/978-1-5225-3147-0.ch008

Silvius, G. (2018). Sustainability Evaluation of IT/IS Projects. In I. Management Association (Ed.), Sustainable Development: Concepts, Methodologies, Tools, and Applications (pp. 26-40). Hershey, PA: IGI Global. doi:10.4018/978-1-5225-3817-2.ch002

Singh, R., Srivastava, P., Singh, P., Upadhyay, S., & Raghubanshi, A. S. (2017). Human Overpopulation and Food Security: Challenges for the Agriculture Sustainability. In R. Singh, A. Singh, & V. Srivastava (Eds.), *Environmental Issues Surrounding Human Overpopulation* (pp. 12–39). Hershey, PA: IGI Global. doi:10.4018/978-1-5225-1683-5.ch002

Singh, R., Srivastava, P., Singh, P., Upadhyay, S., & Raghubanshi, A. S. (2017). Human Overpopulation and Food Security: Challenges for the Agriculture Sustainability. In R. Singh, A. Singh, & V. Srivastava (Eds.), *Environmental Issues Surrounding Human Overpopulation* (pp. 12–39). Hershey, PA: IGI Global. doi:10.4018/978-1-5225-1683-5.ch002

Srivastava, N. (2017). Climate Change Mitigation: Collective Efforts and Responsibly. In I. Management Association (Ed.), Natural Resources Management: Concepts, Methodologies, Tools, and Applications (pp. 64-76). Hershey, PA: IGI Global. doi:10.4018/978-1-5225-0803-8.ch003

Stanganelli, M., & Gerundo, C. (2017). Understanding the Role of Urban Morphology and Green Areas Configuration During Heat Waves. *International Journal of Agricultural and Environmental Information Systems*, *8*(2), 50–64. doi:10.4018/IJAEIS.2017040104

Stewart, M. K., Hagood, D., & Ching, C. C. (2017). Virtual Games and Real-World Communities: Environments that Constrain and Enable Physical Activity in Games for Health. *International Journal of Game-Based Learning*, *7*(1), 1–19. doi:10.4018/IJGBL.2017010101

Stone, R. J. (2017). Modelling the Frequency of Tropical Cyclones in the Lower Caribbean Region. In W. Ganpat & W. Isaac (Eds.), *Environmental Sustainability and Climate Change Adaptation Strategies* (pp. 341–349). Hershey, PA: IGI Global. doi:10.4018/978-1-5225-1607-1.ch013

Syed, A., & Jabeen, U. A. (2018). Climate Change Impact on Agriculture and Food Security. In A. Eneanya (Ed.), *Handbook of Research on Environmental Policies for Emergency Management and Public Safety* (pp. 223–237). Hershey, PA: IGI Global. doi:10.4018/978-1-5225-3194-4.ch012

Tam, G. C. (2017). The Global View of Sustainability. In *Managerial Strategies and Green Solutions for Project Sustainability* (pp. 1–24). Hershey, PA: IGI Global. doi:10.4018/978-1-5225-2371-0.ch001

Tam, G. C. (2017). Understanding Project Sustainability. In *Managerial Strategies and Green Solutions for Project Sustainability* (pp. 110–139). Hershey, PA: IGI Global. doi:10.4018/978-1-5225-2371-0.ch005

Tam, G. C. (2017). Perspectives on Sustainability. In *Managerial Strategies and Green Solutions for Project Sustainability* (pp. 53–76). Hershey, PA: IGI Global. doi:10.4018/978-1-5225-2371-0.ch003

Tang, M., & Karunanithi, A. T. (2018). Visual Logic Maps (vLms). In *Advanced Concept Maps in STEM Education: Emerging Research and Opportunities* (pp. 108–149). Hershey, PA: IGI Global. doi:10.4018/978-1-5225-2184-6.ch005

Taşçıoğlu, M., & Yener, D. (2018). The Value and Scope of GIS in Marketing and Tourism Management. In S. Chaudhuri & N. Ray (Eds.), *GIS Applications in the Tourism and Hospitality Industry* (pp. 189–211). Hershey, PA: IGI Global. doi:10.4018/978-1-5225-5088-4.ch009

Tianming, G. (2018). Food Security and Rural Development on Emerging Markets of Northeast Asia: Cases of Chinese North and Russian Far East. In V. Erokhin (Ed.), *Establishing Food Security and Alternatives to International Trade in Emerging Economies* (pp. 155–176). Hershey, PA: IGI Global. doi:10.4018/978-1-5225-2733-6.ch008

Tiftikçigil, B. Y., Yaşgül, Y. S., & Güriş, B. (2017). Sustainability of Foreign Trade Deficit in Energy: The Case of Turkey. In N. Ray (Ed.), *Business Infrastructure for Sustainability in Developing Economies* (pp. 94–109). Hershey, PA: IGI Global. doi:10.4018/978-1-5225-2041-2.ch005

Tiwari, S., Vaish, B., & Singh, P. (2018). Population and Global Food Security: Issues Related to Climate Change. In I. Management Association (Ed.), Climate Change and Environmental Concerns: Breakthroughs in Research and Practice (pp. 41-64). Hershey, PA: IGI Global. doi:10.4018/978-1-5225-5487-5.ch003

Toujani, A., & Achour, H. (2018). A Data Mining Framework for Forest Fire Mapping. In I. Management Association (Ed.), Information Retrieval and Management: Concepts, Methodologies, Tools, and Applications (pp. 771-794). Hershey, PA: IGI Global. doi:10.4018/978-1-5225-5191-1.ch033

Tripathy, B., & K., S. B. (2017). Rough Fuzzy Set Theory and Neighbourhood Approximation Based Modelling for Spatial Epidemiology. In I. Management Association (Ed.), *Public Health and Welfare: Concepts, Methodologies, Tools, and Applications* (pp. 1257-1268). Hershey, PA: IGI Global. doi:10.4018/978-1-5225-1674-3.ch058

Trukhachev, A. (2017). New Approaches to Regional Branding through Green Production and Utilization of Existing Natural Advantages. In I. Management Association (Ed.), Advertising and Branding: Concepts, Methodologies, Tools, and Applications (pp. 1758-1778). Hershey, PA: IGI Global. doi:10.4018/978-1-5225-1793-1.ch081

Tsobanoglou, G. O., & Vlachopoulou, E. I. (2017). Social-Ecological Systems in Local Fisheries Communities. In G. Korres, E. Kourliouros, & M. Michailidis (Eds.), *Handbook of Research on Policies and Practices for Sustainable Economic Growth and Regional Development* (pp. 306–316). Hershey, PA: IGI Global. doi:10.4018/978-1-5225-2458-8.ch026

Tsou, M. (2010). Geographic Information Retrieval and Text Mining on Chinese Tourism Web Pages. *International Journal of Information Technology and Web Engineering*, *5*(1), 56–75. doi:10.4018/jitwe.2010010104

Tuydes-Yaman, H., & Karatas, P. (2018). Evaluation of Walkability and Pedestrian Level of Service. In I. Management Association (Ed.), Intelligent Transportation and Planning: Breakthroughs in Research and Practice (pp. 264-291). Hershey, PA: IGI Global. doi:10.4018/978-1-5225-5210-9.ch012

Uddin, S., Chakravorty, S., Ray, A., & Sherpa, K. S. (2018). Optimal Location of Sub-Station Using Q-GIS and Multi-Criteria Decision Making Approach. *International Journal of Decision Support System Technology*, *10*(2), 65–79. doi:10.4018/IJDSST.2018040104

Uddin, S., Chakravorty, S., Sherpa, K. S., & Ray, A. (2018). Power Distribution System Planning Using Q-GIS. *International Journal of Energy Optimization and Engineering*, *7*(2), 61–75. doi:10.4018/IJEOE.2018040103

Uzun, F. V. (2018). Natural Resources Management. In A. Eneanya (Ed.), *Handbook of Research on Environmental Policies for Emergency Management and Public Safety* (pp. 1–21). Hershey, PA: IGI Global. doi:10.4018/978-1-5225-3194-4.ch001

V, M., Agrawal, R., Sharma, V., & T.N., K. (2018). Supply Chain Social Sustainability and Manufacturing. In I. Management Association (Ed.), *Technology Adoption and Social Issues: Concepts, Methodologies, Tools, and Applications* (pp. 226-252). Hershey, PA: IGI Global. doi:10.4018/978-1-5225-5201-7.ch011 van der Vliet-Bakker, J. M. (2017). Environmentally Forced Migration and Human Rights. In C. Akrivopoulou (Ed.), *Defending Human Rights and Democracy in the Era of Globalization* (pp. 146–180). Hershey, PA: IGI Global. doi:10.4018/978-1-5225-0723-9.ch007

Vaskov, A. G., Lin, Z. Y., Tyagunov, M. G., Shestopalova, T. A., & Deryugina, G. V. (2018). Design of Renewable Sources GIS for ASEAN Countries. In V. Kharchenko & P. Vasant (Eds.), *Handbook of Research on Renewable Energy and Electric Resources for Sustainable Rural Development* (pp. 1–25). Hershey, PA: IGI Global. doi:10.4018/978-1-5225-3867-7.ch001

Vázquez, D. G., & Gil, M. T. (2017). Sustainability in Smart Cities: The Case of Vitoria-Gasteiz (Spain) – A Commitment to a New Urban Paradigm. In L. Carvalho (Ed.), *Handbook of Research on Entrepreneurial Development and Innovation Within Smart Cities* (pp. 248–268). Hershey, PA: IGI Global. doi:10.4018/978-1-5225-1978-2.ch012

Wahab, I. N., & Soonthodu, S. (2018). Geographical Information System in Eco-Tourism. In S. Chaudhuri & N. Ray (Eds.), *GIS Applications in the Tourism and Hospitality Industry* (pp. 61–75). Hershey, PA: IGI Global. doi:10.4018/978-1-5225-5088-4.ch003

Weiss-Randall, D. (2018). Cultivating Environmental Justice. In *Utilizing Innovative Technologies to Address the Public Health Impact of Climate Change: Emerging Research and Opportunities* (pp. 110–143). Hershey, PA: IGI Global. doi:10.4018/978-1-5225-3414-3.ch004

Weiss-Randall, D. (2018). Cultivating Resilience. In *Utilizing Innovative Technologies to Address the Public Health Impact of Climate Change: Emerging Research and Opportunities* (pp. 204–235). Hershey, PA: IGI Global. doi:10.4018/978-1-5225-3414-3.ch007

Weiss-Randall, D. (2018). Climate Change Solutions: Where Do We Go From Here? In *Utilizing Innovative Technologies to Address the Public Health Impact of Climate Change: Emerging Research and Opportunities* (pp. 236–268). Hershey, PA: IGI Global. doi:10.4018/978-1-5225-3414-3.ch008

Whyte, K. P., List, M., Stone, J. V., Grooms, D., Gasteyer, S., Thompson, P. B., . . . Bouri, H. (2018). Uberveillance, Standards, and Anticipation: A Case Study on Nanobiosensors in U.S. Cattle. In I. Management Association (Ed.), Biomedical Engineering: Concepts, Methodologies, Tools, and Applications (pp. 577-596). Hershey, PA: IGI Global. doi:10.4018/978-1-5225-3158-6.ch025

Wulff, E. (2017). Data and Operational Oceanography: A Review in Support of Responsible Fisheries and Aquaculture. In P. Diviacco, A. Leadbetter, & H. Glaves (Eds.), *Oceanographic and Marine Cross-Domain Data Management for Sustainable Development* (pp. 303–324). Hershey, PA: IGI Global. doi:10.4018/978-1-5225-0700-0.ch013

Yener, D. (2017). Geographic Information Systems and Its Applications in Marketing Literature. In S. Faiz & K. Mahmoudi (Eds.), *Handbook of Research on Geographic Information Systems Applications and Advancements* (pp. 158–172). Hershey, PA: IGI Global. doi:10.4018/978-1-5225-0937-0.ch006

About the Contributors

Ramesh C. Poonia has rich experience of 14+ years as an academician. At present he is an Associate Professor of Computer Science at Amity Institute of Information Technology, Amity University Rajasthan. He has received his PhD in Computer Science from the Banasthali University - Rajasthan. His research interests include Vehicular Networks, Wireless & Mobile Networks, Network Protocol Evaluation, Network Simulation & Modeling, and Internet Congestion Control. He has published more than 22 papers and books; some of them have been published in referred journals/conference proceedings as: IEEE Trans. on Computers, IEEE International Conference on Signal Processing and Communication Engineering Systems, ACM - International Conference Proceedings Series (ICPS), International Journal of Wireless & Mobile Networks – AIRCC, IGI Global, Elsevier, Springer and Inderscience. Dr. Poonia is serving as senior member, technical program committee member, member of editorial board/reviewer board of various renowned national and international conferences/journals.

Xiao-Zhi Gao received his B.Sc. and M.Sc. degrees from the Harbin Institute of Technology, China in 1993 and 1996, respectively. He earned the D.Sc. (Tech.) degree from the Helsinki University of Technology (now Aalto University), Finland in 1999. Dr. Gao has published more than 290 technical papers on refereed journals and international conferences. His current research interests are nature-inspired computing methods (e.g., neural networks, fuzzy logic, evolutionary computing, swarm intelligence, and artificial immune systems) with their applications in optimization, data mining, control, signal processing, and industrial electronics. Dr. Gao is an Associate Editor of the Applied Soft Computing, International Journal of Machine Learning and Cybernetics, Journal of Intelligent Automation and Soft Computing, International Journal of Innovative Computing, Information and Control, and International Journal of Swarm Intelligence and Evolutionary Computation. He also serves in the editorial boards of a few international journals. Dr. Gao is the General Chair of the 2005 IEEE Mid-Summer Workshop on Soft Computing in Industrial Applications. He is an invited plenary speaker at the 2014 International Workshop on Synchro-phasor Measurements for Smart Grid, 2006 International Workshop on Nature Inspired Cooperative Strategies for Optimization, and 2001 NATO Advanced Research Workshop on Systematic Organization of Information in Fuzzy Systems.

Linesh Raja is currently working as Assistant Professor at Amity Institute of Information Technology, Amity University Rajasthan, India. He earned the Ph.D. in computer science from Jaipur National University, India Before that he has completed his Masters and Bachelor degree from Birla Institute of Technology, India. Dr. Linesh has published several research papers in the field of wireless communication, mobile networks security and internet of things in various reputed national and international journals. He has chaired various sessions of international conferences. Currently he is the editor of Handbook of Research on Smart Farming Technologies for Sustainable Development, IGI Global. At the same time he is also acting as a guest editor of various reputed journal publishing house, such as Taylor & Francis, Inderscience and Bentham Science. He is the member of ACM and founder member of ACM Jaipur chapter.

Sugam Sharma obtained his PhD in Computer Science from Iowa State University, USA in 2013. Presently, he is working in Center for Survey Statistics and Methodology, Iowa State University. His research interests include Big Data, Cloud Computing, Multidimensional Databases, GIS, Sensor Network, Smart Home, and Food Informatics. He has published several research papers in high impact factor venues and also serves as an editor for numerous prestigious journals. Previously, he also has worked as a software consultant at NAVTEQ Traffic Inc., and System Analyst at Ames National Laboratory, USA. Dr. Sharma also holds MS in Computer Science from Jackson State University, USA and BE in Computer Science from Roorkee, India. Lately, Dr. Sharma's published work on hunger got international media attention and he was interviewed by several new and media outlets. KCCI Des Moines also aired his short interview on March 13, 2018 in 9:00 PM segment.

Sonali Vyas is serving as academician and research person from past 6+ years. Currently she is working as Assistant Professor of Computer Science at Amity Institute of Information Technology, Amity University Rajasthan. She was awarded PhD in Computer Science from the Jaipur National University – Rajasthan. Her research area is Database Virtualization, Data Mining and Big Data Analytics. At present she is supervising three PhD research scholars in the field of Big Data Analytics and Data Mining in IoT. She has published research papers in referred journals/conference proceedings as: National Conference on Computational and Mathematical

Sciences, JGRCST, CSIT-TCI 2015, IEEE Delhi Section, International Journal of Innovative Research in Computer and Communication Engineering and many more. She is also acting as a guest editor in Elsevier International Journal of Computer and Electrical Engineering (IJCEE) special issue. She is also acting as reviewer and member of various renowned national and international conferences/journals.

* * *

Pankaj Agarwal is working with Amity University Rajasthan as Assistant Professor since 2011. He completed his M. Tech. in mechanical engineering from Jagganath University, Jaipur. He has 10 years of experience in the field of teaching and research. He has published 9+ research papers and book articles of international repute. He is currently supervising several DST, Rajasthan project.

Vaibhav Bhatnagar is a Research Scholar in Amity Institute of Information Technology, Lecturer in S.S. Jain Subodh PG College.

Smita Chaudhry is Professor & Director, Institute of Environmental Studies, Kurukshetra University, Kurukshetra, Haryana, India, actively involved in its establishment; 1991-2007, served as Assistant and Associate Professor in Department of Bioscience & Biotechnology, Banasthali Vidyapeeth University, Rajasthan. She received her post graduation and doctoral from Kumaun University, Nainital, India. She has done pioneer work in restoration of landslides of the Himalayas, published in reputed Journals. She has 27 years research experience & completed various research projects and guided Ph.D. and also a large number of M.Sc and M. Tech students in various fields of Restoration Ecology, Bioremediation and Phytoremediation, Carbon Sequestration & Climate Change, Impact Analysis, Landuse/Land cover changes, Agroecology, Hydropower & Environmental Sustainability, Energy Auditing & Management. She has more than 60 international and national publications in addition to project reports. She is the member of various International and National Societies and has organized a number of seminars and workshops. She has also chaired sessions in National and International Seminars. She is Fellow of National Institute of Ecology, New Delhi.

Santosh R. Durugkar is pursuing Ph.d in computer science and engineering from Amity university Rajasthan, Jaipur. He is currently working wireless sensor networks and it future application for Irrigation systems. He has completed M.E (Info Tech) from University of Pune, Pune.

Gonçalo Manuel Fonseca is a Portuguese Mechanical Engineer. His primary professional pursuits are sustainable development and hardware projects. After the faculty, he has been able to work in different industries. In 2014, he had the possibility to work on a pioneering way to harvest biomass using autonomous vehicles. In 2016 he was defining ways to optimize machine process in the paper production process at the major Portuguese paper and pulp producer. From 2017, he works on the design and construction of equipment for offshore renewable energy industry. His goals include promoting good practices for sustainable energy production and consumption all over the world.

Shivani Garg, Assistant Professor at Institute of Environmental Studies, Kurukshetra University Kurukshetra since 2015. She has completed M.Tech (Energy and Environmental Management) and M.Sc.(Botany) from Kurukshetra University. Her main area research is connected with wastewater treatment, wastewater, energy management, biogas studies, waste management and environmental pollution. She has also presented the papers in National and International Seminars.

Dharm Singh Jat received his ME and PhD in Computer Science and Engineering. He is a Professor with the Department of Computer Science of Namibia's University of Science and Technology. From 1990 to 2014, he was with College of Technology and Engineering, Udaipur, India. He has guided more than 4 PhD and 23 master students and his interests span the areas of multimedia communications, wireless technologies, mobile communication systems, video transmission over wired-wireless networks, Network security and Internet of things. He has been the recipient of more than 14 prestigious awards, such as CSI Chapter Pattern, Distinguished Academic Achievement, Eminent Engineering Personality, Significant Contribution, Best Technical Staff, Outstanding Services, IEI Fellow, Senior Member IEEE and Distinguished ACM Speaker. He Formed Special Interest Group on Wireless Networks of Computer Society of India and designated Convenor and chairman, SIG-WNs by CSI. He is sole author of 03 Books and an editor of 13 books and has authored/co-authored over 135 research publications in peer-reviewed reputed journals, and conference proceedings.

Richa Kothari is currently Professor & Dean, School of Sciences in ITM University, Gwalior. She is also Head, School of Pharmacy, Dept. of Chemistry and Environmental Science, ITM University. She has joined as an Assistant Professor in the Department of Chemistry, BIMR, Gwalior in 2001 and senior lecturer in Department of Chemistry, ITM University, Gwalior in 2008. She earned her Doctorate in Chemistry from Jiwaji University, Gwalior. She is recipient of Best Teacher Award in BIMR and ITM University, Gwalior. She has guided students at Ph.D. and M.Sc.

level. She has completed three major research projects funded by DRDE, Gwalior, MPCST, Bhopal and MP Biotechnology Council. She has 07 patients to her credit. She has published 22 research papers and three books. She has successfully completed one UGC and three PCI inspections of School of Pharmacy. She has contributed to the development of B.Tech., B.Sc. M.Sc. B.Pharma., M. Pharma as member of Board of studies of School of Engineering and Technology and School of Pharmacy, School of Science in ITM University, Gwalior. Dr Kothari has organized various national and international conference/seminars/workshops. Dr Richa Kothari is professor in Department of chemistry, ITM University Gwalior.

Anton Limbo currently a research scholar in the Department of Computer Science, Namibia University of Science and Technology.

Chucknorris Garikayi Madamombe is currently a research scholar in the Department of Computer Science, Namibia University of Science and Technology. He is a technology and business visionary with executive and hands on experience in automating multimillion dollar project enterprises. Strong success record in creating robust IT solutions and proven ability to align IT to the business vision while managing costs and risk. Deep understanding of business process re-engineering and optimization. Proven ability to manage various projects of both SMEs and larger enterprises. Strong management and leadership skills. Strong programming/software development background and web applications development using: ASP.NET, Razor web pages, C#, CSS3, HTML & JavaScript. Vast experience in Microsoft SQL Server Database, MySQL database and Oracle Database. Ability to build complex internet applications using Oracle Forms and Reports 11g. Excellent programming skills using PL/SQL language. Experience in lecturing Business Administration Modules, Information Technology and Web development modules.

Deepak Panwar is working with Amity University Rajasthan as Assistant Professor since 2011. He has completed his M. Tech. and PhD in CSE from Gautam Buddha University Uttar Pradesh. He has 8 years of experience of teaching and research in field of nature inspired algorithm . He has published 8+ research papers and several book chapters of international repute.

Paulo Peças has a graduation (1991) and a doctorate (2004) in Mechanical Engineering at Instituto Superior Técnico (Universidade de Lisboa). Since 1998, he is an assistant professor at Instituto Superior Técnico in the Manufacturing Technologies and Industrial Management scientific area, being responsible for programmes in Production Management and Industrial Management. During the last ten years has participated and integrated coordination teams of R&DT projects in collaboration with the industry, at both national and European levels. The research activity has focused the development of models to support decision-making, comparative analysis and best-practices implementation in the scope of Life Cycle Engineering and lean & agile manufacturing. These models foster the build-up and implementation of solutions for productivity and competitiveness increasing in several industrial sectors. He has published and edited scientific books, published 22 research papers in International Journals and 6 chapters in scientific books, 60 publications in International Conferences Proceedings and 23 dissemination articles in national magazines.

G. L. Saini is working with Amity University Rajasthan as Assistant Professor since 2011. He completed his M. Tech. In CSE from Rajasthan Technical University and perusing PhD from Amity University Rajasthan. He has 8 years of experience in the field of development, programming language teaching and research. He has published 8+ research papers in journals of international repute.

Charu Singh is currently doing Internship from Sat-Com (PTY) Ltd, Namibia. She published 4 papers in refered journal and conference proceedings. Her research interest in VAD and background noise suppression Algorithms, Wireless Communication, and DSP based application for reliable communication.

Vijander Singh received his MCA degree from University of Rajasthan in 2005 and M.Tech. Computer Science & Engineering from Rajasthan Technical University Kota in the year 2009. He has been awarded his Ph.D. degree from Banasthali University in Computer Science & Engineering in 2017. He has qualified UGC-NET in Computer Science subject and GATE-2018 in Computer Science & Engineering subject. He has 12 years of experience in teaching and research. He has published more than 20 research papers in national and international journals and attended several conferences and workshops. Dr. Vijander is a member of ACM, IET, technical program committee member, member of editorial board/reviewer board of various renowned international conferences & journals. He has also organised several workshops at national level. His areas of interest are Networking, MANET and VANET. Presently, he is working as Assistant Professor at Amity University Rajasthan, India.

Claus Grøn Sørensen is a Senior Scientist/Head of Research Unit in the Operations Management division of the Department of Engineering, and Head of Centre on Smart Farming. He holds a PhD in Production and Operations Management. He has over 25 year experience in production and operations management, decision analysis, information modeling, system analysis, and simulation and modeling of technology application in agriculture. Research topics include resource analyses and optimizations, whole farm analyses and optimizations, the feasibility of introducing robotic systems in agriculture, development of management information systems and smart applications (e.g. FP7, FutureFarm, AgriFood2; IoF2020). He has participated (as project coordinator, WP leader, and partner coordinator) in multiple international projects. He is the author of more than 350 articles in peer reviewed Journals, conference proceedings, and scientific reports. He is currently the President of the European Society of Agricultural Engineers (EugAgEng), and past-chairman for CIGR (International Commision of Agricultural Engineering) Section V on System Management.

Umamaheswari Subbian received her MCA degree from Bharathidasan University, Trichy, India, M.Phil degree and Ph.D. degree from Bharathiar University, Coimbatore, India. She is currently working as Associate Professor, Department of Computer Science, Dr. G. R. Damodaran College of Science, Coimbatore, India. Her current research interests include web programming, cloud computing, mobile databases, Internet of Things, Data Analytics and mobile ad hoc networks. She has publications in peer reviewed journals and International Conferences. She is a Member of IEEE. She has been offered with the Summer Research Fellowship by Indian Academy of Sciences, Bangalore during 2015 and Fellowship to attend the India School of Internet Governance (inSIG 2016) during 2016. She has also been awarded with IETF Fellowship during March 2018.

Index

N

P

S

T

U

V

W

www.ingramcontent.com/pod-product-compliance
Lightning Source LLC
LaVergne TN
LVHW061546290926
849162LV00004B/923